G. Bruschi A. Borghetti (Eds.)

Cellular Aspects of Hypertension

With 50 Figures and 34 Tables

Springer-Verlag

Berlin Heidelberg New York
London Paris Tokyo
Hong Kong Barcelona
Budapest

Dr. Giacomo Bruschi
Dr. Alberico Borghetti

Istituto di Clinica Medica e Nefrologia,
Facolta di Medicina e Chirurgia, Università degli Studi di Parma,
Via A. Gramsci, 14, I-43100 Parma, Italy

ISBN 978-3-662-00985-7   ISBN 978-3-662-00983-3 (eBook)
DOI 10.1007/978-3-662-00983-3

Library of Congress Cataloging-in-Publication Data
Cellular aspects of hypertension/G. Bruschi, A. Borghetti (eds). p. cm.
  ISBN 978-3-662-00985-7
  1. Hypertension—Cytopathology. I. Bruschi, G. (Giacomo), 1946–.
II. Borghetti, Alberico. [DNLM: 1. Calcium—metabolism. 2. Cell
Membrane—metabolism. 3. Hypertension—metabolism. 4. Muscle,
Smooth, Vascular—physiology. WG 340 C393]
RC685.H8C378 1991 616.1'3207—dc20 DNLM/DLC 91-4839

© Springer-Verlag Berlin Heidelberg 1991
Softcover reprint of the hardcover 1st edition 1991

Typesetting: Best-set Typesetter Ltd., Hong Kong
25/3130-543210 – Printed on acid-free paper

Table of Contents

List of Contributors*

Afargan, M. *193*[1]
Aharonovitz, O. *193*
Armanini, D. *237*
Arosio, E. *227*
Astarie, C. *219*
Barakeh, J. *79*
Bergmann, C. *123*
Blaustein, M.P. *97*
Bohr, D.F. *209*
Borghetti, A. *69*
Borrione, A.C. *149*
Bosson, D. *237*
Bruschi, G. *69*
Bruschi, M.E. *69*
Canessa, M. *181*
Cannell, M.B. *79*
Cinotti, G.A. *157*
Contney, S.J. *51*
Cowley, A.W. *51*
Devynck, M.-A. *219*
Erne, P. *91*
Friedman, S.M. *169*
Fritschka, E. *245*
Furspan, P.B. *209*
Gairard, A. *123*
Garay, R. *257*
Garthoff, B. *135*

Gruber, K. *257*
Guzzo, P. *227*
Hermsmeyer, K. *91*
Hirth-Dietrich, C. *135*
Horie, R. *25*
Hvarfner, A. *111*
Kauser, K. *51*
Kazda, S. *135*
Khalil, R. *79*
Knorr, A. *135*
Kotelevtsev, Y.V. *175*
Kwan, C.Y. *61*
Kuhnle, U. *237*
Lauciello, C. *227*
Lechi, A. *227*
Lechi, C. *227*
Le Quan Sang, K.-H. *219*
Livne, A.A. *193*
Luckhaus, G. *135*
Mantero, F. *237*
Marche, P. *201*
Markus, S. *193*
Martella, L. *237*
Mazeaud, M. *219*
Menè, P. *157*
Milan, E. *237*
Minuz, P. *227*

---

*The addresses of the authors are given on the first page of each contribution.
[1] Page on which contribution begins.

Introduction

In the last two decades, investigations at the cellular level have progressively gained ground in the context of hypertension research. This choice of approach is due to some extent to the build up of know-how that molecular and cellular biology have been producing at a continuous rate. As the contents list of this volume shows, a large mass of work has been directed to gaining some insight into pathogenetic mechanisms. The pathogenesis of primary hypertension has been progressively categorized as a distinct biological problem, not amenable to the theoretical models that proved successful in understanding the nature of secondary forms of hypertension. At the same time, great efforts have been made to simplify this problem by sorting out, if possible, a few crucial mechanisms from the network of contributory factors in the regulation of blood pressure. The idea that what is to be sought is a primary structural and/or functional fault in arterial muscle has met with widespread acceptance. The strength of this argument lies in the fact that peripheral vascular resistance is increased in all forms of hypertension and, in turn, the diameter of resistance vessels is the dominant factor in the computation of total peripheral resistance. On the basis of this, cardiovascular structural adaptation was proposed as a positive feedback mechanism tending to maintain hypertension, once begun, whatever the initiating factor is. The presence of altered "reactivity" of hypertensive vessels has also been often asserted, but is still a subject of debate. Progress in this area is expected from better understanding of the inner mechanisms of single vascular smooth muscle cells. Different control systems have been identified and characterized and now comprise a host of transducing and regulatory proteins: membrane receptors and lipases, G-proteins, calcium channels, calciproteins, cyclicnucleotide kinases, C kinases, and myosin kinase. The hierarchy that links these elements of the contractile cascade is still insufficiently understood; nonetheless, drugs that attack these mechanisms, like adrenoceptor and calcium channel blockers, are the basis of everyday therapy and new agents are constantly being developed. The progress of antihypertensive treatment in the future will depend even more on the manipulation of specific biochemical systems.

Blood cell investigations have grown as a subsection of hypertension research and are particularly well represented in this book. Blood cells have been considered as carriers of general biochemical mechanisms (such as ion transport systems) or models of excitable, more representative tissues (e.g., the platelets). Blood cells are available in unlimited amounts and are easy to manipulate. Therefore, if they were to have an even modest defect, it should be possible to detect it at least under specially designed experimental conditions. The idea of finding some blood test for the classification and management of essential hypertension, though attractive, still seems quite ambitious. Nevertheless, if found, it would have such a clinical impact that the continued search for it seems justified.

Parma, October 1991   Giacomo Bruschi

Part I
Function and Structure of Vascular Tissue

Structure and Function of Resistance Vessels in Hypertension

M. J. Mulvany

Department of Pharmacology, Aarhus University, Universitetsparken 240, DK-8000 Aarhus C, Denmark

Introduction

The resistance vessels play a major role in the pathogenesis of hypertension, for by definition it is they that are responsible for the increased peripheral resistance and thus the increased blood pressure. However, it is still much debated how much the increased resistance is due to altered function or structure of the resistance vessels, and how much to increased levels of activation. The aim of this paper is twofold: first, to review briefly some of the abnormalities of resistance vessels in hypertension, and, second, to discuss the cellular basis for these abnormalities and how this may account for the difficulties in normalising resistance vessel properties with antihypertensive treatment.

Haemodynamics in Essential Hypertension

At least three abnormalities of the haemodynamics in essential hypertension seem incontrovertible. First, the total peripheral resistance (TPR) is increased under resting conditions [15]. Second, the pressor response – the increase in peripheral resistance caused by infusion of agonists – is increased [7,8]. Third, as Folkow [10] has shown, the TPR is increased under conditions of complete relaxation (and even though a recent report has questioned this [28], the vast majority of studies have supported Folkow's original finding). In principle, the increased resting TPR can be explained in terms of alteration of either the resistance vessels or the level of activation, while the increased pressor response is most easily explained in terms of alterations in the structure or function of the resistance vessels alone. However, the increased TPR under totally relaxed conditions cannot be explained in terms of altered function, and this must therefore indicate that there is altered structure.

The haemodynamic evidence shows therefore that resistance vessel structure is altered in hypertension, but this does not exclude the possibility that

G. Bruschi A. Borghetti (Eds.)
Cellular Aspects of Hypertension
© Springer-Verlag Berlin · Heidelberg 1991

resistance vessel function or neurohumoral activation is also altered. Nor does it say what kind of structural alteration is present: decreased lumen or rarefication [12] could both give increased TPR under relaxed conditions. To examine this question, there is little alternative to taking the vessels out and examining them under in vitro conditions, and our work in this field has used vessels both from the two rat models of hypertension (the spontaneously hypertensive rat (SHR) and Goldblatt 1-kidney, 1-clip renal hypertensive rats) and from patients with essential hypertension.

Methods

In the rat, vessels have been taken from a variety of vascular beds [16]. In all cases, the dilated internal diameters of these vessels was 150–200 μm, and the vessels were thus small enough to contribute to the control of peripheral resistance [5,6]. It should be emphasised, however, that the vessels we have examined are only representative for the proximal part of the resistance vasculature. The more distal part of the resistance vasculature – the arterioles and precapillary sphincters – contribute just as much to the peripheral resistance, although their function may differ. In the human work, sub-cutaneous biopsies, taken under local anaesthesia from otherwise healthy volunteers, have been used [3].

The work has been based on the use of our myograph technique [20], in which short segments of the arteries were threaded on to fine wires attached to a force transducer and micrometer, respectively. The micrometer could be used to stretch the vessel as required, while the force transducer could be used to measure the response of the vessels to agonists added to the chamber in which the vessels were suspended. The whole arrangement was mounted on the stage of a microscope, permitting direct measurements of the vessel dimensions: lumen diameter and media thickness [22]. These dimensions were measured under standardised conditions corresponding to the vessel's being relaxed and exposed to a transmural pressure of 100 mmHg. The dimensions measured are thus a structural parameter, not influenced by any active tone from the smooth muscle cells.

Structure and Function of Resistance Arteries

Data from our group have shown previously (see Mulvany [16] for review) that in the SHR the media:lumen ratio of the resistance vessels we have examined is increased in all the vascular beds investigated. This is also the conclusion drawn by most investigators concerning the proximal resistance vessels of the SHR (whether studied in vivo, in vitro or histologically [11,17,19]), although in the more distal vessels this does not seem to be the case – a conclusion we can support, as I will show later.

The situation regarding the functional properties of the resistance vessels is, however, less clear-cut. In SHR mesenteric vessels, we and others have evidence for increased noradrenaline sensitivity under conditions where the neuronal uptake of noradrenaline is inhibited with, for example, cocaine [23,30]. Another functional difference concerns the calcium sensitivity, which is also increased in SHR vessels from the mesenteric bed [21]. These functional differences do not, however, extend to all vascular beds [24]. Furthermore, in cross-breeding experiments, little correlation between vascular function and blood pressure is found [18]. On the other hand, a consistent finding has been the increased reactivity – the increased maximum response – but this can be fully explained in terms of the altered structure. Thus, the rat studies suggest to us that the major difference in the resistance vessels of the SHR is structural rather than functional. In renal hypertensive rats, the mesenteric resistance vessels have increased media:lumen ratios, but have little alteration in their functional characteristics [13].

In the human studies, evidence for structural change has also been found, such that the media:lumen ratio of proximal resistance vessels from patients with essential hypertension is increased [3]. Increased media:lumen ratios in other forms of human hypertension: pre-eclampsia (even though the pressure had only been raised for a few weeks [1]), and in hypertension associated with uraemia [2]. However, the functional properties of these vessels do not seem to differ, in that the sensitivity of the vessels to noradrenaline, serotonin, vasopressin and angiotensin is normal [3]. The only difference in the responses to these agonists is an increase in the maximum response, an increase which can be entirely explained in terms of the increased media:lumen ratio of the vessels. Thus, as in the SHR, it seems that altered structure plays a more important role than function in essential hypertension as well. It should, however, be noted that, as in the SHR, the sensitivity of proximal resistance vessels to cocaine is also increased, possibly reflecting the abnormal noradrenaline turnover reported by Esler and colleagues [9] in essential hypertension.

Modes of Altered Vascular Structure

The above data therefore suggest that the media:lumen ratio of the proximal resistance vessels is increased, but this does not necessarily mean that the lumen is reduced, or that there has been growth. Figure 1 shows different ways in which the media:lumen ratio of a vessel can increase: first, by adding material onto the outer surface (no change in lumen but increased vascular mass); second, by redistributing the vascular material (no change in vascular mass, but reduced lumen). On the other hand, if material is added such that the wall is extended (no change in wall thickness), then the media:lumen ratio can decrease even though there is vascular growth. Figure 1 demonstrates, therefore, that the finding of an increased media:lumen ratio does not in itself

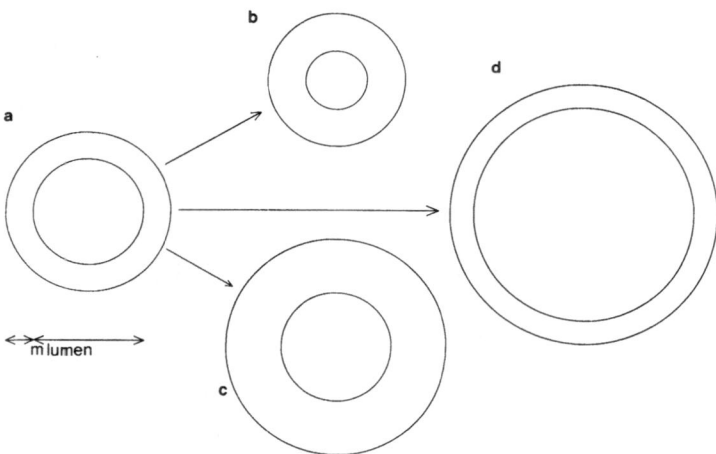

Fig. 1a-d. Different modes of structural change of small arteries. **a** The control situation, showing the media cross-section (*lumen*, internal diameter, *m*, media thickness). **b** The media volume is unchanged, the lumen diameter reduced and the media:lumen ratio increased. **c** The media volume is increased with a maintained lumen diameter. **d** The media volume is increased by the same amount as in **c**, the lumen diameter is increased and the media:lumen ratio is decreased

provide information about lumen diameter. More particularly, with the current interest in growth factors, it is perhaps important to remember that increased media:lumen ratio is not in itself evidence of growth.

It might be thought that it would be relatively easy to distinguish between the possibilities shown in Fig. 1: one could perhaps just take a vessel each from a hypertensive and a normotensive individual and compare them. However, that presupposes that the vessels are comparable, which is not necessarily the case: as a number of investigators have shown, the architecture of the vasculature differs in normotensive and hypertensive individuals [12,27], so that it is not possible to choose vessels which are precisely comparable. Nor is random selection the answer, for random selection has a greater chance of selecting larger vessels than smaller vessels. Similarly, there is no guarantee in the more proximal resistance vasculature that vessels will be comparable. Therefore there is a sampling problem.

Relations Between Media Cross-sectional Area and Lumen

In an attempt to circumvent this problem, we have plotted media cross-sectional area (i.e. volume per unit length) against lumen diameter. The idea is that, although there will inevitably be variation in the location of the vessel which is being examined, the relation between the media cross-sectional area

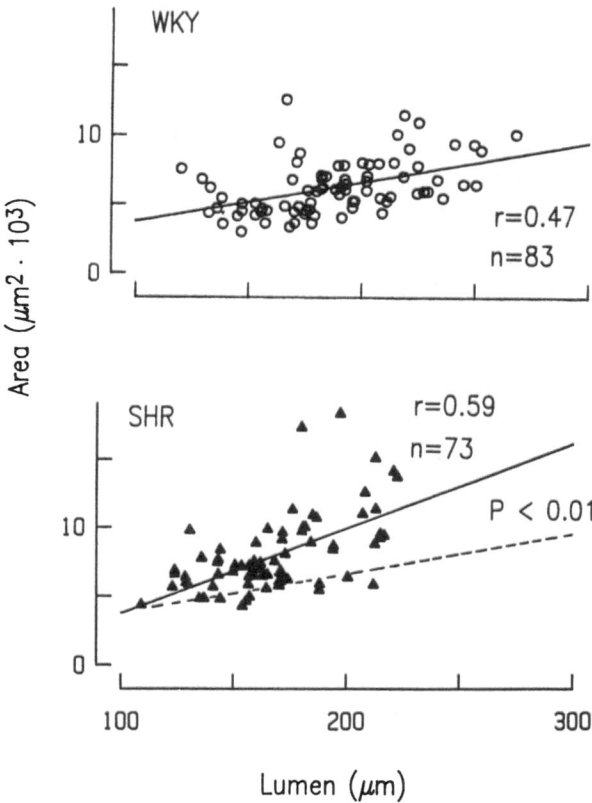

Fig. 2. Relation of cross-sectional area of the media to normalised lumen diameter in the mesenteric resistance vessels of 83 Wistar-Kyoto rats (*top panel*) and 73 spontaneously hypertensive rats (*lower panel*). The *solid lines* are regression lines. In the *lower panel*, the *stippled line* shows the regression line for the WKY data redrawn from the upper panel. Data from several series of experiments as reported by [16]

and lumen should be a fixed characteristic of each animal concerned. The following gives the results of analyses of the relation between media cross-section and lumen for three types of hypertension: SHR, renal hypertension in Wistar-Kyoto (WKY) rats, and human essential hypertension.

Figure 2 summarises data from a number of different experimental series, and shows that, for the more proximal resistance arteries, the media cross-sectional area for a given lumen was larger in the SHRs. This was not the case for the more distal vessels: the data suggest that for vessels distal to 100-μm vessels, there is little difference in media:lumen ratio. Furthermore, even regarding the proximal resistance vessels, these data do not provide direct evidence for growth, for it all depends on the relation between lumen diameters of the vessels from the SHRs and WKYs. On average, the lumen diameter of the SHR vessels was less than that of the WKY vessels, by about 10%. Although, as indicated, there is no guarantee that this result is not clouded by a sampling error, this actually agrees well with the haemodynamic data, which also suggests that the lumen diameter of SHR vessels is on average around 10% less than that of WKY vessels [11]. On this basis, a 200-μm WKY vessel, for example, corresponds to a 180-μm SHR vessel, which then implies that at this level there is vascular growth of around 20%. This would therefore suggest that in the mesenteric vessels, at least, the increased media:lumen ratio is due to encroachment of the media into the lumen (as suggested by Folkow [10]). However, in other vascular beds, the situation may differ: Baumbach and Heistad [4], for example, have reported that in the cerebral circulation, the increased media:lumen ratio of the proximal resistance vessels is due to remodelling, although that result must also depend on whether the vessels they examined are strictly comparable.

A similar analysis of vessels taken from essential hypertensives [3] showed the same general features as in the SHR vessels. In the larger vessels, for a given lumen, the media cross-sectional area was increased; in vessels distal to 100-μm vessels there seems to be little difference. However, the difference between the slopes was more modest, and if again it is assumed (based on haemodynamic data) that a 180-μm vessel from a hypertensive individual corresponds to a 200-μm vessel from a normotensive individual, then there seemed to be little difference in media mass. Here, then, the increased media:lumen ratio would seem to be due primarily to a redistribution of material, but again everything depends on the assumptions concerning the comparability of vessels. Nevertheless, it should be pointed out that similar conclusions were drawn many years ago by Short [29] in post-mortem studies of the human mesentery.

The case for renal hypertension is shown in Fig. 3. Again, this plots the media cross-sectional area against lumen, but this time for mesenteric resistance arteries from 1-kidney, 1-clip renal hypertensive rats. Again, the difference in media cross-sectional area is seen in the more proximal vessels; again, no difference is apparent in 100-μm vessels. However, in this case there is a very clear distinction between the characteristics, and there would have to be a very large difference in the lumen diameter of comparable vessels if there was not to be any growth.

We can thus now summarise what I believe can be said about vascular structure in hypertension. There is overwhelming evidence that the media:lumen ratio of the more proximal vessels is increased, and this can account for the increased pressor response. Whether there is also growth depends on the assumptions one can make about the lumen diameter. If, as the haemodynamic data from the relaxed vasculature indicate, vessels in hypertensives have lumen diameters which are around 10% less than comparable vessels from normotensives, then there is little doubt that there is growth (increased media cross-sectional area) in the SHR and renal hypertensive rats, but the position in humans is less clear-cut.

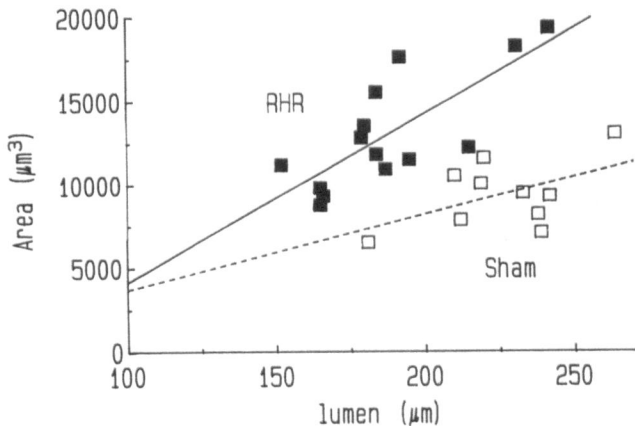

Fig. 3. Relation of cross-sectional area of the media to normalised lumen diameter of mesenteric resistance vessels from 1-kidney, 1-clip Goldblatt renal hypertensive rats (*RHR*) and from sham-operated controls (*Sham*). The *solid lines* are regression lines. From data reported by [13]

Cellular Basis

The cellular basis for these findings is currently being investigated. In the SHR, the size of the vascular smooth muscle cells is not increased [25,26]. Thus, to the extent that there is an increased amount of vascular smooth muscle in the resistance vessels (see above), there is hyperplasia. In the renal hypertensive rats, on the other hand, the smooth muscle cells of the mesenteric resistance vessels were larger: there was cellular hypertrophy, not hyperplasia [13]. As regards the human data, preliminary results show that – as in the SHRs – the size of the vascular smooth muscle cells in subcutaneous resistance arteries from essential hypertensives is not increased (N. Korsgaard, personal communication), suggesting that here, too, spontaneous hypertension is not associated with cellular hypertrophy. Whether there is hyperplasia is not clear, for – as discussed above – it is not clear to what extent there is any increase in the quantity of vascular smooth muscle. Part of the increased media:lumen ratio of resistance arteries of essential hypertensives is therefore likely to be due to reorganisation of existing vascular smooth muscle cells rather than synthesis of new cells.

These results suggest that the mechanisms behind the altered media:lumen ratio seen in spontaneous hypertension and in renal hypertension differ. One can speculate that the reason for this difference could lie in renal hypertension's being associated with an increased load on the individual smooth muscle cells, and that the response to increased load is cellular hypertrophy (as in skeletal muscle). On this basis, one might expect that anti-hypertensive treatment which reduces the load on the vascular smooth muscle would result in cellular hypotrophy, and this indeed seems to be the case [14]. In SHRs which had been treated for 20 weeks to reduce blood pressure, vascular smooth muscle volume was significantly correlated with mean blood pressure. If this result also applies to human essential hypertension, then it suggests an interesting corollary. Although vascular smooth muscle morphology appears to be normal in essential hypertension, antihypertensive treatment which normalises blood pressure also results in abnormal vascular smooth muscle morphology. This possibility requires further investigation.

Conclusion

The available evidence shows that the resistance vasculature in hypertension is altered, but that the vascular smooth muscle appears normal. These alterations appear to be primarily structural in nature, the altered media:lumen ratio perfectly balancing the new requirements. To this extent, the increased peripheral resistance can be fully explained in terms of altered arrangement or number of the smooth muscle cells within the resistance vessels. The fact that antihypertensive treatment does not appear – in the rats – to be able to reduce the number of cells, but rather to make them abnormally small, suggests that present antihypertensive therapies may not be ideal. Perhaps it would be better to find methods of preventing vascular smooth muscle replication.

References

1. Aalkjaer C, Johannsen P, Pedersen EB, Rasmussen A, Mulvany MJ (1984) Characteristics of resistance vessels in preeclampsia and normotensive pregnancy. J Hypertens 2 (S3):183–185
2. Aalkjaer C, Pedersen EB, Danielsen H, Fjeldborg O, Jespersen B, Kjaer T, Srensen SS, Mulvany MJ (1986) Morphological and functional characteristics of isolated resistance vessels in advanced uraemia. Clin Sci 71:657–663
3. Aalkjaer C, Heagerty AM, Petersen KK, Swales JD, Mulvany MJ (1987) Evidence for increased media thickness, increased neuronal amine uptake, and depressed excitation-contraction coupling in isolated resistance vessels from essential hypertensives. Circ Res 61:181–186
4. Baumbach GL, Heistad DD (1988) Cerebral circulation in chronic arterial hypertension. Hypertension 12:89–95
5. Bohlen HG (1986) Localization of vascular resistance changes during hypertension. Hypertension 8:181–183
6. Davis MJ, Ferrer PN, Gore RW (1986) Vascular anatomy and hydrostatic pressure profile in the hamster cheek pouch. Am J Physiol 250:H291–H303
7. Doyle AE, Fraser JRE (1961) Vascular reactivity in hypertension. Circ Res IX: 755–761
8. Duff RS (1956) Adrenaline sensitivity of peripheral blood vessels in human hypertension. Br Heart J 19:45–52
9. Esler M, Jennings G, Korner P, Willett I, Dudley F, Hasking G, Anderson W, Lambert G (1988) Assessment of human sympathetic nervous system activity from measurements of norepinephrine turnover. Hypertension 11:3–20

10. Folkow B (1956) Structural myogenic, humoral and nervous factors controlling peripheral resistance. In: Harington M (ed) Hypotensive drugs. Pergamon, London, pp 163–174
11. Folkow B (1982) Physiological aspects of primary hypertension. Physiol Rev 62: 347–504
12. Hutchins PM, Darnell AE (1974) Observation of a decreased number of small arterioles in spontaneously hypertensive rats. Circ Res 34/35:I161–I165
13. Korsgaard N, Mulvany MJ (1988) Cellular hypertrophy in mesenteric resistance vessels from renal hypertensive rats. Hypertension 12:162–167
14. Korsgaard N, Christensen J, Mulvany J (1991) Basic Res Cardiol
15. Lund-Johansen P (1983) Haemodynamics in early essential hypertension – still an area of controversy. J Hypertens 1:209–213
16. Mulvany MJ (1987a) Vascular structure and smooth muscle contractility in experimental hypertension. J Cardiovasc Pharmacol 10 [Suppl 6]:s79–s85
17. Mulvany MJ (1987b) The Fourth Sir George Pickering Memorial Lecture. The structure of the resistance vasculature in essential hypertension. J Hypertens 5:129–136
18. Mulvany MJ (1988) Resistance vessel structure and function in the etiology of hypertension studied in F2-generation hypertensive-normotensive rats. J Hypertens 6:655–663
19. Mulvany MJ, Aalkjaer C (1990) Structure and function of small arteries. Physiol Rev 70:921–961
20. Mulvany MJ, Halpern W (1977) Contractile properties of small arterial resistance vessels in spontaneously hypertensive and normotensive rats. Circ Res 41:19–26
21. Mulvany MJ, Nyborg NCB (1980) An increased calcium sensitivity of mesenteric resistance vessels in young and adult spontaneously hypertensive rats. Br J
22. Mulvany MJ, Hansen PK, Aalkjaer C (1978) Direct evidence that the greater contractility of resistance vessels in spontaneously hypertensive rats is associated with a narrowed lumen, a thickened media, and an increased number of smooth muscle cell layers. Circ Res 43:854–864
23. Mulvany MJ, Aalkjaer C, Christensen J (1980) Changes in noradrenaline sensitivity and morphology of arterial resistance vessels during development of high blood pressure in spontaneously hypertensive rats. Hypertension 2:664–671
24. Mulvany MJ, Nilsson H, Nyborg N, Mikkelsen E (1982) Are isolated femoral resistance vessels or tail arteries good models for the hindquarter vasculature of spontaneously hypertensive rats. Acta Physiol Scand 116:275–283
25. Mulvany MJ, Baandrup U, Gundersen HJG (1985) Evidence for hyperplasia in mesenteric resistance vessels of spontaneously hypertensive rats using a 3-dimensional disector. Circ Res 57:794–800
26. Owens GK, Schwartz SM, McCanna M (1988) Evaluation of medial hypertrophy in resistance vessels of spontaneously hypertensive rats. Hypertension 11:198–207 Pharmacol 71:585–596
27. Schmid-Schonbein GW, Firestone G, Zweifach BW (1986) Network anatomy of arteries feeding the spinotrapezius muscle in normotensive and hypertensive rats. Blood Vessels 23:34–49
28. Schulte KL, Braun J, Meyer-Sabellek W, Wegscheider K, Gotzen R, Distler A (1988) Functional versus structural changes of forearm vascular resistance in hypertension. Hypertension 11:320–325
29. Short D (1966) Morphology of the intestinal arterioles in chronic human hypertension. Br Heart J 28:184
30. Whall CW, Myers MM, Halpern W (1980) Norepinephrine sensitivity, tension development and neuronal uptake in resistance arteries from spontaneously hypertensive and normotensive rats. Blood Vessels 17:1–15

Vascular Smooth Muscle Cell in Hypertension: Dissecting Out Cause and Effect

J. D. Swales

Department of Medicine, Clinical Sciences Building, Leicester Royal Infirmary
GB–Leicester, Great Britain

Introduction

A large number of phenotypic differences has been demonstrated between the spontaneously hypertensive rat (SHR) and control normotensive strains. Even more differences have been demonstrated between patients with essential hypertension and normotensive healthy subjects. This review will concentrate on the hypothesis that there is a disturbance of cell membrane function in genetic hypertension and will further consider the nature of the association between abnormalities of cell membrane function and blood pressure.

Following the initial demonstration of increased erythrocyte sodium in some hypertensive subjects, a number of transmenbrane flux pathways and other membrane functions has been shown to be disturbed both in essential hypertension and in the SHR (Tables 1, 2). Some of these findings are still controversial, and in some cases (e.g. erythrocyte sodium [1] and erythrocyte sodium potassium co-transport) [2] abnormalities in different populations have been in opposite directions. There are four possible explanations for associations with blood pressure, explanations which are by no means mutually exclusive. These are:

1. Relationships are spurious and attributable to unrecognised confounding factors.
2. Abnormalities are secondary to hypertension.
3. Abnormalities are linked to mechanisms producing hypertension without being directly responsible for hypersion. In particular, abnormalities may be linked genetically or phenotypically.
4. Abnormalities may reflect changes in relevant tissues which produce hypertension, i.e. they are pathogenetic.

G. Bruschi A. Borghetti (Eds.)
Cellular Aspects of Hypertension
© Springer-Verlag Berlin · Heidelberg 1991

Table 1. Essential hypertension: positive reports on ion handling

Variable	Cell	Change
Intracellular sodium	Erythrocyte	Increased/decreased[a]
Transmembrane sodium flux	Erythrocyte	Increased
Ouabain-sensitive rubidium uptake	Erythrocyte	Increased
Sodium efflux (plasma-incubated) rate constant	Erythrocyte	Increased
Ouabain-sensitive sodium efflux rate constant	Erythrocyte	Decreased
Sodium–potassium co-transport	Erythrocyte	Decreased (Paris)
Sodium–potassium co-transport	Erythrocyte	Increased (Boston)
Lithium–sodium countertransport	Erythrocyte	Increased
Calcium binding	Erythrocyte	Decreased
Calcium ATPase	Erythrocyte	Decreased
Cell-membrane fluidity	Erythrocyte	Decreased
Ouabain-sensitive sodium efflux rate constant	Leucocyte	Decreased
Intracellular sodium	Leucocyte	Increased
Intracellular sodium	Lymphocyte	Increased
Intracellular exchangeable calcium	Adipocyte	Increased
Intracellular pH	Platelet	Increased
Sodium–proton exchange	Platelet	Increased
Calcium content	Platelet	Increased

Modified from [58].
[a] Increased in some studies, decreased in others [1].

Spurious Relationships Between Abnormalities and Hypertension

Initial studies of erythrocyte sodium content [3], sodium – potassium co-transport [4] and lithium – sodium countertransport [5] reported dramatic differences between values obtained from hypertensive and normotensive subjects, with little overlap between the two. Later reports either failed to confirm the earlier observations or showed a much less marked disturbance. Inadequate matching of hypertensive and normotensive populations can have important consequences. In particular, the effects of weight, age, race and sex need to be taken into account. It is well established, for instance, that erythrocyte sodium is increased in black populations. In an interesting recent study [6] Smith et al. compared erythrocyte ion transport and sodium content in black and white hypertensives. Although there was no difference in erythrocyte sodium in hypertensives as opposed to normotensives, black people had significantly higher sodium levels than white. Disproportionate numbers of black patients in the hypertensive group would have created a spurious elevation in erythrocyte sodium in hypertension, and likewise excessive numbers of blacks in the control group would have created an apparent reduction in erythrocyte sodium levels in hypertension.

Bramley et al. demonstrated increased cellular sodium in unmatched hypertensive patients [7]. However, after correction for age and sex, differences between controls and hypertensives disappeared. Cooper et al. [8]

Table 2. Genetic hypertension in the rat: positive reports on ion handling

Variable	Model	Tissue	Change
Sodium and potassium permeability	SHR	Erythrocyte	Increased
Ouabain-resistant net potassium	SHR	Erythrocyte	Increased
Net sodium efflux	SHR	Erythrocyte	Decreased
	Sabra hypertensive	Erythrocyte	Decreased
Net potassium influx	Lyon hypertensive	Erythrocyte	Increased
Calcium binding	SHR	Erythrocyte	Decreased
Calcium binding	SHR	Adipocyte	Decreased
Cell-membrane fluidity	SHR	Erythrocyte	Decreased
Sodium content and ouabain-sensitive sodium efflux	SHR	Thymocyte	Decreased
Net sodium efflux	SHR	Tail artery	Increased
Sodium and potassium	SHR	Arterial wall	Increased
Ouabain-insensitive sodium efflux	SHR	Arterial smooth muscle	Increased
Calcium binding	SHR	Arterial wall (subcellular fraction)	Decreased
Calcium ATPase	SHR	Erythrocyte	Decreased
		Aorta	Decreased
		Aortic smooth muscle	Decreased
Na^+, K^+-ATPase	SHR	Arterial smooth muscle	Increased
Intracellular calcium pool	SHR	Adipocyte	Increased
Intracellular pH	SHR	Lymphocyte	Increased
			Decreased
Intracellular pH	SHR	Resistance vessels	Increased
Intracellular pH	SHR	Platelets	Increased
			Decreased

Modified from [58].

examined lithium – sodium countertransport in several distinct populations ranging from black school-children to adults recruited from routine health screening programmes. Although there was a highly significant correlation between countertransport and blood pressure in the group of 448 individuals, this lost statistical significance when age and weight were entered into the equation in a step-wise linear regression analysis.

In an even larger study, Hunt et al. obtained blood samples from 2091 individuals in Utah [9]. Weak positive associations between sodium – lithium countertransport and both weight and blood pressure were observed, but the most striking finding was an association between plasma lipids and sodium – lithium countertransport: triglycerides were positively associated whilst high density lipoprotein (HDL)2 and HDL3 cholesterol was negatively correlated.

When multivariate analysis was performed for lipids, weight and blood pressure, the association between triglycerides and total and HDL cholesterol on one hand and countertransport on the other remained, as did the association between weight and countertransport. However, no relationship between blood pressure and countertransport remained when blood pressure was taken as a continuous variable [9,10]. It should be emphasised, however, that there was still a relationship between sodium – lithium countertransport and the presence or absence of hypertension (defined qualitatively by an arbitrary cut-off point), suggesting a possible genetic association between sodium – lithium countertransport and predisposition to hypertension. In another study dietary weight reduction was associated with a fall in sodium – lithium countertransport as well as in blood pressure [11].

Physical activity has not been systemically examined. However, in two studies exercise training has produced a highly significant fall in sodium – lithium countertransport [12,13]. In one of these studies this was associated with an increase in HDL cholesterol, although there was no change in blood pressure [12]. In the other study no change in plasma lipids could be demonstrated [13].

Abnormalities as Secondary Effects

Many of the abnormalities in organ function (e.g. in the kidneys) which can be demonstrated in essential hypertension are the consequence rather than the cause of increased perfusion pressure [14]. This possibility has rarely been considered in studies of electrolyte transport in hypertension. Nevertheless, alterations in, for instance, circulating catecholamines may influence transport processes. Thus, elevated noradrenaline can inhibit sodium transport in leucocytes but stimulate it in other tissues [15]. The effects that dietary advice often given to hypertensive subjects may have on cell transport systems has not been examined, although the close association between, for instance, sodium – lithium countertransport and plasma triglycerides [9] suggests that this requires consideration.

Genetic and Phenotypic Linkage of Abnormalities

Even in in-bred strains of rat such as the SHR, genetic finger-printing [16] has indicated a large number of differences far in excess of the two to four genes that have been calculated to be responsible for hypertension in this strain [17]. The genetic fingerprinting technique applied does not indicate the proteins for which these sequences code, and it seems likely that many, if not most, are non-coding sequences. However, these results suggest that there are genetically linked features in this strain that are not related to blood pressure, but are an abnormality fortuitously bred into the strain. Cross-breeding studies are necessary to show associations between blood pressure and genotypic or phenotypic abnormalities [18,19]. The concept that abnormalities of ionic handling reflect changes in the membrane lipid environment which contribute to blood pressure through other second messenger systems (see below) postulates such a phenotypic linkage.

Abnormalities as Pathogenetic

The hypothesis that an ionic abnormality gives rise to hypertension directly is clearly attractive, but before it can be accepted the other three possibilities have to be excluded. Existing theories of pathogenesis have related disturbances in ion handling to alterations in smooth muscle sodium, calcium or hydrogen ions regulating tone or growth.

Theories of Pathogenesis

Two hypotheses have linked hypertension to increased intracellular sodium. Tobian et al. [20] originally suggested that retention of sodium and fluid within the arterial wall resulted in a reduction in luminal area and an increase in resistance to flow. However, whilst increased sodium content could be demonstrated in the large artery wall from hypertensive patients and animals, it seems likely that in most cases these were secondary to increased wall mass as a result of elevated blood pressure. It is also difficult to see how waterlogging of the resistance vessel wall could result in increased maximal contraction, which is a feature of hypertension.

Blaustein proposed that inhibition of a putative sodium – calcium exchange in vascular smooth muscle led to an increase in second messenger calcium [21]. Increased smooth muscle sodium was related to elevated concentrations of a sodium transport inhibitor, produced as a result of sodium retention [22]. Evidence for inhibition of an arterial sodium pump has been demonstrated in several rat models of hypertension associated with volume expansion [23]. Whilst it seems that such a sodium transport inhibitor may well be an important regulator of sodium excretion, it has still not been characterised chemically, and the association between inhibition of the sodium – potassium pump and hypertension in the predominant genetic forms of hypertension (e.g. SHR and essential hypertension) is not persuasive [24]. The absence of evidence for sodium retention in essential hypertension and the uncertainty as to whether the sodium – calcium exchange mechanism operates over physiological concentrations in vascular smooth muscle has prevented general acceptance of the Blaustein hypothesis [25].

Other hypotheses have related increased vascular tone to abnormalities of transmembrane monovalent cation flux giving rise to smooth muscle membrane depolarisation, increased calcium influx and enhanced vasomotor tone [26,27]. A pathogenic role has been attributed to decreased smooth muscle calcium binding in decreasing membrane stability in this context [27]. Some workers have linked hypertension to smooth muscle sodium proton antiport activity. It has been postulated that increased activity, which has been demonstrated in the platelets from hypertensive patients [28] and in cultured vascular smooth muscle of SHR, gives rise to cell alkalinisation, which may be a stimulus for growth [28,29]. In support of this hypothesis, intracellular pH is elevated in SHR resistance vessels [30].

In the face of a large volume of frequently contradictory data and a proliferation of hypotheses, I believe the best approach is to test current hypotheses in the context of two questions. Firstly, is there an underlying disturbance of cell membrane function in hypertension? Secondly, what are the primary tissue abnormalities that elevate blood pressure? Unless we can define what abnormality it is that needs explanation (in particular whether the abnormality is one of growth or tone), we shall not be able to test existing hypotheses adequately.

There is now strong evidence that there is an association between physicochemical properties of cell membrane lipids and blood pressure. The Utah study has shown that circulating lipids are associated with both blood pressure and ion fluxes [9]. Other studies have demonstrated a reduction in the unsaturated fat component of the cell membrane from erythrocytes and platelets in humans and SHR [31–33]. This may be associated with decreased membrane fluidity which has been demonstrated in erythrocytes both in humans and in SHR [34,35]. The manner in which alterations in membrane lipids may influence more than one transport system has been shown in an in vitro study carried out by Brand and Whittam [36]. Erythrocyte membrane is more permeable to the nitrate anion than to chloride. Substituting nitrate for chloride increases permeability of the erythrocyte to monovalent cations, probably as a result of rearrangement of membrane lipids: thus, nitrate reproduces one of the flux abnormalities most commonly reported in hypertension, i.e. an increase in passive sodium fluxes. This induced change in membrane properties also gives rise to a reduction in outward sodium and potassium cotransport and a failure of the active sodium pump to respond to its increased intracellular sodium load. Likewise, modification of the lipid component of semipurified preparations of sodium – potassium ATPase modifies its activity [37], whilst sodium and potassium co-transport can be altered both in vitro and in vivo by changing the erythrocyte membrane lipid composition [38,39]. We have recently obtained direct evidence that the membrane environment is responsible for decreased membrane-bound calcium ATPase activity in SHR erythrocytes [40]. Both basal and calmodulin-stimulated activity were depressed in membrane preparations but normal when the enzyme was solubilized. Similar observations have been made in human erythrocytes [41].

Over the last few years it has been recognised that cell membrane phospholipids perform an important second messenger role in the regulation of contraction, growth and neurotransmitter release [42]. Hydrolysis of the inner lamina lipid phosphoinositol to water-soluble inositol triphosphate is an important second messenger system for the control of intracellular calcium and probably mediates the calcium response to such stimuli as angiotensin II, vasopressin and endothelin. In addition, the other product of phosphoinositol hydrolysis, diacylglyercol, performs another role as an activator of the trans-membrane sodium – hydrogen ion exchange system, which may play an important role in cell differentiation and growth. The phosphoinositol cycle also regulates basal arterial sodium pumping [43]. Recent reports have described abnormalities of phosphoinositide turnover in the early stages of hypertension in SHR and in humans [43,44]. This system therefore provides a possible link between global disturbance of cell membrane function with an influence on several ionic pathways and hypertension. They do not provide the opportunity to identify the cellular processes which are responsible for this association.

One of the most fruitful approaches to addressing this issue is to ask the second question stated above: What is the tissue abnormality which we are attempting to explain?

Tone vs Structure

The proximate cause of hypertension in most clinical forms and experimental models is an increase in peripheral resistance. Two processes could contribute to this increased resistance. These are (1) increased vasoconstrictor tone and (2) structural changes associated with an increased wall to lumen ratio [45]. There are, however, clear-cut qualitative differences between animal models and humans. Studies of isolated resistance vessel beds and of single resistance vessels from several experimental models of hypertension support the view that the vasoconstrictor response to certain agonists is specifically enhanced. Differentiating between specific vascular hypersensitivity and increased reactivity due to structural changes is much more difficult in humans. Recently, however, the myographic technique developed by Mulvany and Halpern has been applied by our group in collaboration with that of Mulvany to the isolated human resistance vessel obtained at skin biopsy [46]. No evidence of an increased constrictor response to any agonist could be demonstrated either in patients with essential hypertension [46] or in relatives of hypertensive patients [47]. On the other hand, there was a 29% increase in wall-to-lumen ratio in the resistance vessels of hypertensives, which, if characteristic of the resistance vessel bed in general, closely matches elevation in blood pressure (32%). Folkow [48] has demonstrated a similar close matching between wall-to-lumen ratio and blood pressure elevation in the chronic stage of SHR. This indicates that in both species, structural hypertrophy of the resistance vessel bed is a sufficient cause of the elevation of blood pressure. From the study of relatives, this appears to be as true of the early stages of hypertension as of established hypertension in humans [47]. The interpretation of disturbances in cell membrane function therefore demands a genetically conditioned abnormality which gives rise to resistance vessel structural changes.

Trophic Response vs Trophic Stimulus

The increased resistance vessel wall-to-lumen ratio observed both in SHR and in essential hypertension could be the result of an enhanced load imposed by enhanced transient pressure spikes in hypertension or by an enhanced smooth muscle trophic response to transient pressor spikes.

There is very strong evidence for an early enhanced neurogenic influence upon haemodynamics in the early stage of hypertension in SHR and in humans [45]. This evidence comprises an increased alerting cardiovascular response in both species and restoration of the haemodynamic pattern of early essential hypertension by a combination of sympathetic and parasympathetic blockade [49]. Other studies have shown increased noradrenaline overflow from the synaptic junction [50], which is most marked in the early phases of essential hypertension [50], and increased sympathetic activity both in SHR [51] and in borderline clinical hypertension [52]. It could be argued, therefore, that the trophic changes seen in the resistance vessels in hypertension are a normal response to a genetic abnormality of autonomic circulatory control. An alternative explanation, however, postulates that the smooth muscle trophic response to a given pressure load is enhanced. Thus, smooth muscle cells from 12-week SHR show an enhanced growth rate compared with control animals [53], and arterial mass is increased at SHR even at birth [54]. Destruction of the sympathetic nervous system either chemically or immunologically has produced discrepant results, with blood pressure being either reduced or lowered to normal levels in SHR [55,56]. In the most extensive study the large artery trophic changes were not influenced, although smaller arterial hyperplasic changes were prevented by neonatal sympathectomy [56]. The interpretation of such studies is rendered difficult by uncertainty about the completeness of sympathectomy and the possibility of secondary induced changes in animals deprived of autonomic activity. It seems probable, however, that increased autonomic activity observed in the early stages of hypertension is at least partly responsible for the trophic resistance vessel changes.

It still remains uncertain how far the undeniable differences between SHR and WKY smooth muscle constitute an additional component in the trophic response. Accordingly, Bund and Heagerty in our department have studied the relationship between pressure and structure in SHR by placing a partially constricting ligature around one external iliac artery of SHR at 5 weeks. As a result, mean arterial pressure in that limb was reduced to that recorded in WKY controls. This procedure prevented resistance vessel structural changes as assessed myographically and indeed normalised structure at both 12 and 24 weeks [57], which would suggest that the primary distrubance is of autonomic regulation of blood pressure rather than the trophic smooth muscle response to pressure.

Conclusion

Dissecting out the significance of altered electrolyte transport requires knowledge of the tissue-based abnormalities with which they are associated. The strong evidence for an overactive sympathetic nervous system in the early stages of genetic hypertension suggests that the putative global disturbance of cell membrane function is acting as a determinant of autonomic activity, perhaps through the regulation of membrane charge or second messenger systems (such as the phosphoinositide systems) within the sympathetic neurones or vasomotor centre. An additional enhanced trophic responsiveness of vascular smooth muscle cannot be excluded, although I would suggest that the current emphasis upon smooth muscle in analysis of the cellular physiology of hypertension may be misplaced.

References

1. Simon G (1989) Is intracellular sodium increased in hypertension? Clin Sci 76: 455–461
2. Canessa M, Bize I, Solomon H, Adragna N, Tosteson DC, Dagher GE, Garay R, Meyer P (1981) Sodium countertransport and co-transport in human red cell function, dysfunction and genes in essential hypertension. Clin Exp Hypertens. 3:783–795
3. Losse, H, Weymeyer H, Wessels F (1960) Wasser- und Elektrolytgehalt von Erythrozyten bei arterieller Hypertonie. Klin Wochenschr 38:393–395
4. Garay RP Meyer P (1979) A new test showing abnormal net Na+ AND K+ fluxes in erythrocytes of essential hypertensive patients. Lancet i:349–353
5. Canessa M, Adragna N, Solomon HS, Connolly TM, Tosteson DC (1980) Increased sodium – lithium countertransport in red cells of patients with essential hypertension. N Eng J Med 302:772–776
6. Smith JB, Wade MB, Fineberg NS, Weinberger MH (1988) Influence of race, sex and blood pressure on erythrocyte-sodium transport in humans. Hypertension 12:251–258
7. Bramley PM, Paulin JM, Millar JA (1986) Intracellular cations and transmembrane cation transport in essential hypertension. The importance of controlled clinical observations. J Hypertens 4:589–596
8. Cooper R, Trevisan M, Ostrow D, Sempos C, Tamler J (1984) Blood pressure and sodium – lithium counter-transport: findings in population-based surveys. J Hypertens 2:467–471
9. Hunt SC, Williams RR, Smith JB, Ash KO (1986) Associations of three erythrocyte cation transport systems with plasma lipids in Utah subjects. Hypertension 8:30–36
10. Williams RR, Hunt SC, Wu LL, Hasstedt SJ, Hopkins PN Ash KO (1988) Genetic and epidemiological studies on electrolyte transport systems in hypertension. Clin Physiol Biochem 6:136–149
11. McDonald AM, Dyer AR, Liu K, Stamler R, Gosch FC, Grim R, Berman R, Stamler J (1988) Sodium – lithium countertransport and blood pressure control by nutritional intervention in mild hypertension. J Hypertens 6:283–292

12. Adragna NC, Chang JL, Morey MC, Williams RS (1985) Effects of exercise on cation transport in human red cells. Hypertension 7:132–139
13. Hespel P, Lijnen P, Fagard R, M'Buyamba-Kabangu JR, Van Hoogf R, Lissens W, Rosseneu M, Amery A (1988) Changes in erythrocyte sodium and plasma lipids associated with physical training. J Hypertens 6:159–166
14. Swales JD (1986) Hypertension and the kidney. In: Lote C (ed) Advances in renal physiology. Croom Helm London, pp 247–296
15. Riozzi A, Heagerty AM, Bing RF, Thurston H, Swales JD (1984) Noradrenaline: a circulating inhibitor of sodium transport. Br Med J 289:1025–1027
16. Samani NJ, Swales JD, Jeffreys AJ, Morton DB, Naftilan AJ, Lindpaintner K, Ganten D, Brammar WJ (1989) DNA finger-printing of spontaneously hypertensive and Wistar-Kyoto rats: implications for hypertension research. J Hypertens 7:809–816
17. Rapp JP (1983) Genetics of experimental and human hypertension. In: Genest J, Kuchel O, Hamet P Cantin M (eds) Hypertension. McGraw-Hill New York pp 582–595
18. Rapp JP, Wang SM, Dean E (1988) A genetic polymorphism in the renin gene of Dahl rats co-segregates with blood pressure. Science 243:542–544
19. Furspan PB, Jokelainen PT, Sing CF, Bohr DF (1987) Genetic relationship between a lymphocyte membrane abnormality and blood pressure in spontaneously hypertensive/prone and Wistar-Kyoto rats. J Hypertens 5:293–297
20. Tobian L, Janecek A, Tomboulian A, Ferreira D (1961) Sodium and potassium in the walls of the arterioles in experimental renal hypertension. J Clin Invest 40:1922–1925
21. Blaustein MP (1977) Sodium ions, calcium ions, blood pressure regulation and hypertension: a reassessment and a hypothesis. Am J Physiol 232:C165–C173
22. De Wardener HE, Clarkson EM (1985) Concept of natriuretic hormone. Physiol Rev 65:658–759
23. Haddy FJ, Overbeck HW (1976)The role of humoral agents in volume-expanded hypertension. Life Sci 19:935–948
24. Swales JD (1983) Abnormal ion transport by cell membranes in hypertension. In: Birkenhager WH Reid JL (eds) Clinical aspects of essential hypertension. Elsevier Amsterdam, pp 239–267 (Handbook of hypertension vol 1)
25. Brading AG Lategan TW (1985) Sodium-calcium exchange in vascular smooth muscle. J Hypertens 3:109–116
26. Bruner CA, Myers JH, Sing CF, Jokelianen PT, Webb RC (1986) Genetic basis for altered vascular responses to ouabain and potassium free solutions in hypertension. Am J Physiol 252:H1276–H1282
27. Rinaldi G, Bohr D (1988) Plasma membrane and its abnormalities in hypertension. Am J Med Sci 295:389–395
28. Livne A, Veitch R, Grinstein S, Balfe JW, Marquez-Julio A, Rothstein A (1987) Increased platelet sodium – hydrogen ion exchange rates in essential hypertension: application of a novel test. Lancet 1:533-536
29. Berk BC, Vallega G, Muslin AJ, Gordon HM, Canessa M, Alexander RW (1989) Spontaneously hypertensive rat vascular smooth muscle cells in culture exhibit increased growth and Na^+/H^+ exchange. J Clin Invest 83:822–829
30. Izzard AS, Heagerty AM (1989) The measurement of internal pH in resistance arterioles: evidence that intracellular pH is more alkaline in SHR than WKY animals. J Hypertens 7:173–180
31. Ollerenshaw JD, Heagerty AM, Bing RF, Swales JD (1987) Abnormalities of erythrocyte membrane fatty acid composition in human essential hypertension. J Human Hypertens 1:9–12
32. Naftilin AJ, Dzau VJ, Loscalzo J (1986) Preliminary observations on abnormalities of membrane structure and function in essential hypertension. Hypertension 8: II-174–II-179

33. Nara Y, Sato T, Mochizuki S, Mano M, Horie R, Yamori Y (1986) Metabolic dysfunction in smooth muscle cells of spontaneously hypertensive rats. J Hypertens 4 [Suppl 3]:105–107
34. Orlov SN, Postnov YV (1982) Ca^{++} binding and membrane fluidity in essential and renal hypertension. Clin Sci 63:281–284
35. Tsuda K, Tsuda S, Minatogawa Y, Iwahashi H, Kido R, Masuyama Y (1988) Decreased membrane fluidity of erythrocytes in cultured vascular smooth muscle cells in spontaneously hypertensive rats: an electron spin resonance study. Clin Sci 75:477–480
36. Brand SC Whittam R (1984) The effect of furosemide on sodium movements in human red cells. J Physiol 348:301–306
37. Kimelberg HK (1975) Alterations in phospholipid dependent ($Na^+ + K^+$) ATPase activity due to fluid lipidity. Biochim Biophys Acta 413:143–156
38. Cooper RA (1977) Abnormalities of cell membrane fluidity in the pathogenesis of disease. N Engl J Med 297:371–377
39. Jackson PA Morgan DB (1982) The relationship between membrane cholesterol and phospholipid and sodium efflux in erythrocytes from healthy subjects and patients with chronic cholestasis. Clin Sci 62:101–107
40. Adeoya AS, Norman RI Bing RF (1989) Erythrocyte membrane calcium adenosine 5-triphosphatase activity in the spontaneously hypertiensive rat. Clin Sci 77:395–400
41. Adeoya AS, Norman RI, Bing RF (1989) Erythrocyte membrane Ca^{2+} Mg^{2+} adenosine 5-triphosphatase activity in essential hypertension. J Hypertens 7:921–922
42. Heagerty AM, Ollerenshaw JD (1987) The phosphoinositide signalling system and hypertension. J Hypertens 5:515–524
43. Heagerty AM, Ollerenshaw JD, Swales JD (1986) Abnormal vascular phosphoinositide hydrolysis in spontaneously hypertensive rat. Br J Pharmacol 89:803–807
44. Riozzi A, Heagerty AM, Ollerenshaw JD, Swales JD (1987) Erythrocyte phosphoinositide metablolism in essential hypertensive patients and their normotensive offspring. Clin Sci 73:29–32
45. Folkow B (1982) Physiological aspects of primary hypertension. Physiol Rev 62:347–504
46. Aalkjaer, C, Heagerty AM, Petersen KK, Swales, JD, Mulvany MJ (1987) Evidence for increased media thickness and increased neuronal amine uptake, but depressed excitation contraction coupling in isolated resistance vessels from essential hypertensives. Circ Res 61:181–186
47. Aalkjaer C, Heagerty AM, Bailey I, Mulvany MJ, Swales JD (1987) Studies of isolated resistance vessels from offspring of essential hypertensive patients. Hypertension 9 [Suppl III]:III-155–III-158
48. Folkow B (1990) The 'structural factor' in hypertension: with special emphasis on the hypertrophic adaptation of the systemic resistance vessels. In: Laragh JH Brenner BM (eds) Hypertension: pathophysiology, diagnosis and management. Raven New York pp 565–581
49. Julius S, Pascual AV, London R (1971) Role of parasympathetic inhibition in the hyperkinetic type of borderline hypertension. Circulation 44:413–418
50. Esler M, Jennings G, Korner P, Blombery P, Burke F, Willett I, Leonard P (1984) Total and organ specific noradrenaline plasma kinetics in essential hypertension. Clin Exp Hypertens 6:507–521
51. Lundin S Thoren P (1982) Renal function and sympathetic activity during mental stress in normotensive and spontaneously hypertensive rats Acta Physiol Scand 115:115–124
52. Anderson EA, Sinkey CA, Lawton WJ Mark AL (1989) Elevated sympathetic nerve activity in borderline hypertensive humans. Evidence from direct intra-neural recordings. Hypertension 14:177–183

53. Yamori Y, Igawa T, Kanebe T, Nara Y, Tagami M (1988) Enhanced growth rate of cultured smooth muscle cells from spontaneously hypertensive rats. Heart Vessels 4:94–99

54. Gray SD (1982) Anatomical and physiological aspects of cardiovascular function in Wistar-Kyoto and spontaneously hypertensive rats at birth. Clin Sci 63:383s–385s

55. Folkow, B, Hallback M. Lundgren Y, Weiss L (1972) The effects of 'immuno-sympathectomy' on blood pressure and vascular 'reactivity' in normal and spontaneously hypertensive rats. Acta Physiol Scand 84:512–513

56. Lee RMKW, Trigle CR, Cheung DWT Coughlin MD (1987) Structural and functional consequence of neonatal sympathectomy on the blood vessels of spontaneously hypertensive rats. Hypertension 10:328–338

57. Bund SJ, Heagerty AM (1989) Resistance vessel morphology and blood pressure development: no evidence for enhanced smooth muscle growth in spontaneously hypertensive rats. J Hypertens 7:927–928

58. Swales JD (1983) Abnormal ion transport by cell membranes in hyertension. In: Birkenhager WH Reid JL (eds) Clinical aspects of essential hypertension Elsevier, Amsterdam, pp 239–266 (Handbook of hypertension, vol 1)

Cellular Mechanisms of Spontaneous Hypertension and Stroke: Role of Vascular Smooth Muscle Cells

Y. Yamori[1], Y. Nara[1], T. Nabika[2], M. Tagami[3], and R. Horie[1]

[1] Department of Pathology, Shimane Medical University, Izumo 693, Japan
[2] Department of Laboratory Medicine, Shimane Medical University, Izumo 693, Japan
[3] Department of Internal Medicine, Sanraku Hospital, Tokyo 101, Japan

Introduction

The development of rat models for hypertension and stroke has enabled us to study cellular mechanisms of spontaneous hypertension and stroke experimentally and to extend our investigation into the pathogenesis of these diseases in humans from cellular aspects. This paper will first review the development of rat models of various cardiovascular diseases and their genetic background, then describe the characteristics of cultured vascular smooth muscle cells (VSMC) from rat models for hypertension and stroke, and finally report on our experimental studies on stroke as well as pathological studies in humans, seeking analogies between rat models and humans in the cellular aspects of stroke, the most prevalent complication of hypertension.

Rat Models of Cardiovascular Diseases

Development of Various Strains

Since the VSMC in our studies were obtained from the aorta of rat models for hypertension and stroke, spontaneously hypertensive rats (SHR) [20,33] and stroke-prone SHR (SHRSP) [21,34,40], the development of these models is briefly overviewed here with some emphasis on the new substrains and their genetic background [35]. In fact, the recent blooming of studies on cellular aspects on hypertension partly stems from our findings of cell membrane or transport abnormalities [43–46] and of the accelerated growth of VSMC in SHR and SHRSP [11,32,50].

Experimental hypertension research has made great progress thanks to the exploitation of SHR, which are now used widely around the world, and SHRSP, selectively bred from SHR by Yamori et al. [40], are unique models which rapidly develop severe hypertension (over 200 mmHg), almost

G. Bruschi A. Borghetti (Eds.)
Cellular Aspects of Hypertension
© Springer-Verlag Berlin · Heidelberg 1991

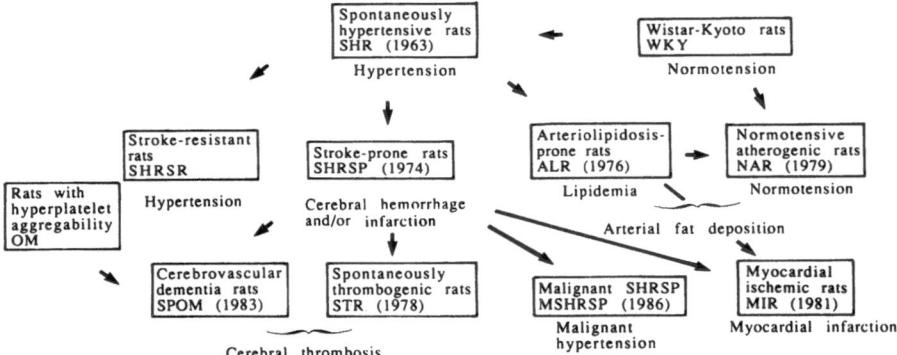

Fig. 1. Development of various rat models of cardiovascular diseases from SHR and related strains. *OM*, Osborne-Mendel strain

100% nowadays dying of typical hemorrhagic or ischemic stroke. We have established various rat models derived from SHR and SHRSP (Fig. 1) and , in particular, have obtained a new strain by cross-breeding between SHRSP and the normotensive OM strain which showed the most accelerated platelet aggregation among 24 inbred rat strains, the strains thus obtained (SPOM), showing both hypertension and accelerated platelet aggregation, develops lacunar stroke due to multiple cerebrovascular thrombosis [35] and may be used as a good model for experimental studies on cerebrovascular dementia, a condition on the increase in the elderly population [38]. Pathologically, the strain is close to spontaneous thrombogenic rats (STR), which were selectively bred from SHRSP on the basis of slower development of severe hypertension during the period of youth (50–100 days after birth) [42]. On the other hand, the substrain of SHRSP which develop severe hypertension at a younger age die mainly of cerebral hemorrhage before even reaching the age of 4 months. Although maintaining such as strain was difficult, Okamoto et al. [22] succeeded in breeding the offspring from SHRSP treated with antihypertensive agents and established MSHRSP which quickly developed severe hypertension at a young age and died of malignant hypertension.

Another model of interest is a hypertensive strain, so-called myocardial ischemic rat (MIR), which frequently develops the clinical symptoms of cardiac failure and dies of hypertensive heart disease in combination with ischemic myocardial lesions due to the narrowing of coronary arteries by the thickened intima and thrombosis [44].

In summary, various rat models for cardiovascular diseases (CVD), derived from SHR or SHRSP are now available for experimental research on cellular aspects of not only hypertension but also its complications such as stroke, cerebrovascular dementia, heart disease, and malignant hypertension. SHR were selectively bred from Wistar-Kyoto (WKY) rats with higher blood pressure, and SHRSP were established by selectively maintaining the offspring of SHR that died of stroke. A genetic disposition to hypertension is the common background of these various, CVD. Since these rats develop hypertension and CVD spontaneously, they are the best models so far for studying genetic pathogenesis and its interaction with environmental factors of CVD which can not be studied in other artificial models of hypertension. Genetic mechanisms and gene-environmental interaction in hypertension and other complications can now be studied experimentally at the cellular level in these models.

Genetic Background of Various Strains and Controls

The genetic characteristics of these models, although they may not always be involved in the genetic pathogenesis, are important in investigations of the genetic pathogenesis and the cellular mechanisms of these CVD. Yamori and Okamoto [39] were the first to find that renal nonspecific esterase isozymes were characteristic of SHR. Matsumoto et al. [15] recently analyzed over 30 loci in SHR and SHRSP substrains as well as control WKY rats and demonstrated that both SHR and WKY had some substrain variabilities. By using esterase isozymes, SHR substrains and WKY animals can be clearly differentiated from each other, and SHRSP differ from stroke-resistant SHR (SHRSR) [33], in the molecular forms of phosphogluconate dehydrogenase and pepsinogen-1 [35]. With advanced techniques for gene analysis becoming available, we may hope that the major genes involved in the genetic cellular mechanisms of hypertension and stroke will be clarified in the near future and utilized as highly reliable predictors for hypertension and stroke and important tools for their primary prevention.

It was recently noted that major histocompatibility genes were mostly similar in SHR and SHRSP but different in the substrains of WKY [15]. SHR, SHRSP and original Japanese WKY animals have commonly the k-type of RT-1 histocompatibility gene but WKY substrains in the United States have the l-type. These two types constitute the major genetic difference, so it is important that the appropriate controls for SHR and SHRSP should be used for further research on genetic and cellular aspects of hypertension and stroke.

Cellular Mechanisms of Spontaneous Hypertension

Principal Hypertension Mechanisms

The early arterial alterations in hypertension observed under a microscope were functional vasoconstriction and dilatation due to increased sympathetic activity [48], and early morphological signs of vascular smooth

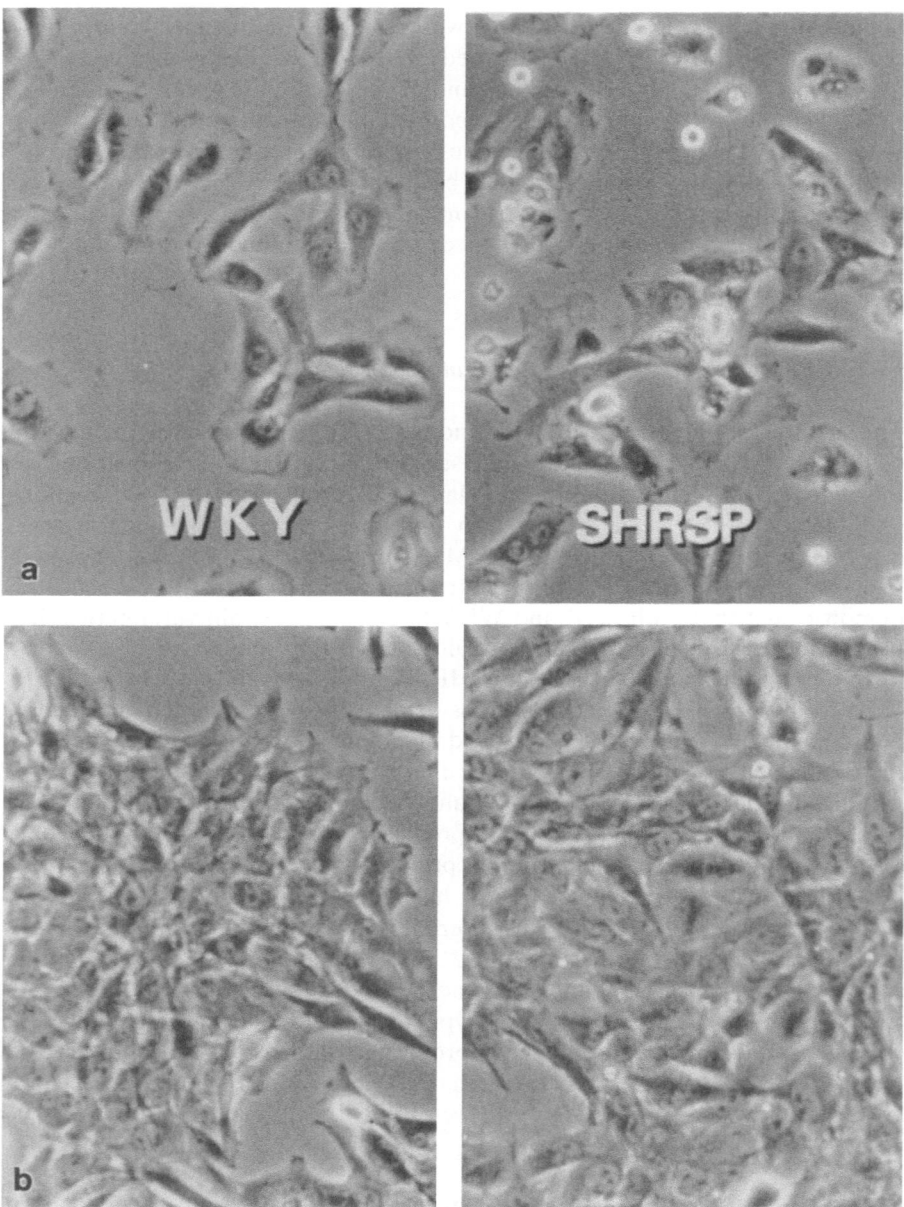

Fig. 2a,b. Time-lapse observation of fixed fields on cultured VSMC from the aorta of WKY rat (*left*) and SHRSP (*right*) on the 2nd day of seeding (**a**) and 24 hours after (**b**)

muscle cell (VSMC) hyperplasia and hypertrophy can be detected light and electron microscopically at the age of 3 months in SHRSP [32]. Early changes in VSMC morphology were recently confirmed by scanning electron microscope in the medial wall of mesenteric arteries; more segmental bulges were noted on the surface of VSMC in SHR [4].

Yamori [30] was the first to demonstrate that arteries were very active in amino acid incorporation in SHR compared with WKY rats. As early as at 4 weeks of age, even before the development of hypertension, and also later, protein synthesis was significantly increased in SHR in mesenteric arteries and aorta [30–32]. This means that the biochemical process of vascular hypertrophy and hyperplasia starts early in the development of hypertension.

VSMC Proliferation Under Culture Conditions

In order to exclude the effect of blood pressure on VSMC proliferation, the growth rates under tissue culture conditions of VSMC obtained from SHR, SHRSP, and WKY by an explant method were compared. The migration of VSMC was first noted to be more active from the explants from the aorta of SHR or SHRSP than from those of WKY, and the growth rates of VSMC from SHR and SHRSP were demonstrated to be faster than those from WKY animals [33,51]. This difference in the growth rate was recently demonstrated by time-lapse videoing of cultured VSMC obtained from SHRSP and WKY (Fig. 2). Filming was started 2–3 days after seeding of VSMC and pictures taken at the rate of 1 shot per minute for 24 h. The objective observation of the process of cell growth for 24 h in a film which less than 1 min to show confirmed the faster growth rate of VSMC from the aorta of SHR.

Statistical analysis of cell numbers proved that VSMC obtained from SHRSP and SHR had a significantly faster rate of growth than those from WKY animals; doubling times of VSMC from 24-week-old (and 12-week-old) SHRSP, SHR and WKY are 24 ± 2 (26 ± 1), 31 ± 4 (30 ± 4) and 51 ± 9 (47 ± 11) h, respectively [47]. The faster growth rate was noted similarly in VSMC obtained from younger (12-week-old) rats. Therefore, it can be concluded that VSMC from hypertensive rats are genetically more active in cell proliferation. This enhanced VSMC growth is considered to accelerate structural vascular changes [6] and thus to contribute to the pathogenesis of genetic hypertension [11,30–32,35,50].

Intracellular Ionic Imbalance of VSMC

Our studies on cultured SMC first demonstrated that Na and K ions were more permeable at the cell membrane of SHRSP and SHR, as indicated by the increased Na influx after the inhibition of Na, K-ATPase by 1 mM ouabain [32,45,46]. Nabika et al, [18] further demonstrated that such an increase in Na permeability affected the growth rate of VSMC: when growth factor (fetal calf serum, FCS) was removed from the culture medium for 48 h, thymidine incorporation was reduced to the minimum stable level, and then the growth factor was again added to the media with different Na concentrations; increased Na concentration of the culture media clearly accelerated the growth rate of smooth muscle cells. Since such an acceleration of cellular kinetics was noted in SHR and SHRSP with cell membrane more permeable to Na, we may speculate that extracellular Na concentration affects the growth of VSMC from hypertensive rats even more.

On the other hand, Matlib et al. [14] isolated membrane fractions from mesenteric arteries of SHR and WKY rats and were able to demonstrate that Na^+-Ca^{2+} exchange tended to be activated in the plasma membrane from SHR which showed a reduction of Na^+, K^+-ATPase activity.

We can now briefly summarize the cellular mechanism of genetic hypertension and its interaction with environmental factors [32,35]. Intracellular Na^+ or K^+ in VSMC tends to increase or decrease in dependence on the permeability or transport abnormality of the cell membrane. Such intercellular electrolyte imbalance tends to elevate cytosolic ionic Ca through Na^+-Ca^{2+} exchange and functionally or structurally increases peripheral vascular resistance. These VSMC are also more excitable in response to neurogenic stimuli or norepinephrine [17,37]. Environmental factors, especially excess salt intake, aggravate this process through inhibition of Na^+, K^+-ATPase by the action of ouabain or digoxin-like natriuretic factors.

Transport and Oncogene Induction of VSMC

Cytoplasmic pH (pH_i) is regulated by several different H^+ transport mechanisms such as Na^+-H^+ and Cl^--HCO^{3-} exchange systems [16]. Some mitogens and growth factors induce cytoplasmic alkalinization by activating Na^+-H^+ exchange [7]. Because of its importance in cell function, a new simple method was developed to monitor pH_i continuously in floating cells by estimation of the fluorescent intensity of 2',7'-bis-carboxyethyl-5,6,carboxyfluorescein (BCECF) [28]. We applied this method for the quantitative analysis of Na^+-H^+ exchange activity in cultured VSMC under physiological conditions and proved that it is enhanced in VSMC from SHRSP compared with those from WKY [12] (Table 1).

Since activation of Na^+-H^+ exchange and subsequent intracellular alkalinization have been postulated as playing an important role in the initiation process of cell proliferation, c-fos mRNA induction was measured 30 min after the addition of FCS to quiescent VSMC, when c-fos was maximally induced (T. Nabida et al., unpublished data). c-fos mRNA induction was not fully inhibited at any reduced concentration of Na^+ suggesting that activation of Na^+-H^+ exchange and subsequent intracellular

Table 1. Na^+-H^+ exchange activity in cultured VSMC from SHRSP and WKY at various extracellular Na^+ concentrations $[Na^+]_o$

Na^+-H^+ exchange	$[Na]_o$ (mM)					
	20	40	60	80	100	120
SHRSP	0.01143 ± 0.0007	$0.0145 \pm 0.0011^*$	$0.0190 \pm 0.0013^{**}$	$0.0204 \pm 0.0013^{**}$	$0.0213 \pm 0.0013^{**}$	$0.0210 \pm 0.0012^{**}$
WKY	0.00831 ± 0.0012	0.0096 ± 0.0015	0.0127 ± 0.0009	0.0144 ± 0.0007	0.0158 ± 0.0006	0.0158 ± 0.0008

Na^+-H^+ exchange activity is expressed in terms of the change in cytoplasmic pH per second.
$^* p < 0.05$; $^{**} p < 0.01$ (difference from WKY).

Table 2. Intracellular pH (pH_i) and Na^+-H^+ exchange in hypertension

Cell type	SHR		Essential hypertension	
	pH_i	Na^+-H^+ exchange	pH_i	Na^+-H^+ exchange
Erythrocyte	–	–	↓ [24]	–
Leucocyte/lymphocyte	↓ [2]	↑ [5]	↑ [19]	↑ [19]
Platelet	–	–	↑ [1]	↑ [13]
Vascular smooth muscle cells	↓ [12] ↑ [3, 10]	↑ [12] ↑ [3, 10]	–	–

alkalinization was not necessary in the induction of *c-fos* mRNA. This was further confirmed by the observation, that inhibition of Na^+-H^+ exchange by amiloride or a more specific inhibitor, 5-*N*-ethyl-*N*-isopropyl amiloride (EIPA), completely blocked FCS-stimulated intracellular alkalinization at the concentration of $10 \mu M$ but showed no inhibitory effect on c-fos mRNA induction.

Since Na^+-H^+ exchange as well as *c-fos* mRNA induction was assumed to be stimulated by protein kinase C activation, the effects of removal of extracellular Na^+ on *c-fos* mRNA induction by 12-*O*-tetradecanoylphorbol 13-acetate (TPA), an activator of protein kinase C, were also observed. Exposure of quiescent VSMC to $100 nM$ TPA caused potent induction of *c-fos* mRNA, which was only particularly inhibited in Na-free buffer. Thus, even protein kinase C activation did not seem to need Na^+-H^+ exchange activation in the induction of *c-fos* mRNA. All these findings indicate that Na^+-H^+ exchange activation and subsequent intracellular alkalinization are not necessary for the induction of this proto-oncogene when cells are stimulated by FCS.

Intracellular pH and Ionic Balance of Cultured VSMC

Although VSMC from SHRSP show growth activation, the pH_i of VSMC in the quiescent state is significantly lower ($p < 0.002$) in SHRSP than in WKY rats: 7.26 ± 0.03 ($n = 12$) vs 7.35 ± 0.03 ($n = 9$). Interesting enough, it was recently reported that the pH_i of lymphocytes from SHR was also lower [2]. Therefore, the observed activation of Na^+-H^+ exchange may be secondary to acidification.

As summarized in Table 2, a lowering of pH_i was also observed in erythrocytes from hypertensive patients [24] but pH_i was reported to be increased in leukocytes [19], and in platelets from some hypertensive [1]. There are discrepancies in the findings for the cell types used. Even in the same cell types, VSMC, alkalinization was reported in cultured (not quiescent) VSMC [3] or dissected resistant arterioles in SHR [10]. Although these available data of pH_i in hypertension are inconsistent, Na^+-H^+ exchange has been consistently reported to be activated in VSMC [3,10,12], and lymphocytes [5] of SHR as well as in leukocytes [19] and platelets [13] of hypertensive patients.

Resnick et al. [25] further reported reduction of intracellular Mg^{2+} in erythrocytes from hypertensive patients concomitantly with acidification. Mg was also reduced in erythrocytes of hypertensive rats, as was first reported by Henrotte et al. [9]. Since Mg may be more generally reduced in cells in hypertension, we observed the effect on blood pressure and survival rate in SHRSP of a Mg-rich diet (0.8%) compared with a control diet containing 0.2% Mg. The Mg-enriched diet slightly lowered blood pressure and markedly improved the survival rate [35,52].

Other than Mg, we have previously reported that Ca, K, dietary fiber, protein, and especially substances containing sulfur amino acids such as taurine and methionine were effective for reducing stroke and hypertension-induced CVD [49].

Interaction between these beneficial factors and genetic mechanisms has been analyzed and is speculated to be as follows [53]: the intracellular Mg level decreases primarily or secondarily under the influence of sympathetic activation and the reduction lowers Na^+, K^+-ATPase activity. Mg and K supplementation activate this Na pump activity and counteract the process of ionic imbalance in hypertension. Protein, its final metabolite, urea, Ca, and K accelerate urinary Na excretion and dietary fiber reduces intestinal absorption of Na, thus attenuating the adverse effect of Na at different levels and mechanisms. Sulfur amino acids attenuate sympathetic activity and also intracellular Ca^{2+} mobilization [51].

Cellular Mechanisms of Stroke

Hypertension and Cerebral Blood Flow

As for the mechanism of stroke, a typical cardiovascular complication of hypertension, it has been morphologically confirmed that arterionecrosis, so-called fibrinoid degeneration, is the major vascular lesion in small intra-cerebral arteries [23,26,27]. It is completely different from the atherosclerotic proliferative changes of coronary arteries in ischemic heart disease.

We noted that the commonest site of stroke was the basal ganglia both in SHRSP and in humans [41]. Our morphological studies on the vascular architecture of the commonest sites of stroke showed that these sites were fed by recurrent arteries which induced reduction of blood flow in hypertension due to hemodynamic mechanisms.

Yamori and Horie [36] reported earlier in 1976 that regional cerebral blood flow was significantly reduced in SHRSP with severe hypertension over 210 mmHg, but not in SHR and WKY rats which developed no stroke spontaneously. Since the improvement of cerebral blood flow by the administration of antihypertensive agent resulted in the prevention of stroke, the reduction of cerebral blood flow is a good predictor for stroke, and also an important pathogenic mechanism of stroke.

Pathology of VSMC in Intracerebral Arteries in SHRSP

In SHRSP the reduction of cerebral blood flow induces early changes in the VSMC of intracerebral arteries, namely, focal degeneration and necrosis, which always occur at the outer layer of the media due to the impairment of oxygen and nutrient supply through the blood–brain barrier [26,27,35]. Therefore, arterionecrosis is caused by local nutritional disturbance. When medial VSMC start to necrose, macrophages invade subintimal space by disrupting the endothelial barrier and induce massive infiltration of large molecular components of plasma, i.e., fibrin. Thus, macrophages play a key role in the development of "fibrinoid necrosis," also working as scavenger cells to take up cell debris and fibrin and return into blood stream. The arterial lumen becomes narrow due to the subintimal fibrin deposition, and is often occluded with thrombosis.

The pathogenetic process of fibrinoid necrosis in hypertension can be summarized as follows. Focal degeneration and necrosis start at the outer layer of the media of perforating arteries, which is exposed to hypoxia and hyponutrition. Macrophages invade the subintima probably through the chemotactic effect of necrotized mass, and induce massive infiltration of fibrin which often results in the occlusion of perforating arteries. This pathological process has been confirmed in humans also.

Immunohistochemical Studies in Human Stroke

In order to test this new hypothesis in humans, we extensively examined stroke victims who underwent autopsy within 5 h after death [35]. In hypertensive patients who had died of cerebral hemorrhage or infarction, the medial wall of the perforating artery was thickened and replaced with collagen. These intracerebral arteries were immunohistochemically examined using monoclonal antibodies against human VSMC actin (HHF35) and human macrophages (HAM56) [8,35]. Anti-human VSMC actin antibody reacted only in a thin layer near the intima, even though the vascular wall appeared thickened (Fig. 3). This means that in humans too, medial degeneration starts at the outer media.

Fig. 3a–d. Thickened intracerebral artery of a 66-year-old hypertensive man with cer- ▷ ebral infarction: **a** hematoxylin-eosin **b** Masson trichrome stain, **c** reaction to anti-human macrophage monoclonal antibody (HAM56), and **d** reaction to anti-human VSMC actin monoclonal antibody (HHF35)

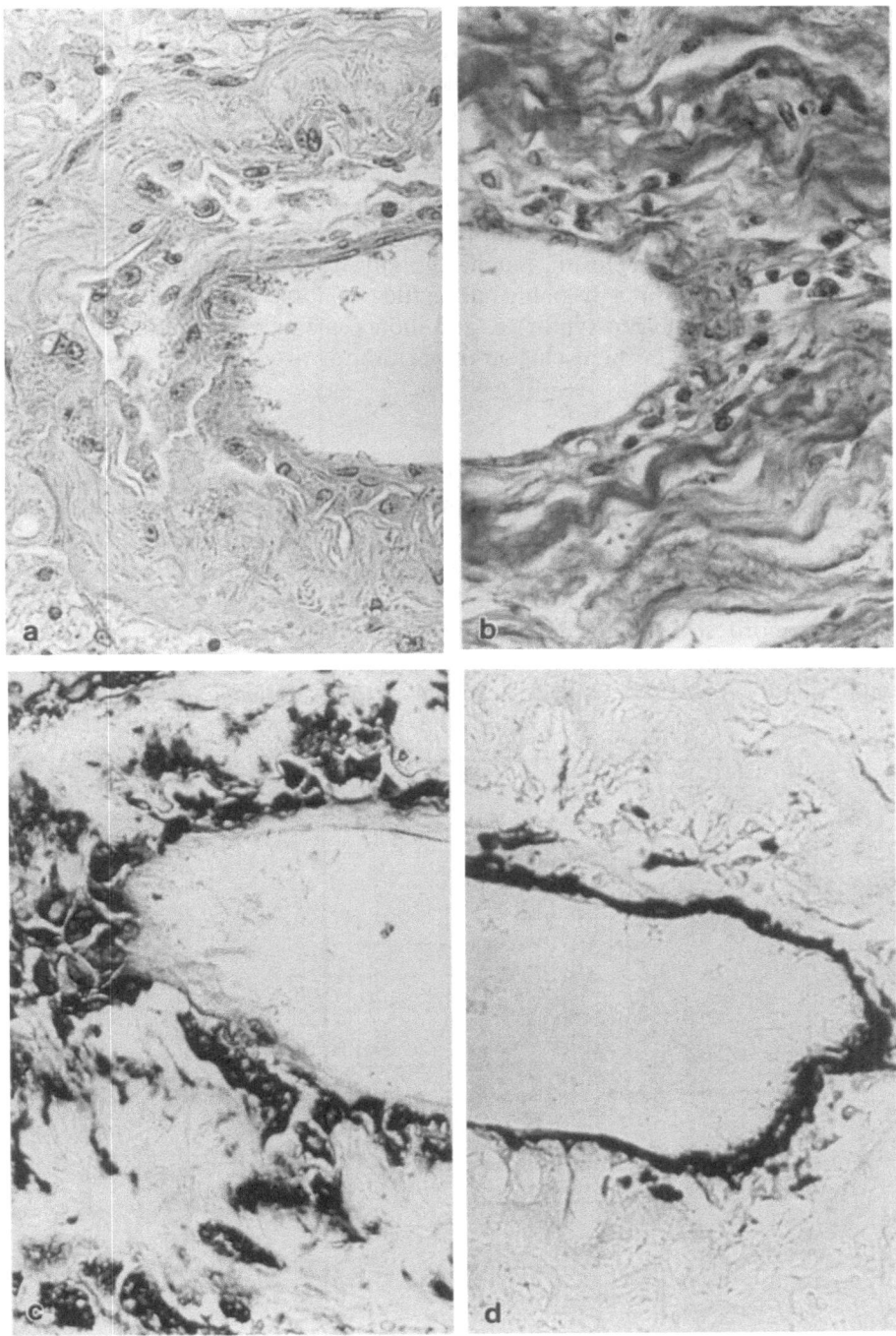

Anti-human macrophage antibody reacted with cells in the subintima and media and demonstrated massive infiltration of macrophages in the process of arterial lesions in hypertension. Thus, a similarity of pathology between stroke in SHRSP and in humans was clearly demonstrated.

From our immunohistochemical analyses of human autopsy cases, a new concept of the pathogenetic process of fibrinoid necrosis in hypertension can be proposed in human also. Focal degeneration and necrosis start at the outer layer of media which is exposed to hypoxia and hyponutrition in hypertension. Macrophages invade the media initially from Virchow-Robin's space and later through intima probably due to the chemotactic effect of necrotized mass, stimulate collagen synthesis, and induce massive infiltration of fibrin which often results in the occlusion of perforating arteries.

Cellular Basis of Stroke

Our further studies on cultured VSMC demonstrated that the effect of hypoxia was greater in cells from SHRSP than in those from normotensive WKY rats [35]. When the oxygen content was reduced from 20% to 5%, there was initially no difference in the change shown by cells from SHRSP and those from WKY, but 20 days later the VSMC from SHRSP showed more severe degeneration whereas those from WKY rats were still active with large intact nuclei. Therefore, we may conclude that vulnerability to hypoxia is also a cellular pathogenic mechanism of stroke.

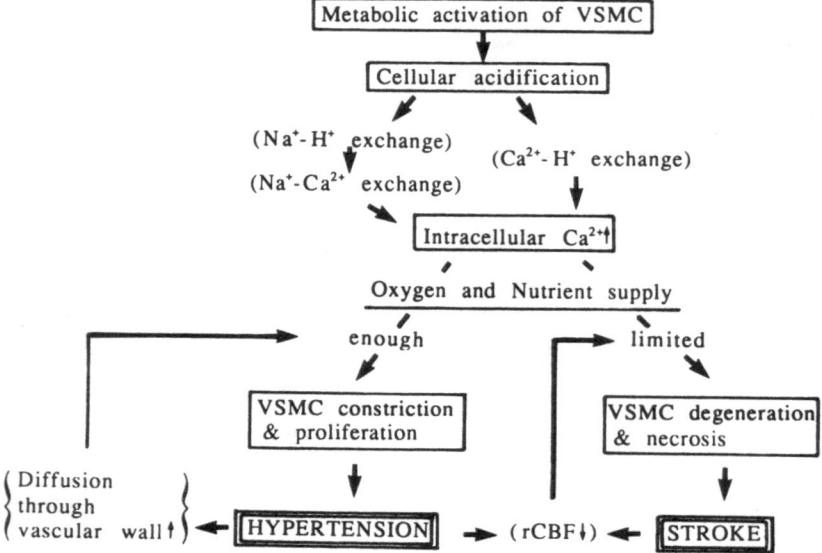

Fig. 4. Possible common cellular mechanisms of hypertension and stroke. *rCBF*, regional cerebral blood flow

These findings are consistent with our previous data indicating that VSMC have a faster growth rate, and thus a more active metabolism, when exposed to a sufficient supply of oxygen and nutrients. In fact, ATPase activity related to energy metabolism is greater in cultured VSMC from SHR and SHRSP than in those from WKY rats [35]. Since a metabolic inhibitor antimycin A attenuates the growth rate of cultured VSMC, particularly those from hypertensive strains [35], metabolic activation of VSMC may be involved in the pathogenic mechanism of hypertension, i.e., accelerated structural changes due to the faster growth of VSMC (Fig. 4).

VSMC in SHR show acidification, possibly due to metabolic activation. This secondarily activates Na^+-H^+ exchange and may presumably also activate Ca^{2+}-H^+ exchange to elevate cytosolic Ca^{2+} as Batlle et al. [2] proposed. Therefore, metabolic activation may enhance constriction of these cells under hypernutritional conditions.

Conclusion

The present findings suggest that metabolic activation of VSMC causes hypertension through the accelerated growth of VSMC when enough oxygen and nutrients are supplied. This can be come a "vicious cycle" because hypertension increases the nutrient supply to VSMC through augmented diffusion in the peripheral circulation. The same mechanism, however, may also contribute to the process of arterionecrosis in the brain. When regional nutrient and oxygen supply to intracerebral VSMC is reduced, especially in hypertension, these cells in SHRSP, whose metabolic demands are greater than those of WKY rats, are considered to be more vulnerable and prone to degenerate and fall into necrosis. Therefore, metabolic activation of VSMC may be a common cellular factor of hypertension and stroke.

References

1. Astarie C, Levenson J, Simon A, Meyer P, Devynck MA (1989) Platelet cytosolic proton and free calcium concentrations in essential hypertension. J Hypertens 7: 485–491
2. Batlle DC, Salah A, Rombola G (1990) Reduced intracellular pH in lymphocytes from the spontaneously hypertensive rat. Hypertension 15:97–103
3. Berk BC, Vallega G, Muslin AJ, Golden HM, Canessa M, Alexander RW (1989) Spontaneously hypertensive rat vascular smooth muscle cells in cultured exhibit increased growth and Na/H exchange. J Clin Invest 83:822–829
4. Chaldakov GN, Nara Y, Horie R, Yamori Y (1989) A new view of the arterial smooth muscle cells and autonomic nerve plexus by scanning electron microscopy in spontaneously hypertensive rats. Exp Pathol 36:181–184
5. Feig PV, D'Occhio MA, Boylan JW (1987) Lymphocyte membrane sodium-protein exchange in spontaneously hypertensive rats. Hypertension 9:282–288

6. Folkow B, Hallback M, Lundgren Y, Sivertsson R, Weiss L (1973) Importance of adaptive changes in vascular design for establishment of primary hypertension, studied in man and in spontaneously hypertensive rats. Circ Res 32 [Supply]:I2–16

7. Glaser L, Whiteley B (1987) Control of ion fluxes by mitogenic polypeptides. Hypertension 10 [Suppl I]:I-27–I-31

8. Gown AM, Tsukada T, Ross R (1986) Human atherosclerosis: immunocytochemical analysis of the cellular composition of human atherosclerotic lesions. Am J Pathol 125:191–207

9. Henrotte JG, Santanomana M, Bourdon R (1985) Concentration en magnesium, calcium et zinc du plasma et des erythrocytes de rats spontanement hypertendus. C R Acad Sci [III] 300:431–436

10. Izzard AS, Heagerty AM (1989) The measurement of internal pH in resistance arterioles: evidence that intracellular pH is more alkaline in SHR than WKY animals. J Hypertens 7:173–180

11. Kanbe T, Nara Y, Tagami M, Yamori Y (1983) Studies of hypertension-induced vascular hypertrophy in cultured smooth muscle cells from spontaneously hypertensive rats. Hypertension 5:887–892

12. Kobayashi A, Nara Y, Nishio T, Mori C, Yamori Y (1990) Increased Na^+-H^+ exchange activity in cultured vascular smooth muscle cells from stroke-prone spontaneously hypertensive rats. J Hypertens 8:153–157

13. Livne A, Balef JW, Veitch R, Marguez-Julio A, Grinstrin S, Rothstein A (1987) Increased platelet Na^+-H^+ exchange rates in essential hypertension: application of a novel test. Lancet I:533–536

14. Matlib MA, Schwartz A, Yamori Y (1985) A Na^+-Ca^{2+} exchange process in isolated sarcolemmal membrane of mesenteric arteries from WKY and SHR rats. Am J Physiol 249:C166–C172

15. Matsumoto K, Yamada T, Natori T, Ikeda K, Yamada J, Yamori Y (1991) Genetic variability in SHR SHRSR, SHRSP and WKY strains. Clin Exp Hypertens A13 (lin press)

16. Moolenar WH, Tertoolen LGJ (1984) The regulation of cytoplasmic pH in human fibroblasts. J Biol Chem 259:7563–7569

17. Mtabaji JP, Kihara M, Yamori Y (1985) Zinc and vascular reacivity in rat mesenteric vessels: possible altered dihomo-γ-linolenic acid metabolism in spontaneously hypertensive rats. Prostaglandins Leucotrienes Med 18:235–243

18. Nabika T, Nara Y, Endo J, Yamori Y (1986) Effects of Na^+ on kinetics of cell proliferation in cultured vascular smooth muscle cells. Hypertension 4 [Suppl 6]:S303–S305

19. Ng LL, Dudley C, Bomford J, Hawley D (1989) Leucocyte intracellular pH and Na/H antiport activity in human hypertension. J Hypertens 7:471–475

20. Okamoto K, Aoki K (1963) Development of a strain of spontaneously hypertensive rats. Jpn Circ J 27:282–293

21. Okamoto K, Yamori Y, Nagaoka A (1974) Establishment of the stroke-prone SHR. Circ Res 34/35 [Suppl 1]:143–153

22. Okamoto K, Yamamoto K, Morita N, Ohta Y, Chikugo T, Higashizawa T, Suzuki T (1986) Establishment and use of M strain of stroke-prone spontaneously hypertensive rat. J Hypertens 4 [Suppl 3]:S21–S24

23. Ooneda G (1970) Pathology of cerebral hemorrhage: focused on vascular lesions. Nippon Byori Gakkai Kaishi 59:27–56

24. Resnick LM, Gupta RI, Sosa RE, Corbett ML, Laragh JH (1987) Intracellular pH in human and experimental hypertension. Proc Natl Acad Sci USA 84:7663–7667

25. Resnick LM, Gupta RK, Gruenspan H, Laragh JH (1988) Intracellular free magnesium in hypertension: relation to peripheral insulin resistance. J Hypertens 6 [Suppl 4]:S199–S201

26. Tagami M, Kubota A, Kitamura J, Nara Y, Yamori Y (1986) Electron microscopic studies of ruptured and occluded arteries in stroke-prone spontaneously hypertensive rats. J Hypertens 4 [Suppl 3]:S413–S415

27. Tagami M, Nara Y, Kubota A, Sunaga T, Maezawa H, Fujino H, Yamori Y (1987) Ultrastructural characteristics of occluded perforating arteries in stroke-prone spontaneously hypertensive rats. Stroke 18:733–740

28. Thomas JA, Buchsbaum RN, Zimmiak A, Racker R (1979) Intracellular pH measurements in Ehrlich ascites tumor cells utilizing spectroscopic probes generated in situ. Biochemistry 18:2210–2218

29. Yamori Y (1974) Contribution of cardiovascular factors to the development of hypertension in spontaneously hypertensive rats. Jpn Heart J 15:194–196

30. Yamori Y (1976a) Interaction of neural and nonneural factors in the pathogenesis of spontaneous hypertension. In: Julius S, Esler M (eds) The nervous system in arterial hypertension. Thomas, Springfield, pp 17–50

31. Yamori Y (1976b) Neural and non-neural mechanisms in spontaneous hypertension. Clin Sci Mol Med 51:431s–434s

32. Yamori Y (1983) Physiopathology of the various strains of spontaneously hypertensive rats. In: Genest J, Kuchel O, Hamet P et al. (eds) Hypertension. McGraw-Hill, Montreal, pp 556–581

33. Yamori Y (1984a) Development of the spontaneously hypertensive rats (SHR) and of various spontaneous rat models, and their implications. In: de Jong W (ed) Experimental and genetic models of hypertension. Elsevier, Amsterdam, pp 224–239 (Handbook of hypertension, vol 4)

34. Yamori Y (1984b) The stroke-prone spontaneously hypertensive rat: contribution to risk factor analysis and prevention of hypertensive diseases. In: de Jong W (ed) Experimental and genetic models of hypertension. Elsevier, Amsterdam, pp 240–255 (Handbook of hypertension, vol 4)

35. Yamori Y (1989) Prediction and preventive pathology of cardiovascular diseases. Acta Pathol Jpn 39:683–705

36. Yamori Y, Horie R (1977a) Developmental course of hypertension and regional cerebral blood flow in stroke-prone spontaneously hypertensive rats. Stroke 8: 456–461

37. Yamori Y, Horie R (1977b) Vascular reactivity in pathological states. In: Carrier O Jr, Shibata S (eds) Factors influencing vascular reactivity. Igaku-Shoin, Tokyo, pp 268–281

38. Yamori Y, Ishino H (1987) Possible nutritional prevention of cerebrovascular dementia in the Japanese elderly: implications of an autopsy study. In: Yamori Y, Lenfant C (eds) Prevention of cardiovascular diseases: an approach to active long life. Elsevier, Amsterdam, pp 179–190

39. Yamori Y, Okamoto K (1970) Zymogram analysis of various organs from spontaneously hypertensive rats. Lab Invest 22:206–211

40. Yamori Y, Nagaoka A, Okamoto K (1974) Importance of genetic factors in hypertensive cerebrovascular lesions: an evidence obtained by successive selective breeding of stroke-prone and -resistant SHR. Jpn Circ J 38:1095–1100

41. Yamori Y, Horie R, Sato M, Handa H (1976) Pathogenetic similarity of stroke-prone SHR and humans. Stroke 7:46–53

42. Yamori Y, Ohta K, Horie R, Ohtaka M, Nara Y, Oosha A (1979) A new model for cerebral thrombosis and its pathogenesis. Jpn Heart J 20 [Suppl I]:343–345

43. Yamori Y, Nara Y, Horie R, Ooshima A (1980) Abnormal membrane characteristics of erythrocytes in rat models and men with predisposition to stroke. Clin Exp Hypertens 2:1009–1021

44. Yamori Y, Kihara M, Nara Y (1982a) Myocardial-ischemic rats (MIR): coronary vascular alteration induced by a lipid-rich diet. Atherosclerosis 42:15–20

45. Yamori Y, Nara Y, Kanbe T, Imafuku H, Mori K, Kihara M, Horie R (1982b) Diversity of membrane abnormalities in spontaneous hypertension. Clin Sci 63 [Suppl 8]:27s–29s
46. Yamori Y, Nara Y, Imafuku H (1984a) Biomembrane abnormalities in spontaneous hypertension. In: Villarreal H, Sambhi MP (eds) Topics in pathophysiology of hypertension. Martinus Nijhoff, Boston, pp 3–13
47. Yamori Y, Igawa T, Tagami M, Kanbe T, Nara Y, Kihara M, Horie R (1984b) Humoral trophic influence on cardiovascular structural changes in hypertension. Hypertension 6 [Suppl III]:III-27–III-32
48. Yamori Y, Ikeda K, Kulakowski EC, McCarty R, Lovenberg W (1985) Enhanced sympathetic-adrenal medullary response to cold exposure in spontaneously hypertensive rats. J Hypertens 3:63–66
49. Yamori Y, Horie R, Nara Y, Tagami M, Kihara, M, Mano M, Ishino H (1987) Pathogenesis and dietary prevention of cerebrovascular diseases in animal models and epidemiological evidence for the applicability in man. In: Yamori Y, Lenfant C (eds) Prevention of cardiovascular diseases: an approach to active long life. Elsevier, Amsterdam, pp 163–177
50. Yamori Y, Igawa T, Kanbe T, Nara Y, Tagami M (1988) Enhanced growth rate of cultured smooth muscle cells from spontaneously hypertensive rats. Heart Vessels 4:94–99
51. Yamori Y, Nara Y, Shimizu S, Mano M, Nabika T (1989a) Common cellular mechanisms in the development of hypertension and atherosclerosis. In: Meyer P, Marche P (eds) Blood cells and arteries in hypertension and atherosclerosis. Raven, New York, pp 233–246
52. Yamori Y, Nara Y, Ikeda K, Tsuchikura S, Eguchi T, Mano M, Horie R, Sugahara T (1989b) Recent advances in experimental studies on dietary prevention of cardiovascular diseases. In: Yamori Y, Strasser T (eds) New horizons in preventing cardiovascular diseases. Elsevier, Amsterdam, pp 1–11
53. Yamori Y, Nara Y, Horie R, Morii F, Nishima T, Nomiyama H, Nomiyama K (1990) Experimental and epidemiological studies on the relation of trace elements in the pathogenesis and prevention of cardiovascular diseases. In: Tomita H (ed) Trace elements in clinical medicine. Springer, Berlin Heidelberg New York, pp 65–75

Hypertrophic Growth of Vascular Smooth Muscle*

G.K. Owens

Department of Physiology, University of Virgina School of Medicine, Charlottesville,
VA 22908, USA

Introduction

Smooth muscle cell (SMC) growth is increased in arteries of hypertensive
patients and animals [1,12,24,30,36]. This increased growth contributes to
development of medial thickening in resistance vessels [1,10] and has been
hypothesized to play an important role in the etiology of the hypertension by
conferring a geometric advantage to the thickened vessel such that vascular
resistance is greater at any given level of smooth muscle activation [9]. In
contrast, in a large conduit vessels, accelerated smooth muscle growth is
believed to represent an adaptive process to minimize changes in wall stress
resulting from increased blood pressure [27,28]. Whereas accelerated growth
of arterial smooth muscle cells is a common feature of virtually all hypertensive models, the cellular nature of the growth response, with regard to the
contribution of cellular hypertrophy versus hyperplasia, appears to vary as a
function of the vascular site examined and/or the hypertensive model studied
[29]. For example, aortic medial thickening in the spontaneously hypertensive
rat (SHR) is due primarily to enlargement of preexisting SMCs or cellular
hypertrophy, and is accompanied by development of polyploidy in a large
fraction of hypertrophic SMCs [29,32,33]. In contrast, medial thickening in
intermediate-size resistance vessels in the SHR and in the aorta of rats made
hypertensive by aortic coarctation is due to SMC hyperplasia [31].
  Whereas the factors that mediate hypertrophic versus hyperplastic
growth of SMC have not been clearly identified, there is good evidence that
the regulatory signals for these growth processes are different, and that the
growth response of a given SMC is a function of the nature of the growth
stimulus rather than an inherent property of the SMC (see review, [29]). As
such, a critical first step in elucidating factors that regulate the increased
growth of SMC in hypertension is to definitively ascertain the nature of the

* This work was supported by Public Health Service Grants P01-HL19242 and R01-HL
  38854 from the National Institutes of Health.

G. Bruschi A. Borghetti (Eds.)
Cellular Aspects of Hypertension
© Springer-Verlag Berlin · Heidelberg 1991

cellular growth response *in the blood vessel and hypertensive model of interest.* This paper will focus on examination of possible mechanisms that regulate hypertrophy of vascular smooth muscle in hypertension, with particular emphasis on examination of the role of contractile agonists as growth mediators and their mechanism of action.

Factors Implicated in the Control of SMC Hypertrophy

There is circumstantial evidence suggesting that SMC hypertrophy may be a response to increased blood pressure or wall stress. Hypertrophy of aortic SMC occurs predominantly after blood pressure has increased [27,28]. Our laboratory [27,28] and others [23] have consistently observed a high correlation between the level and duration of blood pressure elevation and the extent of vascular SMC hypertrophy. Normalization of blood pressure in hypertensive animals by treatment with antihypertensive drugs is effective in preventing further development of SMC hypertrophy and hyperploidy as well as in reversing SMC hypertrophy, but does not reverse changes in ploidy, at least over a treatment period of several months [11,28]. Additional studies have shown that protection of vascular beds by partial ligation of upstream supply vessels prevents the development of medial hypertrophy associated with hypertension [4]. Wolinsky [47,48] demonstrated that vessel wall thickness in both hypertensive and normotensive animals, and indeed the number of lamellar units in aortas from different species, correlates closely with wall stress, indicating that vessel wall stress may be regulated by vascular growth and remodeling. Glagov and co-workers [22,42], demonstrated that mechanical stretch increased overall protein synthesis as well as connective tissue synthesis in cultured SMCs, although their studies did not address the question whether these changes were associated with cellular hypertrophy. While these studies support a possible role for mechanical factors in mediation of SMC hypertrophy, almost nothing is known regarding how mechanical stresses might be transduced into a growth signal. One intriguing possibility is that stretch-activated ion channels could be involved in the transduction pathway. However, this is speculative, and it is clear that further studies are needed in this important area.

Whereas mechanical factors are likely to be involved in control of SMC hypertrophy, results of drug treatment studies in the SHR indicate that development of SMC hypertrophy is not simply a response to increased blood pressure [11,28]. For example, we observed that captopril was more effective than hydralazine in preventing increases in aortic SMC content, medial smooth muscle weight, and SMC polyploidy in SHR, although the drugs induced nearly identical blood pressure lowering [28]. These observations suggest that factors other than blood pressure per se are involved in the SMC hypertrophic response and imply a possible role for angiotensin II in the growth response. While the SHR is not, on the basis of measurements of circulating renin [45], considered to be a renin-dependent form of hyper-tension, there is increased vascular renin activity in the SHR compared to the WKy rat [3,14], and the efficacy of captopril in reducing blood pressure in SHR is thought to be due to inhibition of local angiotensin II formation [2]. Indeed, there is considerable evidence for the existence of a local vas-cular renin-angiotensin system [8]. Captopril also significantly reduced SMC size and medial smooth muscle content in WKy animals, suggesting that angiotensin II may play some role in the regulation of vascular smooth muscle mass in normotensive strains as well.

Mechanisms Whereby Contractile Agonists Induce Smooth Muscle Cell Hypertrophy

There are a variety of mechanisms whereby angiotensin II could influence vascular SMC growth. It might stimulate growth indirectly through release of neuropeptides [2] or polypeptide growth factors from non-SMC, as well as through alterations in mechanical forces. Alternatively, it may stimulate SMC growth directly. In results consistent with this, we have demonstrated that angiotensin II [16] as well as arginine vasopressin [15] induce increased protein synthesis and cellular hypertrophy in cultured rat aortic SMC made quiescent in a defined serum-free medium, but had no mitogenic activity either alone or in combination with serum or platelet-derived growth factor (PDGF). Our results in rat aortic SMC have been confirmed by several other groups including Scott-Burden et al. [40] and Berk et al. [5]. Campbell-Boswell and Robertson [6] also found that angiotensin II treatment induced cellular hypertrophy in human aortic SMC in medium containing high concen-trations of serum, although these investigators also reported that angiotensin II induced increases in number of SMCs. However, proliferative effects were extremely modest, and there is agreement that angiotensin II is at best a very weak mitogen for cultured SMC as compared to polypeptide growth factors such as platelet-derived growth factor (PDGF) [38] or fibroblast growth factor (FGF) [17]. One must consider, however, that the growth response of SMCs to angiotensin II may vary as a function of the differ-entiated state of the cell. As such, cultured SMCs as well as intimal SMCs, both of which are phenotypically modulated [7,13,34], may show altered growth responses to angiotensin II as compared to fully differentiated SMCs. This might explain observations by Powell et al. [37] that the converting enzyme inhibitor cilazapril was extremely potent in inhibiting *proliferation* of intimal SMCs following balloon catheter-induced vascular injury, although it is unclear from these studies whether angiotensin II directly stimulated SMC proliferation or whether effects were indirect. The proliferative growth response to angiotensin II observed in some culture studies but not others may also be due to differences in the differentiational status of the SMCs employed. Under the conditions used in our studies, rat aortic SMCs express high levels of smooth-muscle-specific contractile proteins including smooth muscle α-actin [34] and smooth muscle myosin heavy chain [39], while the explant-derived SMCs used by Campbell-Boswell et al. [6] were most likely highly modulated and expressed little, if any, of these differentiated proteins [7]. Thus, the proliferative response observed in some studies may reflect increased growth responsiveness of SMCs that are highly modulated due to loss of growth suppressor mechanisms associated with cellular differentiation. In any event, further studies are needed to determine the effect of the SMCs' differentiated state on contractile agonist-induced growth responses, and the relevance of findings in cultured SMCs to regulation of hypertrophy in vivo.

Relatively little is known regarding how angiotensin II and arginine vasopressin stimulate increased protein synthesis [29]. We (unpublished observations) and Berk et al. [5] have demonstrated that angiotensin II-induced increases in protein synthesis can be blocked with α-amanitin or actinomycin D. Although these data suggest that there is a transcriptional dependent event in the response, they provide no definitive insight as to what this might be, and are subject to criticisms regarding non-specific actions of these drugs. Angiotensin II has been shown to stimulate marked increases in expression of the protooncogenes c-fos [19,20,43] and c-myc [25], raising the possibility that the effects of angiotensin II on growth are mediated by the protein products of these genes. However, while these are clearly growth-related genes, the function of their protein products in normal cells have not been clearly defined [18]. Furthermore, there is some evidence suggesting that angiotensin II-induced changes in fos can be dissociated from increases in protein synthesis. Taubman et al. [43] found that while phorbol ester-induced downregulation of protein kinase C nearly totally blocked angiotensin II induction of fos, it had no effect on angiotensin II-induced increases in protein synthesis. While one cannot rule out the possibility that the small increase in fos that occurs in phorbol-pretreated cells is sufficient to mediate the protein synthetic response, there is at present no direct evidence supporting a major role for fos in the angiotensin II-induced growth response. These observations also suggest that angiotensin II-induced increases in protein synthesis are not dependent on protein kinase C, although the downregulation of protein kinase C in these experiments was only about 50%.

Scott-Burden et al. [40] suggested that angiotensin II-induced increases in protein synthesis may be mediated by alterations in S_6 kinase activity. These investigators demonstrated that angiotensin II activates an S_6 kinase activity in cultured rat aortic SMCs. Phosphorylation of the ribosomal S_6 protein is an early response of many cells to growth factors, and there is evidence that it may play a role in growth factor-induced increases in protein synthesis [44]. However, observations that activation of S_6 kinase activity in these cells was nearly blocked by amiloride [40], an inhibitor of Na^+/H^+ exchange, whereas angiotensin II-induced increases in protein synthesis were unaffected by treatment with dimethylamiloride or by prevention of angiotensin II-induced intracellular alkalinization using Na^+ free media [5] are inconsistent with this hypothesis.

There is some evidence suggesting that increases in intracellular calcium ion concentration may play an important role in mediation of angiotensin II-induced hypertrophy. Berk et al. [5] demonstrated that angiotensin II-induced increases in protein synthesis were markedly inhibited by chelation of intracellular Ca^{2+} with quin 2. However, increasing intracellular Ca^{2+} with ionophore had little effect on protein synthesis, suggesting that Ca^{2+} may play a permissive rather than a regulatory role in control of SMC hypertrophy. This would not be surprising, since protein synthesis involves many Ca^{2+}-dependent processes. While further studies of the role of Ca^{2+} in regulation of hypertrophy are needed, they will require development of markers of hypertrophy other than protein synthesis.

Dzau and co-workers [25,26] presented evidence suggesting that angiotensin II-induced increases in protein synthesis occurred secondary to increases in PDGF A-chain expression by SMCs. These investigators found that angiotensin II increased expression of PDGF A-chain mRNA but not PDGF B-chain in cultured rat aortic SMCs. Increases were dependent on protein synthesis, were evident within 6 h following angiotensin II stimulation, and were accompanied by an increase in PDGF secretion. Since no B-chain PDGF mRNA was detectable, these observations raise the interesting possibility that angiotensin II may increase protein synthesis by stimulating production of PDGF AA homodimer. However, PDGF AA has been reported to stimulate proliferation, not hypertrophy, in a variety of cultured cells [21], including SMCs (D. Bowen-Pope, personal communication). Thus, either multiple control processes must be operating that convert a PDGF AA-induced proliferative response to a hypertrophic response, or the rat aortic SMC cultures employed in these studies are not mitogenically responsive to PDGF AA. Indeed, there is some evidence to support the former: Scott-Burton et al. [41] and Dzau et al. (personal communication) have demonstrated that angiotensin II increased expression of active TGF-β in cultured rat aortic SMCs, while we have demonstrated that TGF-β-induced growth inhibition in serum containing medium is associated with development of SMC hypertrophy [35].

A recurring weakness in the preceding studies is that it is often difficult, if not impossible, to determine whether early changes in second messenger systems or protoocogene expression are causal, permissive, or irrelevant to the chronic changes in protein synthesis and metabolism that lead to cellular hypertrophy. Thus, we have employed an alternative strategy to explore the possible cellular and molecular mechanisms whereby contractile agonists induce SMC hypertrophy, involving first identifying the major proteins that accumulate in hypertrophic cells (i.e., the initial emphasis is on identifying structural rather than regulatory proteins), and then determining the molecular mechanisms that contribute to their accumulation. Table 1 sum-

Table 1. Possible mechanisms whereby angiotensin II and other contractile agonists increase protein synthesis

---

A. Post-transcriptional/translational controls:
  1. Increase in the rate or efficiency of translation
  2. Increase in mRNA stability
  3. Mobilization of a pool of previously untranslated mRNA
  4. Increased number of protein copies per mRNA molecule
B. Transcriptional controls:
  1. Increased transcription of rRNAs resulting in increased protein synthetic capability
  2. Increased transcription of mRNAs

---

marizes some of the possible molecular mechanisms that might be involved. Note that any one or a combination of these processes could be operative.

Utilizing a combination of one- and two-dimensional gel electrophoretic analyses, we demonstrated that angiotensin II and arginine vasopressin-induced hypertrophy of rat aortic SMCs is associated with widespread but selective increases in cellular proteins. Increases in the content of certain proteins, including a number of smooth muscle-specific contractile proteins such as smooth muscle α-actin, smooth muscle tropomyosin, and smooth muscle myosin heavy chain, far exceeded overall increases in cellular protein content [46]. Our results also demonstrated that increases in smooth muscle α-actin were accompanied by increases in mRNA content, demonstrating that increases in actin content were not mediated solely at the translational level. It is not known whether increases in smooth muscle α-actin mRNA are due to increased gene transcription or to increased mRNA stability. If, however, as expected, changes are due to increased transcription, it should then be possible to identify the contractile agonist responsive elements of the smooth muscle α-actin gene and to utilize these DNA regulatory elements as a means to isolate and characterize important regulatory molecules involved in contractile agonist-induced SMC hypertrophy.

References

1. Aalkjaer C, Heagerty AM, Petersen KK, Swales JD, Mulvany MJ (1987) Evidence for increased media thickness, increased neuronal amine uptake, and depressed excitation – contraction coupling in isolated resistance vessels from essential hypertensives. Circ Res 61(2):181–186
2. Antonaccio M, Kerwin L (1981) Pre- and postjunctional inhibition of vascular sympathetic function by captopril in SHR. Hypertension 3:54–62
3. Asaad M, Antonaccio M (1982) Vascular wall renin in spontaneously hypertensive rats: potential relevance to hypertension maintenance and antihypertensive effect of captopril. Hypertension 4:487–493
4. Berecek K, Bohr D (1977) Structural and functional changes in vascular resistance and reactivity in the deoxycorticosterone acetate (DOCA-) hypertensive pig. Circ Res 40 [Suppl I]:146–152

5. Berk BC, Vekshtein V, Gordon HM, Tsuda T (1989) Angiotensin II-stimulated protein synthesis in cultured vascular smooth muscle cells. Hypertension 13(4): 305–314
6. Campbell-Boswell M, Robertson A (1981) Effects of angiotensin II and vasopressin on human smooth muscle cells in vitro. Exp Mol Pathrol 35:265–276
7. Chamley-Campbell J, Campbell G, Ross R (1979) The smooth muscle cell in culture. Physiol Rev 59:1–61
8. Dzau V (1986) Significance of the vascular renin-angiotensin pathway. Hypertension 8:553–559
9. Folkow B (1979) Constriction-distension relationshios of resistance vessels in normo- and hyper-tension. Clin Sci 57:23s–25s
10. Folkow B (1982) Physiological aspects of primary hypertension. Physiol Rev 62: 347–504
11. Freslon J, Giudicelli J (1983) Compared myocardial and vascular effects of captopril and dihydralazine during hypertension development in spontaneously hypertensive rats. Br J Pharmacol 80:533–543
12. Furuyama M (1962) Histometrical investigations of arteries in reference to arterial hypertension. Tohoku J Exp Med 76:388–414
13. Gabbiani G, Kocher O, Bloom W, Vandekerckhove J, Weber K (1984) Actin expression in smooth muscle cells of rat aortic intimal thickening, human atheromatous plaque, and cultured rat aortic media. J Clin Invest 73:148–152
14. Garst J, Koletsky S, Weisenbaugh P, Hadady M, Matthew D (1979) Arterial wall renin and renal venous renin in the hypertensive rats. Clin Sci Mol Med 56:41–46
15. Geisterfer A, Owens G (1989) Arginine vasopressin induced hypertrophy of cultured rat aortic smooth muscle cells. Hypertension 14:413–420
16. Geisterfer A, Peach MJ, Owens GK (1988) Angiotensin II induces hypertrophy, not hyperplasia of cultured rat aortic smooth muscle cells. Circ Res 62:749–756
17. Gospodarowicz D, Hirabayashi K, Giguere L, Tauber JP (1981) Factors controlling the proliferative rate, final cell density, and life span of bovine vascular smooth muscle cells in culture. J Cell Biol 89:568–578
18. Kaczmarek L (1986) Protooncogene expression during the cell cycle. Lab Invest 54(4):365–376
19. Kawahara Y, Kariya K, Araki S, Fukuzaki H, Takai Y (1988) Platelet-derived growth factor (PDGF)-induced phospholipase C-mediated hydrolysis of phosphoinositides in vascular smooth muscle cells: different sensitivity of PDGF- and angiotensin II-induced phospholipase C reactions to protein kinase C-activating phorbol esters. Biochem Biophys Res Commun 156(2):846–854
20. Kawahara Y, Sunako M, Tsuda T. Fukuzaki H, Fukumoto Y, Takai Y (1988) Angiotensin II induces expression of the *c-fos* gene through protein kinase C activation and calcium ion mobilization in cultured vascular smooth muscle cells. Biochem Biophys Res Commun 150(1):52–59
21. Kazlauskas A, Bowen PD, Seifert R, Hart, CE, Cooper JA (1988) Different effects of homo- and heterodimers of platelet-derived growth factor A and B chains on human and mouse fibroblasts. EMBO J 7(12):3727–3735
22. Leung D, Glagov S, Mathews M (1977) A new in vitro method for studying cell responses to mechanical stimulation: different effects of cyclic stretching and agitation on smooth muscle cell biosynthesis. Exp Cell Res 109:285–298
23. Lichtenstein A, Brecher P, Chobanian A (1986) Effects of hypertension and its reversal on the size and DNA content of rat aortic smooth muscle cells. Hypertension 8:1150–1154
24. Mulvany M, Baandrup U, Gundersen H (1985) Evidence for hyperplasia in mesenteric resistance vessels of spontaneously hypertensive rats using a three-dimensional disector. Circ Res 57:794–800

25. Naftilan AJ, Pratt RE, Dzau VJ (1989) Induction of platelet-derived growth factor A-chain and *c-myc* gene expressions by angiotensin II in gulture rat vascular smooth muscle cells. J Clin Invest 83:1419–1424
26. Naftilan AJ, Pratt RE, Eldridge CS, Lin HL, Dzau VJ (1989) Angiotensin II induces *c-fos* expression in smooth muscle via transcriptional control. Hypertension 13: 706–711
27. Owens G (1985) Differential effects of antihypertensive therapy on vascular smooth muscle cell hypertrophy, hyperploidy, and hyperplasia in the spontaneously hypertensive rat. Circ Res 56:525–536
28. Owens GK (1987) Influence of blood pressure on development of aortic medial smooth muscle hypertrophy in spontaneously hypertensive rats. Hypertension 9(2): 178–187
29. Owens G (1989) Control of hypertrophic versus hyperplastic growth of vascular smooth muscle cells. Am J Physiol 26:H1755–1765
30. Owens GK (1989) Growth response of aortic smooth muscle cells in hypertension. In: Blood vessel changes in hypertension: structure and function. Lee MKW (ed) CRC Press, Boca Raton, pp 45–63
31. Owens G, Reidy M (1985) Hyperplastic growth response of vascular smooth muscle cells following induction of acute hypertension in rats by aortic coarctation. Circ Res 57:695–705
32. Owens G, Schwartz S (1982) Alterations in vascular smooth muscle mass in the spontaneously hypertensive rat. Role of cellular hypertrophy, hyperploidy, and hyperplasia. Circ Res 51:280–289
33. Owens G, Rabinovitch P, Schwartz S (1981) Smooth muscle cell hypertrophy versus hyperplasia in hypertension. Proc Natl Acad Sci USA 78:7759–7763
34. Owens G, Loeb A, Gordon D, Thompson M (1986) Expression of smooth muscle specific *a*-isoactin in cultured vascular smooth muscle cells: relationship between growth and cytodifferentiation. J Cell Biol 102:343–352
35. Owens G, Geisterfer A, Yang Y, Komoriya A (1988) Transforming growth factor beta-induced growth inhibition and cellular hypertrophy in cultured vascular smooth muscle cells. J Cell Biol 107:771–780
36. Owens GK, Schwartz SM, McCanna M (1988) Evaluation of medial hypertrophy in resistance vessels of spontaneously hypertensive rats. Hypertension 11(2):198–207
37. Powell JS, Clozel J, Muller KM, Kuhn H, Hefti F, Hosang M, Baumgartner H (1989) Inhibitors of angiotensin-converting enzyme prevent myointimal proliferation after vascular injury. Science 245:186–188
38. Ross R (1987) Platelet-derived growth factor. Annu Rev Med 38:71–79
39. Rovner A, Murphy R, Owens G (1986) Expression of smooth muscle and non-muscle myosin heavy chains in cultured vascular smooth muscle cells. J Biol Chem 261: 14740–14745
40. Scott BT, Resink TJ, Baur U, Burgin M, Buhler FR (1988) Amiloride sensitive activation of S_6 kinase by angiotensin II in cultured vascular smooth muscle cells. Biochem Biophys Res Commun 151(1):583–589
41. Scott BT, Resink TJ, Hahn AP, Buhler FR (1989) Differential stimulation of growth related metabolism in cultured smooth muscle cells from SHR and WKY rats by combinations of EGF and LDL. Biochem Biophys Res Commun 159(2):624–632
42. Sottiurai V, Kollros P, Glagov S, Zarins C, Mathews M (1983) Morphologic alteration of cultured arterial smooth muscle cells by cyclic stretching. J Surg Res 35:490–497
43. Taubman MB, Berk BC, Izumo S, Tsuda T, Alexander RW, Nadal GB (1989) Angiotensin II induces *c-fos* mRNA in aortic smooth muscle. Role of Ca2+ mobilization and protein kinase C activation. J Biol Chem 264(1):526–530
44. Thomas G, Perez-Martin J, Siegmann M, Otto A (1982) The effect of serum, EGF, PGF_{2a} and insulin on S_6 phosphorylation and the initiation of protein and DNA synthesis. Cell 30:235–242

45. Trippodo N, Frolich E (1981) Similarities of genetic (spontaneous) hypertension: man and rat. Circ Res 48:309–319
46. Turla MB, Thompson MM, Corjay, MH, Owens GK (1991) Mechanisms of angiotensin II and arginine-vasopressin-induced increases in protein synthesis and content in cultured rat aortic smooth muscle cells. Circ Res 68:288–299
47. Wolinsky H (1970) Response of the rat aortic media to hypertension. Circ Res 26:507–607
48. Wolinsky H, Glagov S (1969) Comparison of abdominal and thoracic aortic medial structure in mammals. Circ Res 25:677–688

Biomechanical and Electrical Responses of Normal and Hypertensive Veins to Short-Term Pressure Increases*

E. Monos[1], K. Kauser[1], S.J. Contney[2], A.W. Cowley Jr.[2], and W.J. Stekiel[2]

[1] Experimental Research Department and Second Institute of Physiology, Semmelweis University of Medicine, Ulloi ut. 78/a, H-1082 Budapest, Hungary
[2] Department of Physiology, Medical College of Wisconsin, Milwaukee, WI 53226, USA

Observation of elevated central venous pressure or systemic filling pressure in various forms of arterial hypertension strongly suggests that systemic veins may participate in inducing and/or maintaining the hypertensive state [1,12]. However, sufficient information is not available to identify the mechanism(s) by which veins could contribute to the development and/or maintenance of chronic arterial pressure elevation [6]. It is plausible to hypothesize that some basic properties of the smooth muscle – such as intrinsic myogenic reactivity – in the systemic veins are altered in the hypertensive state, for then even a relatively small enhancement of stretch-induced intrinsic tone of the vessel wall could contribute significantly to the hemodynamic changes observed in arterial hypertension. Augmentation of this response may lead to the increase of central cardiopulmonary blood volume, postcapillary resistance, and the reduction of the pressure-buffer capacity of the venous system. This hypothesis is encouraged by our recent data demonstrating that there is an enhanced pressure-induced myogenic tone in isolated, small ($100-150$ μm internal diameter) gracilis arteries from reduced renal mass (RRM) rats – a model of volume-expanded hypertension – relative to non-RRM controls [13]. In accordance with these data, results recently published by Mulvany [10] suggest that increased intrinsic oscillatory activity of the mesenteric resistance vessels of spontaneous hypertensive rats plays a part in the development of high blood pressure.

The current view in the literature, based on reports studying the intrinsic behavior of systemic veins and finding little or no pressure-induced myogenic response, is that myogenic activity plays an insignificant role in the overall ability of the systemic veins to regulate capacity [14]. One of the possible causes of the failure, however, could be the vulnerable thin-walled nature of

* This work was supported by National Heart, Lung, and Blood Institute Research Grant HL-29587, by National Science Foundation Grant INT-8908904 (USA), and by OTKA-1/1314/1988 (Hungary).

G. Bruschi A. Borghetti (Eds.)
Cellular Aspects of Hypertension
© Springer-Verlag Berlin · Heidelberg 1991

these vessels, especially of the smaller veins, and the methodological difficulties encountered in making such measurements successfully [5,11].

Thus, the purpose of this study was to characterize quantitatively the intrinsic myogenic properties of systemic veins from normal rats using precise biomechanical and electrophysiological measurements during in vitro short-term pressure loads, and to define changes in this myogenicity after chronic exposure of veins in vivo to gravitationally elevated pressure and after hemodynamic loading caused by volume-expanded hypertension.

Materials and Methods

In Vitro *Short-Term Pressure Load*

The distal part of the isolated great saphenous vein from anesthetized Sprague-Dawley (SD) rats of 300–400 g body weight was chosen for the studies. This is a superficial, nearly straight vessel with only a few side branches draining predominantly skeletal muscle beds (outside diameter: 400–500 μm at 1 mmH$_2$O). The cylindrical vein segments, cannulated at both ends and extended to their in vivo length, were perfused (0.2 ml/min) and superfused (4 ml/min) with modified, carbonated Krebs-Henseleit solution (KHS) in a specially designed double-walled tissue chamber (25 ml) at 37°C. Intraluminal pressure was regulated by height adjustment of the outflow tubing and set at constant levels between 0 and 20 mmHg by 5 mmHg stepwise increases. Biomechanical parameters of the vessels (active and passive strains, stresses, distensibility, etc.) were calculated from vessel diameter (measured via videomicroscopy), wall thickness and intraluminal pressure using standard mechanical equilibrium equations. To determine the active, smooth muscle-related responses of the veins, differences between the total biomechanical response (vessels perfused with normal KHS) and the passive response (relaxed vessels perfused with calcium-free, magnesium-rich KHS) were compared. Membrane potential (E_m) of the venous smooth muscle cells was measured with intracellular glass microelectrodes (tip diameter: ~0.1 μm; tip resistance: 60–100 MΩ) by repeated impalements from the adventitial side of the vessel.

In Vivo *Gravitationally Elevated Venous Pressure Load*

In order to increase the hind limb venous pressure chronically without surgical manipulations, single animals were kept in a tube-like tiltable plastic cage at a 45° angle above the horizontal level for 2 weeks. The animals could eat, drink, and walk up and down in a head-up tilted position without turning around. This position resulted in a steady two-fold increase in the mean venous pressure measured in the femoral vein (2.9 ± 0.2 mmHg vs. 5.9 ± 0.2 mmHg, in conscious animals) without significantly influencing the arterial pressure. Rats kept in identical cages in horizontal position for 2 weeks served as parallel controls. After the 2-week period of tilt or horizontal restriction, rats were anesthetized and saphenous vein segments dissected from them for the in vitro biomechanical and electrophysiological studies, which followed a protocol similar to the one described above.

In an additional series of measurements, in situ obtained E_m values of saphenous venous smooth muscle cells were compared with those obtained from brachial veins in six of the tilted and six of the horizontally restricted rats. The measurements were repeated under local neural blockade by tetrodotoxin (TTX) administered via KHS superfusion.

Further details of the above methods can be found in the literature [7,8].

Sustained Hemodynamic Load by Volume-Expanded Hypertension

The model of volume-expanded arterial hypertension was developed using RRM rats. The model consisted of normal SD rats that had undergone a two-stage 75% surgical reduction of the renal mass under pentobarbital anesthesia. In the first stage 50% of the left kidney (two polar regions) was removed, then after a 2-week recovery period, a right-sided total nephrectomy was performed. After 1 week recovery, these RRM rats were maintained on a 4% NaCl diet with ad libitum water for 6 weeks [13]. We found that saphenous vein pressure was elevated in these chronically hypertensive rats compared to the control age-matched group (1 mmHg vs. 1.75 ± 0.21 mmHg, under anesthesia). At the end of the 6th week, animals were anesthetized and in a similar way to the previous protocols the carefully dissected saphenous veins were studied in vitro.

Results

Pressure-Induced Myogenic Responses of Saphenous Veins from Normal Rats

Isolated perfused and superfused saphenous vein segments exhibited substantial myogenic tone generation to acute pressurization. Perfusion with relaxing KHS resulted in a significantly larger increase of diameter in response to changes in pressure compared to the increase measured with normal KHS. Elevation of the intraluminal pressure induced significant active tangential isometric stress (13.6 ± 5.4 dyn/cm^2 at a normalized midradius of 1.5 mm) and active isobaric tangential strain (max.: $-13.1 \pm 0\%$ in response to the first 5 mmHg pressure step). The pressure-induced activation of the venous smooth muscle decreased the lumen capacity by 32% at 5 mmHg and 27% at 20 mmHg relative to vessels in relaxing KHS.

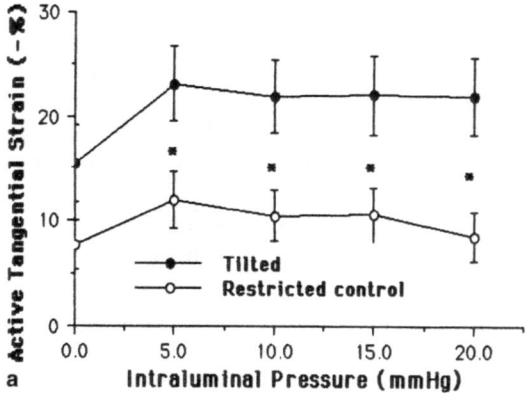

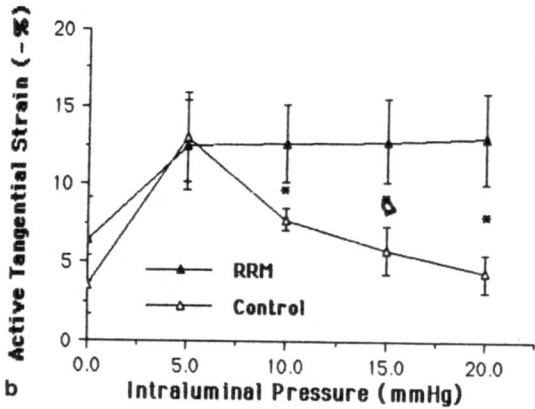

Fig. 1a,b. Active tangential isobaric strain responses (mean ± SEM) to step increases in intraluminal pressure from studies with saphenous veins of tilted rats (**a**) and of RRM rats (**b**). * denotes significant differences ($p < 0.05$) between the response of vessels from tilted and restricted control animals as well as between those obtained from RRM and age-matched controls

Pressure-Induced Electrophysiological Responses In Vitro

A one-step elevation of intraluminal pressure in the in vitro perfused and superfused venous segments from a control level of 2.2 ± 0.1 mmHg to 7.6 ± 0.4 mmHg at constant perfusion rate resulted in a significant, maintained, and reversible 12.9 ± 1.2 mV depolarization during the 1-h observation period. Increase of the intraluminal pressure to 15 ± 0.8 mmHg induced a proportionally larger (18.6 ± 0.9 mV) depolarization in another series of experiments [7].

Myogenic Responses of Saphenous Veins from Tilted Rats: Changes in Biomechanical and Electrical Properties

After a 2-week gravitational venous pressure elevation due to tilting, mean external diameter (but not wall thickness) of in vitro perfused and superfused vein segments ($n = 8$) was significantly larger (by 10%–30%) than that of the restricted controls ($n = 8$). In another series of measurements, the tilted veins ($n = 7$) exhibited a substantially augmented active strain response: the active strain induced by short-term step elevations of intraluminal pressure was twice as large in veins isolated from tilted rats as in those from restricted controls ($n = 5$; Fig. 1a). Mean incremental distensibility of veins from tilted rats was significantly smaller (by approximately 40%) than that of the control vessels at the 5–10 mmHg intraluminal pressure range.

The long-term tilt resulted in changes also in the resting E_m of the saphenous venous smooth muscle both in vitro and in situ. Thus, the smooth muscle cells of isolated veins from tilted rats were significantly hyperpolarized relative to the restricted controls (-58.2 ± 0.9 mV vs. 51.8 ± 0.3 mV) at normal perfusion pressure of 2.3 ± 0.23 mmHg. The mean E_m (-61.3 ± 2.3 mV) obtained in situ from neurally blocked saphenous venous smooth muscle cells of anesthetized tilted animals was also significantly greater in magnitude than in the nontilted controls (-53.5 ± 0.6 mV). Such differences in E_m were not found in brachial venous smooth muscle cells in situ. In the tilt technique used for these measurements [8], the brachial vein is exposed to a chronic pressure reduction rather than elevation.

Pressure-Induced Changes in the Biomechanical Properties of Saphenous Veins from RRM Rats

A 6-weeks period of sustained arterial hypertension concomitant with an elevated venous pressure caused a significant increase in the acute pressure-induced myogenic response of saphenous veins from RRM rats ($n = 5$) compared to the myogenic response of age-matched normal controls ($n = 3$). Fig. 1b shows the active isobaric tangential strain values as a function of the intraluminal pressure. The RRM saphenous veins exhibited a significantly higher response over the 5–20 mmHg pressure range than the age-matched parallel control group. Wall thickness of the RRM veins was increased ($0.225 \pm 0.015 \times 10^{-2}$ cm vs. $0.187 \pm 0.004 \times 10^{-2}$ cm at 10 mmHg) relative to the controls.

Discussion

This is the first demonstration that pressure-induced myogenic tone becomes more pronounced in veins after exposure to chronic pressure elevation. The magnitude of this myogenic response seems to be sufficient to participate effectively in the venous capacity control and may have an important role in pathophysiological situations. Our data support the previously proposed hypothesis that stretch-induced active venous smooth muscle tone may be sufficient in magnitude to significantly modulate both total peripheral resistance and cardiac output, the latter via an enhancement of venous return [7,8]. The results of these studies indicate that both gravitationally elevated venous pressure and volume expansion result in a significantly increased venous myogenic response to short-term, stepwise pressurization in vitro. This observation complements previous findings proving that sustained pressure load can result in an enhanced myogenic response of muscle arteries from rats modelling different types of hypertension [13]. However, further studies are needed to explain the significance of the increased wall thickness of the veins from RRM rats in their augmented active strain response to acute pressure.

Although the cellular mechanisms by which stretch can activate smooth muscle cells are not clearly understood, our electrophysiological results provide evidence that intravascular pressure alters venous smooth muscle transmembrane potential. This finding is in agreement with data demonstrating that pressure elevation in isolated cerebral arteries caused a membrane depolarization of $1.45\,mV/10\,mmHg$ in the smooth muscle cells [3]. The slope of this relation was directly dependent on external Ca^{2+} concentration, suggesting that this cation may play a role in membrane depolarization [4]. Several laboratories recently demonstrated single ionic channels in diverse cell types that are activated by stretch [2,9]. Although these channels are probably not specific for Ca^{2+}, its entry through them would provide an explanation for the relation between increases in vascular cell muscle membrane strain, increases in transmembrane Ca^{2+} flux, decreases in transmembrane potential magnitude, and increases in vascular muscle active stress.

The results indicating that the 2-week gravitational pressure load produced a significant hyperpolarization of the venous smooth muscle membrane are different when compared with the acute pressure response causing depolarization. Membrane alterations participating in this relative hyperpolarization are not yet identified. It is possible that sustained pressure elevation results in increased Na^+/K^+ pump activity (see [8]).

The enhanced acute pressure-induced active strain of the venous segments after long-term pressure loads suggests the existence of an active capacity adaptation process developing in the vessel wall in response to, or in association with, elevated venous pressure.

In conclusion, using in vitro perfused-superfused vein segments extended to their in vivo length and loaded by short-term pressure steps, venous biomechanical and electrophysiological responses can be successfully examined. This technique has the potential to eliminate methodological difficulties underlying previous misconceptions concerning the existence and possible physiological role of venous myogenic tone in regulating vascular capacitance. Applying this method, substantial pressure-dependent myogenic tone was found in rat saphenous vein, which was augmented in vessels previously exposed to sustained pressure elevation in vivo. This provides evidence supporting the role of pressure-induced myogenic tone in the regulation of venous function under both normal and pathological conditions.

References

1. Cowley AW Jr, Barber WJ, Lombard JH, Osborn JL, Liard JF (1986) Relationship between body fluid volumes and arterial pressure. Fed Proc 45:2864–2870
2. Guharay F, Sachs F (1984) Stretch-activated single ion channel currents in tissue-cultured embryonic chick skeletal muscle. J Physiol (Lond) 352:687–701
3. Harder DR (1984) Pressure-dependent membrane depolarization in cat middle cerebral artery. Circ Res 55:197–202
4. Harder DR, Gilbert R, Lombard JH (1987) Vascular muscle cell depolarization and activation in renal arteries on elevation of transmural pressure. Am J Physiol 253 (Renal Fluid Electrolyte Physiol 22):F778–F781
5. Johnson PC (1980) The myogenic response. In: Bohr DF, Somlyo AP, Sparks HP Jr (eds) Cardiovascular system. Vascular smooth muscle. American Physiological Society, Bethesda, pp 409–442 (Handbook of physiology, vol II, sect 2)
6. Monos E (1989) Control mechanisms of the veins. Proc IUPS XVII:130
7. Monos E, Contney SJ, Cowley AW Jr, Stekiel WJ (1989) Electrical and mechanical responses of rat saphenous vein to short-term pressure load. Am J Physiol 256 (Heart Circ Physiol 25):H47–H55
8. Monos E, Contney SJ, Cowley AW Jr, Stekiel WJ (1989) Effect of long-term tilt on mechanical and electrical properties of rat saphenous vein. Am J Physiol 256 (Heart Circ Physiol 25):H1185–H1191
9. Morris CE, Sigurdson WJ (1989) Stretch-inactivated ion channels coexist with stretch-activated ion channels. Science 243:807–809
10. Mulvany MJ (1988) Possible role of vascular oscillatory activity in the development of high blood pressure in spontaneously hypertensive rats. J Cardiovasc Pharmacol 12 [Suppl 6]:S16–S20
11. Rothe CF (1983) Venous system: physiology of the capacitance vessels. In: Bohr DF, Somlyo AP, Sparks HP Jr (eds) The cardiovascular system. Peripheral circulation and organ blood flow. American Physiological Society, Bethesda, pp 397–452 (Handbook of physiology, vol III, sect 2, part 1)
12. Safar ME, London GM (1985) Venous system in essential hypertension. Clin Sci 69:497–504
13. Stekiel WJ, Myers K, Monos E, Lombard J (1988) Stretch-dependent tone in small mesenteric and gracilis muscle arteries from spontaneous (SHR) and volume-expanded hypertensive rats. In: Halpern W et al. (eds) Resistance arteries. Perinatology Press, Ithaca, pp 342–350
14. Vanhoutte PM, Janssens WJ (1978) Local control of venous function. Microvasc Res 16:196–214

Part II
Calcium and Circulatory Control

Dysfunction of Ca^{2+}-pump of Vascular Muscle Membranes: An Important Etiological Factor in Hypertension*

C.Y. Kwan

Department of Biomedical Sciences, McMaster University, Health Sciences Centre,
1200 Main Street West, Hamilton, Ontario, Canada, L8N 3Z5

Membrane Abnormalities in Hypertension

An increase in total peripheral resistance associated with sustained hypertension is thought to be contributed by the hyperresponsiveness of arteries, particularly at the level of small arteries and arterioles. Both structural and functional changes of blood vessels have long been recognized as contributing factors to the increased vascular responsiveness in animals with experimental hypertension [4]. Over the past decade, derangement of membrance functions in cardiovascular tissues has been repeatedly shown to be closely associated with the development of primary hypertension [13,14]. These membrane abnormalities included the handling (e.g., binding, active transport, exchange and/or passive flux) of ions (e.g., H^+, Na^+, K^+, Cl^-, and/or Ca^{2+}), the interactions (e.g., affinity, density, and/or modulation) between membrane receptors and physiologically relevant vasoactive ligands (e.g, angiotensin II, catecholamines, and/or atrial natriuretic peptides), and the properties of membrane-associated anzymes (e.g., those responsible for ion transport, receptor coupling, signal transduction, or other unknown functions probably involving cellular proliferation). The nature of membrane abnormalities varied considerably depending on the mode and duration of hypertension, the vasculature selected, and the methodology employed.

It is evident that a wealth of information about membrane abnormalities in hypertension has accumulated in recent years. Figure 1 schematically summarizes a number of common membrane properties of vascular smooth muscle which are being investigated in hypertension. Some membrane dysfunctions in vascular smooth muscle may account for the altered vascular reactivity leading to elevated peripheral resistance in sustained hypertension, and some may simply be the consequence of elevated blood pressure.

* The author's work cited in this communication was supported by the Heart and Stroke Foundation of Ontario and the Medical Research Council of Canada.

G. Bruschi A. Borghetti (Eds.)
Cellular Aspects of Hypertension
© Springer-Verlag Berlin · Heidelberg 1991

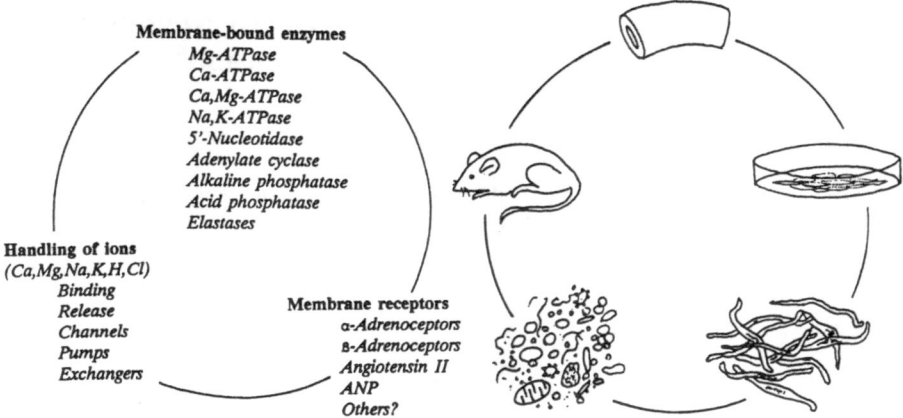

Fig. 1. The multiplicity of vascular membrane abnormalities in hypertension and the experimental approaches (the use of whole animal, isolated vascular segments, cultured cells, isolated single cells and subcellular membranes) commonly used to study these changes. Readers are referred to recent reviews (see [13]) for more detailed discussions on these cellular and subcellular aspects of membrane changes in hypertension

Membrane abnormalities have also been found in tissues and cell types that are not necessarily confined to the cardiovascular system. Conceivably, genetic factors may contribute to the multiplicity of membrane disorders in the development of primary hypertension. It is beyond any doubt that the membrane basis of hypertension has become a rapidly developing area of research in this field. Space prohibits a comprehensive account here of the immense quantity of available information, which in fact has recently been reviewed in a two-volume book specifically devoted to this topic [13]. I will therefore, confine my discussion to a limited scope by presenting only the evidence for diminished Ca^{2+} pump activity in vascular muscle (mostly based on previous studies from this laboratory using the subcellular membrane approach), the consequence of such a Ca^{2+} handling defect on the contractile abnormalities, and its significance to the etiology of hypertension.

Deficiency in Ca^{2+} Pump or in Ca^{2+} Leak?

The pivotal role of Ca^{2+} in the control of contractile function of vascular smooth muscle makes the regulation of Ca^{2+} by vascular smooth muscle cells a probable primary target for dysfunction in hypertension [29,31,33]. The direct recording of elevated cytosolic Ca^{2+} concentration in vascular muscle cells from SHR by using the fluorescent Ca^{2+} indicator technique [5,34] or NMR spectroscopy [9], and the lack of change in Ca^{2+} sensitivity in the contractile function of skinned vascular muscle strips of SHR [27], suggest that contractile abnormalities reside in the membrane systems that regulate the cytosolic concentration of Ca^{2+}. The direct approach using isolated microsomal membrane vesicles [12] from vascular smooth muscle of large and small arteries from hypertensive rats has consistently demonstrated diminished ATP-supported Ca^{2+} transport [10–12] compared with normotensive controls, which could not be accounted for by differences in the relative proportion of membrane vesicles of different orientations [22] or in membrane leakiness to Ca^{2+} [11]. This finding is consistent with the hypothesis of a deficiency in the vascular muscle Ca^{2+} pump (the Ca^{2+} transport ATPase) in hypertension.

Although increased Ca^{2+} leak across the vascular muscle cell membranes in intact muscle strips or cells from SHR is evident on the basis of $^{45}Ca^{2+}$ fluxes [3] and functional studies [28], this may be a consequence of other intrinsic changes due to faulty interactions, such as membrane stabilization by high extracellular Ca^{2+} [2] or membrane destabilization by Ca^{2+} depletion [6]. The possible mechanism for Ca^{2+} entry across the cell membrane as a compensatory response to depletion of the intracellular Ca^{2+} store, previously proposed in vascular smooth muscle [35] and recently shown to occur in non-excitable as well [17,30], has not been adequately considered in vascular muscle in hypertension. This aspect will be discussed in a later section. It appears that membrane changes in Ca^{2+} leak secondary to other intrinsic changes occurring in the intact cell system may have not been readily manifested in isolated membrane vesicles. The same argument may be applicable to the role of Na^+/Ca^{2+} exchange in hypertension. Although supportive evidence has been presented in functional studies of intact vascular tissue [1,25], Na^+/Ca^{2+} activity measured with sarcolemmal membrane vesicles isolated from vascular muscle showed no significant difference between spontaneously hypertensive (SHR) and Wistar-Kyoto (WKY) rats [26].

Cause or Consequence of Hypertension?

Several lines of evidence suggest that the deficient Ca^{2+} pump in vascular muscle membranes from hypertensive rats cannot be interpreted solely as a direct consequence of elevated blood pressure. *First of all*, this Ca^{2+} handling defect was shown to precede the onset of detectable hypertension in our SHR colony [18]. *Secondly*, normalization of blood pressure in SHR by long-term hydralazine treatment, which does not correct the causal mechanism and only serves to control hypertension, failed to correct such a defect in the Ca^{2+} pump [15]. However, in experimental hypertension induced by administration of deoxycorticosterone (DOC) in rats on a high-salt diet, the change in Ca^{2+} pump activity corresponded to the changes in blood pressure: in rats whose blood pressure returned to normal after DOC was withdrawn, the Ca^{2+} pump defect was no longer present, but the defect persisted in rats whose hypertension persisted despite withdrawal of DOC [19]. *Thirdly*, a similar decrease in ATP-supported Ca^{2+} transport was observed in microsomes of cultured aortic smooth muscle cells from adult SHR compared to those from age-matched WKY under similar culture conditions [10]. *Finally,* diminished Ca^{2+} pump activity in vascular muscle membrane was also found in different models of experimental hypertension [20], with the exception of salt-induced hypertension in Dahl salt-sensitive rats [23]. These findings suggest that such a defect, commonly associated with most forms of hypertension, is sufficient but may not be necessary to cause hypertension, nor is this membrane defect a consequence of hypertension per se. *Furthermore,* in spontaneous hypertension, a similar Ca^{2+} pump defect occurred in microsomes obtained from nonvascular smooth muscles [21] and in membranes isolated from nonsmooth muscle tissues such as brain, platelets, erythrocytes, adipocytes, hepatocytes, and cardiac muscle (see other chapters in this volume). Since vascular and nonvascular smooth muscles have fundamentally similar membrane properties for the regulation of Ca^{2+} distribution, this result can be interpreted to mean that the Ca^{2+} handling defect in vascular smooth muscle, as in nonvascular smooth muscle, is not a direct consequence of elevated blood pressure. Such a defect in nonvascular smooth muscle (e.g., vas deferens) membranes, however, was not found in DOC-salt hypertension [16]. This is in agreement with the finding of altered contractile function of vas deferens in spontaneous hypertension but not in DOC-salt hypertension [23,32]. Some the above findings also suggest that in SHR, the deficit in Ca^{2+} pump may be genetically programmed such that it occurs in many tissues and is not necessarily confined to the cardiovascular system. These results provide additional support for the hypothesis that a widespead membrane defect forms the basis for the etiology of primary hypertension. Changes in Ca^{2+} pump using microsomal membranes isolated from vascular and nonvascular smooth muscles of rats with different models of hypertension are summarized in Table 1.

Defect in Plasma Membrane or in Endoplasmic Reticulum?

Although our studies using highly purified plasmalemma-enriched membrane vesicles from large and small arteries strongly suggest that vascular muscle cell membrane is a major site for reduced Ca^{2+} pump activity associated with hypertension, a possible defect in endoplasmic reticulum Ca^{2+} pump still cannot be ruled out ont the basis of subcellular membrane studies due to the intrinsic difficulty in isolating a sufficient quantity of purified endoplasmic reticulum from the vascular muscle in the rat model. In fact, it is not known whether Ca^{2+} sequestration occurs in all the endoplasmic reticulum membranes or is confined to a selected fraction of them. However, in theory, reduced uptake of cytosolic Ca^{2+} by endoplasmic reticulum due to a deficient endoplasmic reticulum Ca^{2+} pump may also lead to an increased cytosolic Ca^{2+} level, if the plasma membrance Ca^{2+} fails to normalize it by effectively extruding the excessive cytosolic Ca^{2+} at any given time. This may provide a

Table 1. Changes in microsomal Ca^{2+} pump of vascular and nonvascular smooth muscle of rats with different models of hypertension

Model of hypertension	Vascular	Nonvascular	Reference
Spontaneous	↓	↓	15, 18, 19, 21
DOC-salt	↓	↔	16, 18, 19
Dahl salt-sensitive	↔	↓, ↔	24
1-kidney, 1-clip	↓	ND	20
2-kidney, 1-clip	↓	ND	20

↓ decreased, ↔ unaltered compared with corresponding normotensive controls, ND not determined.

plausible explanation for the increased vascular tone [28] and the decreased rate of relaxation [32].

As mentioned earlier, depletion of Ca^{2+} from the intracellular store will lead to compensatory influx of Ca^{2+} such that the depleted Ca^{2+} stores can be refilled. If the intracellular store Ca^{2+} pump becomes defective in hypertension and cannot fully refill the stores after an agonist-induced release of Ca^{2+} it can be anticipated that the stores will have a reduced amount of Ca^{2+} available for subsequent release, thus remaining in a state of partial depletion. Consequently, one would expect the compensatory Ca^{2+} entry to be enhanced. To test this hypothesis, we recently studied the effects of temporal depletion of intracellular Ca^{2+} in Ca^{2+}-free, EGTA-containing medium [7] on phenylephrine-induced contractile response (presumably due to intracellular Ca^{2+} release) of aorta from SHR and WKY and the effects of subsequent repletion of Ca^{2+} in the presence of the agonist (presumably due to Ca^{2+} entry). Figure 2A shows that the contractile responses of rat aortic rings to the maximal dose of phenylephrine in Ca^{2+}-free, EGTA-containing medium diminished with increasing period of exposure to the Ca^{2+}-free medium. This was accompanied by increasing responses induced by subsequent addition of Ca^{2+} such that the masimal response remained unaltered. Figure 2B shows how, although the maximal responses of aortas from adult SHR and WKY to phenylephrine remained unaltered with respect to the exposure periods in Ca^{2+}-free, EGTA-containing medium, the temporal decrease of contractile response to phenylephrine in Ca^{2+}-free medium was substantially samller in SHR. Hano et al. [8] has also reported a reduced vascular response to norepinephrine and caffeine in SHR compared to WKY in Ca^{2+}-free medium. Figure 2C indicates that the rate of relaxation of aorta to phenylephrine-induced contraction in Ca^{2+}-containing medium was considerably slower in SHR than in WKY. These results are consistent with the hypothesis that the internal Ca^{2+}-pump (presumably on the endoplasmic reticulum) is also defective in hypertension and that the increased leakiness (manifested as increased Ca^{2+} influx) is a compensatory response to the partial emptiness of the internal stores due to reduced Ca^{2+} pump activity.

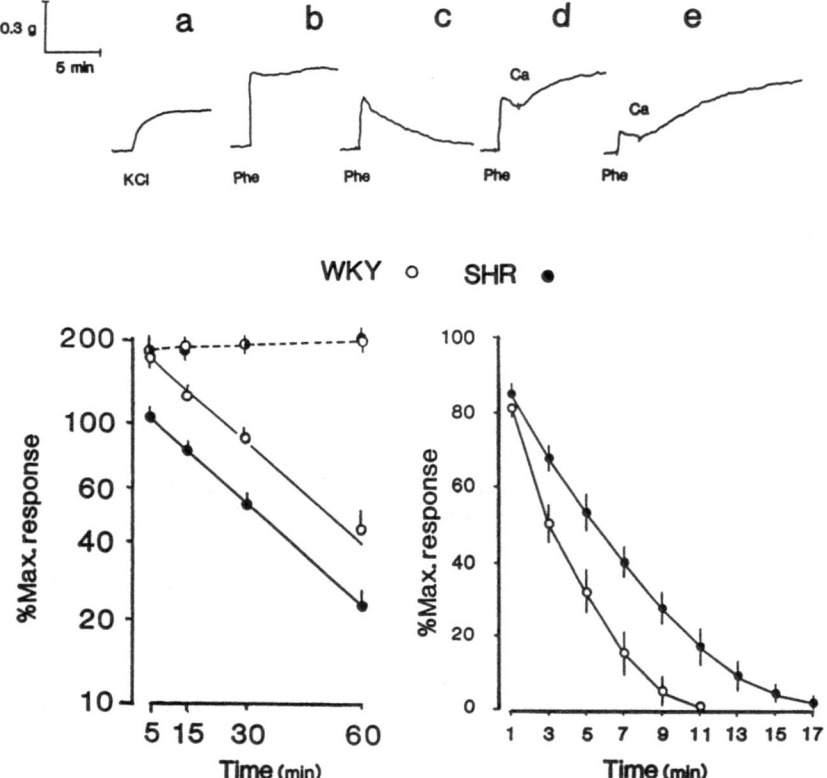

Fig. 2. (*Top*) Typical tracings from a representative experiment showing the contractile responses of aortic rings from WKY to 100 mM KCl (*a*), $10^{-5}M$ Phe in Ca^{2+}-containing medium (*b*), and Ca^{2+}-free medium containing $50 \mu M$ EGTA (*c*). Tracings *d* and *e* represent the responses to $10^{-5}M$ Phe in Ca^{2+}-free medium followed by addition of 2.5 mM (final concentration) Ca^{2+} after the plateau phasic responses. Tracing *e* was obtained after the aortic rings had been incubated in Ca^{2+}-free, EGTA-containing medium for 30 min, whereas tracings *c* and *d* were obtained after 5 min preincubation in Ca^{2+}-free medium containing $50 \mu M$ EGTA. (*Lower left*) Decay of aortic responses of SHR (*closed circles*; *n* = 8) and WKY (*open circles*; *n* = 14) to $10^{-5}M$ Phe as a function of preincubation period in Ca^{2+}-free medium containing $50 \mu M$ EGTA (*solid line*) and after subsequent addition of 2.5 mM Ca^{2+} (*dotted lines*). Contractile response to 100 mM KCl was taken as 100%. (*Lower right*) Rate of loss of contractile responses of aortas of SHR (*closed circles*) and WKY (*open circles*) to $10^{-5}M$ Phe (*n* = 8) in Ca^{2+}-containing Krebs' solution following washout of agonist after the plateau responses. All paired data points except for those obtained 1 min after washout of Phe were significantly different from each other. The plateau response to $10^{-5}M$ Phe prior to washing was taken as 100%

References

1. Blaustein MP (1988) Sodium/calcium exchange and the control of contractility in cardiac muscle and vascular smooth muscle. J Cardiovasc Pharmacol 12 [Suppl 5]: S56–S62
2. Bruner CA, Webb RC, Bohr DF (1989) Vascular reactivity and membrane stabilizing effect of calcium in spontaneously hypertensive rats. In: Aoki K, Frohlich ED (eds) Calcium in essential hypertension Academic, Tokyo
3. Cauvin C, Johns A, Yamamoto MK, Hwang O, Gelband C, van Breemen C (1989) Ca^{2+} movements in vascular smooth muscle and their alterations in hypertension. In: Kwan CY (ed) Membrane abnormalities in hypertension, vol I CRC Press, Boca Raton,
4. Daniel EE, Kwan CY, Lee RMKW, Smeda J (1984) Early structural changes in precapillary vessels in hypertension and their relationship to functional changes. J Cardiovasc Pharmacol 6:S671–S682
5. Erne P, Hermsmeyer K (1989) Intracellular vascular muscle Ca^{2+} modulation in genetic hypertension. Hypertension 14:145–151
6. Guan YY, Kwan CY (1989) Abnormal contractile responses of aortas from spontaneously hypertensive rats to Ca^{2+} after depletion of Ca^{2+} in Ca^{2+}-free medium. Am J Hypertens 2:643–646
7. Guan YY, Kwan CY, Daniel EE (1988) Tthe effects of EGTA on vascular smooth muscle contractility in calcium-free medium. Can J Physiol Pharmacol 66:1053–1056
8. Hano T, Kuchii M, Baba A, Nishio I, Masuyama Y (1987) Ca mobilization from the intracellular Ca store in spontaneously hypertensive rats. J Cardiovasc Pharmacol 10 [Suppl 10]:S72–S73
9. Jelicks L, Gupta RK (1990) NMR measurement of cytosolic free calcium, free magnesium, and intracellular sodium in the aorta of the normal and spontaneously hypertensive rat. J Biol Chem 265:1394–1400
10. Kwan CY (1985) Dysfunction of calcium handling by smooth muscle in hypertension. Can J Physiol Pharmacol 63:366–374
11. Kwan CY (1986) Calcium membrane interactions in smooth muscle in relation to hypertension: a subcellular membrane approach. In: Aoki K (ed) Essential hypertension: calcium mechanisms and treatment. Springer Berlin Heidelberg New York, pp 135–144
12. Kwan CY (1987) Preparation of smooth muscle plasma membranes: a critical evaluation. In: Kidwai AM (ed) Sarcolemmal biochemistry, CRC Press, Boca Raton,
13. Kwan CY (ed) (1989) Membrane abnormalities in hypertension, vols I and II. CRC Press, Boca Raton,
14. Kwan CY, Daniel EE (1981) Biochemical abnormalities of venous plasma membrane fraction isolated from spontaneously hypertensive rats. Eur J Pharmacol 75:321–324
15. Kwan CY, Daniel EE (1982) Arterial muscle membrane abnormalities of hydralazine-treated spontaneously hypertensive rats. Eur J Pharmacol 82:187–190
16. Kwan CY, Grover AK (1983) Membrane abnormalities occur in vascular smooth muscle but not in non-vascular smooth muscle from rats with deoxycorticosterone-salt induced hypertension. J Hypertens 1:257–265
17. Kwan CY, Putney JW Jr (1990) Uptake and intracellular sequestration of divalent cations in resting and methacholine-stimulated mouse lacrimal acinar cells. J Biol Chem 265:678–684
18. Kwan CY, Belbeck L, Daniel EE (1979) Abnormal biochemistry of vascular smooth muscle plasma membrane as an important factor in the initiation and maintenance of hypertension in rats. Blood Vessels 16:259–268
19. Kwan CY, Belbeck L, Daniel EE (1980) Abnormal biochemistry of vascular smooth muscle plasma membrane isolated from hypertensive rats. Mol Pharmacol 17: 137–140

20. Kwan CY, Belbeck L, Daniel EE (1980) Characteristics of arterial plasma membrane in renovascular hypertension in rats. Blood Vessels 17:131–140
21. Kwan CY, Grover AK, Sakai Y (1982) Abnormal biochemistry of subcellular membranes isolated from non-vascular smooth muscles of spontaneously hypertensive rats. Blood Vessels 19:273–283
22. Kwan CY, Grover AK, Daniel EE (1984) On the ouabain-sensitive potassium activated p-nitrophenyl phosphatase activity of vascular muscle plasma membranes. Arch Int Pharmacodyn Ther 272:245–255
23. Kwan CY, Sakai Y, Daniel EE (1984) On the abnormalities of contractile responses of rat vasa deferentia. Clin Exp Hypertens [A]6:1257–1265
24. Kwan CY, Triggle CR, Grover AK, Daniel EE (1985) Subcellular membrane properties in vascular and nonvascular smooth muscles of Dahl hypertensive rats. J Hypertens 4:49–55
25. Lategan TW (1980) Sodium/calcium exchange and hypertension. In: Kwan CY (ed) Membrane abnormalities in hypertension, vol I. CRC Press, Boca Raton,
26. Matlib MA, Schwartz A, Yamori Y (1985) A Na^+-Ca^{2+} exchange process in isolated sarcolemmal membranes of mesenteric arteries from WKY and SHR rats. Am J Physiol 249:C166–C172
27. Mrwa U, Guth K, Haist C, Troschka M, Herrmann T, Wojciechowski T, Gagelmann M (1986) Calcium requirement for activation of skinned vascular smooth muscle from spontaneously hypertensive (SHRSP) and normotensive control rats. Life Sci 38: 191–196
28. Noon JP, Rice PJ, Baldessarini RJ (1978) Calcium leakage as a cause of high resting tension in vascular smooth muscle from spontaneously hypertensive and normotensive rats. Proc Natl Acad Sci USA 65:1605–1607
29. Postnov YV, Orlov SM (1985) Ion transport across plasma membrane in primary hypertension. Physiol Rev 65:904–945
30. Putney JW Jr (1986) A model for receptor-regulated calcium entry. Cell Calcium 7:1–12
31. Robinson BP (1984) Altered calcium handling as a cause of primary hypertension. J Hypertens 2:453–460
32. Sakai Y, Kwan CY, Daniel EE (1984) Contractile responses of vasa deferentia from rats with genetic and experimental hypertension. J Hypertens 2:631–638
33. Sprenger KBG (1985) Alteration of cellular calcium metabolism as primary cause of hypertension. Clin Physiol Biochem 3:208–218
34. Sugiyama T, Yoshizumi M, Takaku F, Urabe H, Tsukahoshi M, Kasuya T, Yazaki Y (1986) The elevation of the cytoplasmic calcium ions in vascular smooth muscle cells in SHR: measurement of the free calcium ions in single living cells by lasermicro-fluorospectromety. Biochem Biophys Res Commun 26:340–345
35. van Breemen C, Lukeman S, Leijten P, Yamamoto H, Loutzenhiser T (1986) The role of the superficial SR in modulating force development induced by Ca^{2+} entry into arterial smooth muscle. J Cardiovasc Pharmacol 8:5111–5115

Changes in Ca^{2+} and Ca^{2+} Sensitivity During Contraction-Relaxation of Arterial Muscle

G. Bruschi, M.E. Bruschi, G. Regolisti, and A. Borghetti

Istituto di Clinica Medica e Nefrologia, Facolta di Medicina e Chirurgica, Università degli Studi di Parma, Via A. Gramsci, 14, I-43100 Parma, Italy

The control of vascular function is a central issue in hypertension, where total peripheral resistance is increased, implying a state of relative vasoconstriction.

At the cellular level of vascular physiology, the calcium ion has long been known to play a cardinal role, but most details began to be understood only with the advent of intracellular Ca^{2+} indicators. This report will focus on the problem of Ca^{2+} measurements in functioning vascular tissues.

Two classes of calcium probes have been applied to the study of intracellular Ca^{2+} in vascular smooth muscle. The first class is that of carboxylate indicators, which have a structure similar to that of EGTA. The second class is that of photoproteins, of which aequorin is the most widely used example.

These two classes of indicators have rather different characteristics, and therefore different advantages and disadvantages. Carboxylic indicators, because of their EGTA-like structure, can produce buffering of, Ca^{2+} transients. On the other hand, they also have some attractive properties, like that of offering, at least in principle, quantitative values of Ca^{2+}, and of reporting an average estimate of Ca^{2+} in the sample under study. On the other side, photoproteins produce no significant buffering, but their signal is difficult to quantify and is also somewhat dominated by local gradients that take place within the preparation.

The first records of intracellular Ca^{2+} signals in vascular tissues were obtained with aequorin during pharmacologic stimulation of the portal vein. Morgan and Morgan [1] observed that stimulation of the vessel with phenylephrine, or high-K depolarization, regularly involved some changes of aequorin light emission. Also, they reported different patterns with different stimuli, like a smooth and continuous rise of the Ca^{2+} signal during high-K stimulation, which differed from the high spikes of light produced by α-adrenergic stimulants like phenylephrine. The latter were followed by a slight but continuous elevation during continuous exposure to this agonist.

However, one problem with aequorin is the difficulty of translating the light signals into Ca^{2+} values, so to correlate the changes in light intensity to

G. Bruschi A. Borghetti (Eds.)
Cellular Aspects of Hypertension
© Springer-Verlag Berlin · Heidelberg 1991

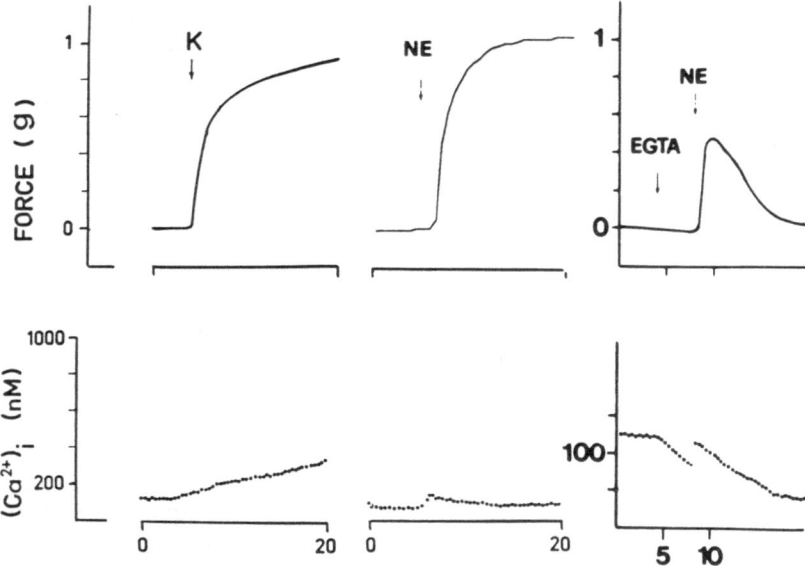

Fig. 1. Representative records of the changes in force (*top*) and intracellular Ca^{2+} (*bottom*) observed in rat aortic smooth muscle during stimulation by 100 mM K (*left*) and 10 μM norepinephrine (*middle*) in 1 mM extracellular Ca^{2+}. The *right panel* shows the response to the same concentration of norepinephrine in the absence of ambient Ca^{2+} (EGTA). Again, the force record is shown above the Ca^{2+} values. Experimental setup and charcteristics of the preparation were as described in [3]. Ca^{2+} values were calculated from 340/380 ratios as reported by Grynkiewicz et al. [2]

the process of force development. It is also difficult to decide whether the patterns observed depend on a homogeneous involvement of the cytosol of single cells, or a homogeneous involvement of every cell in the muscle, again because aequorin is quite a local Ca^{2+} indicator.

Some years ago, fura-2 was synthesized [2] and it became clear that this indicator could be applied to obtain Ca^{2+} measurements even in contracting tissues, including vascular smooth muscle. For example, we assembled a system [3] in which a ring of aortic smooth muscle could be prepared for tension measurements under a resting stretch and simultaneously submitted to fluorescence measurements, as a flat double sheet of tissue onto which exciting light could be directed, and from which fluorescent light could be collected. It was then possible to ask the question whether stimulation of aortic smooth muscle was associated with changes of intracellular Ca^{2+}.

The observation was that a rise in Ca^{2+} actually was associated with contraction, especially during high-K stimulation, but often, and quite consistently with norepinephrine, the increase in Ca^{2+} was quite modest, sometimes even undetectable (Fig. 1). It was also evident that in some circumstances vasoconstriction occurred even if the Ca^{2+} signal did not rise but fell; for example in Fig. 1 (right panel), when norepinephrine was applied

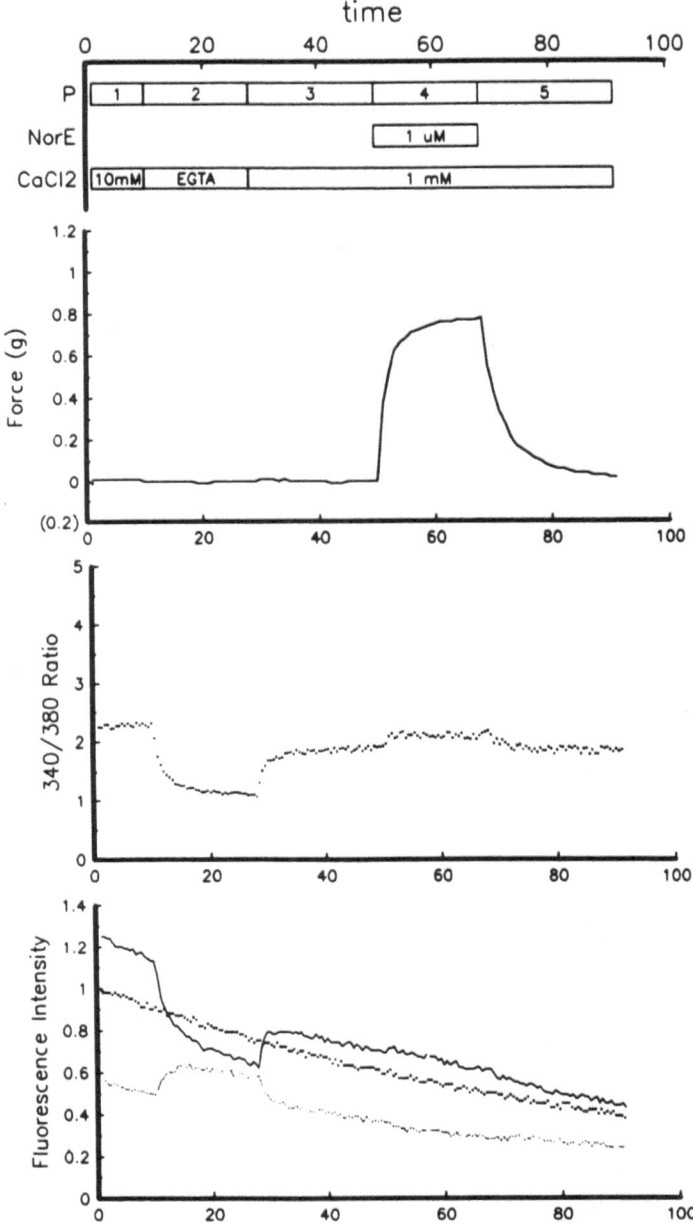

Fig. 2. Simultaneous measurements of force and fura-2 fluorescence in a rat aortic smooth muscle ring. The 340/380 ratio in the *middle panel* is representative of Ca^{2+}. The *bottom panel* shows the fluorescence intensity at single excitation wavelengths (340, 359, 380 nm). In the first 50 min only changes in extracellular Ca^{2+} were induced. Then 1 μM norepinephrine was added, in the presence of 1 mM ambient Ca^{2+}. Abbreviations: *P*, experimental period (sequentially numbered), *CaCl^2*; concentration of extracellular Ca^{2+}; *NorE*, concentration of norepinephrine

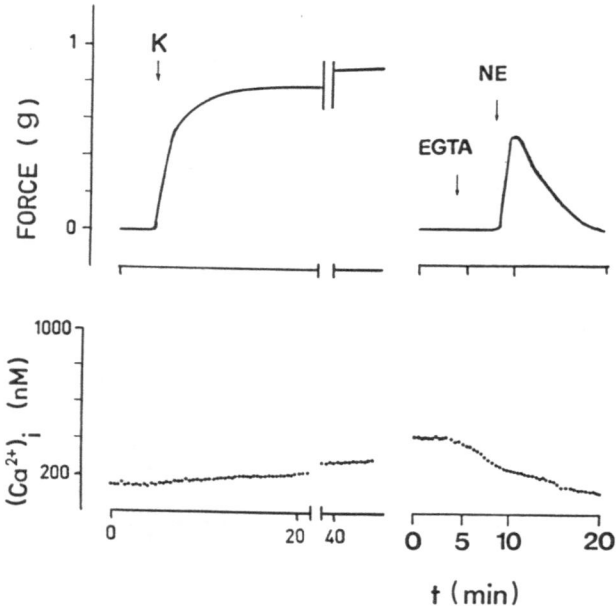

Fig. 3. In muscles loaded with 1–2 mM fura-2, high-K depolarization (80 mM, *left*) was carried out in normal Ca^{2+} (1 mM) medium, while norepinephrine (10 μM, *right*) was applied in Ca^{2+}-free medium

in the absence of extracellular Ca^{2+}. In spite of an immediate transient elevation, intracellular Ca^{2+} is below the baseline during the time of stimulation, while the force record stands above the baseline, at least for some minutes.

Sometimes changes in intracellular Ca^{2+} occurred without vasoconstriction, as in the Fig. 2 experiment. During the first period, in the absence of any stimulus, but with ample changes in extracellular Ca^{2+} (10 mM, then no Ca^{2+} with EGTA, then 1 mM Ca^{2+} again), considerable changes in intracellular Ca^{2+} are induced without any concomitant change in force. Immediately after, in the same muscle, stimulation with 1 μM norepinephrine is associated with the usual force response and a modest elevation of the fura-2 signal. Actually the 60-min Ca^{2+} level, corresponding to excitation (in 1 mM external Ca^{2+}), is somewhat lower than the initial level (10 mM external Ca^{2+}), corresponding to zero force.

An important question concerns the buffering effects of the indicator, and to check this point one can decide deliberately to increase the loading (and so the buffering) of fura-2 and assess the consequences. In Fig. 3, for example, norepinephrine stimulation is carried out in heavily loaded cells (about 1 mM intracellular fura-2), in the absence of extracellular Ca^{2+} (right panel); no change in intracellular Ca^{2+} occurs, except a continuous fall; nevertheless a force transient takes place. In the left panel of Fig. 3, high-K

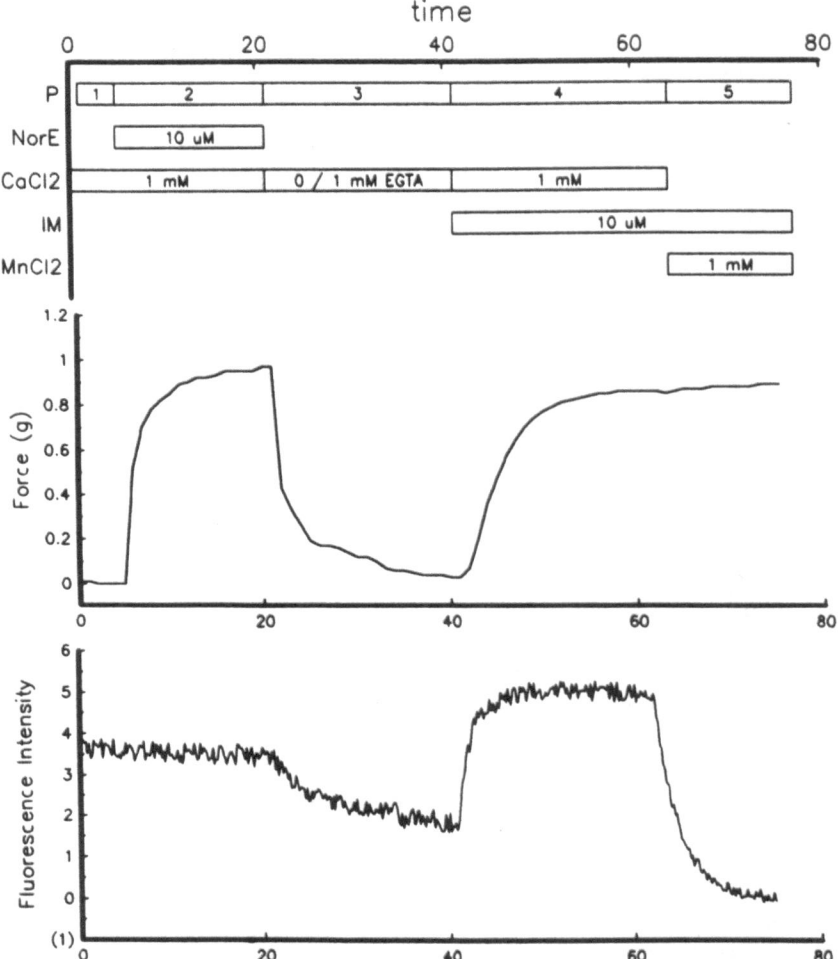

Fig. 4. After loading with $100\,\mu M$ extracellular ($9.4\,mM$ intracellular) quin-2, a ring of rat aortic smooth muscle was exposed to norepinephrine (period 2) and ionomycin (periods 4–5). The *lower panel* represents the fluorescence intensity excited continuously at a single wavelength (340 nm excitation, 490 nm emission). Abbreviations as in Fig. 2; *IM*, ionomycin; *MnCl²*, manganese chloride

stimulation is applied in cells loaded with about $2\,mM$ fura-2; the Ca²⁺ rise is somewhat flattened, but contraction develops quite normally.

An extreme degree of buffering can be realized with another indicator, quin-2, because quin-2 is hydrolyzed and trapped by the cells up to concentrations of several-millimolars. In the example illustrated in Fig. 4, the loading with quin-2 was actually around 10 mM, and after stimulation by a maximal dose of norepinephrine the fluorescent Ca²⁺ signal is practically flat; however, force development is essentially normal.

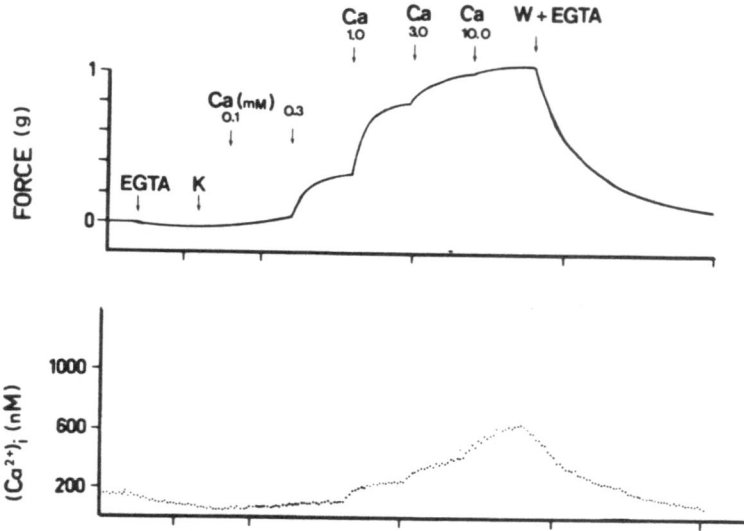

Fig. 5. Cumulative force response of aortic smooth muscle to cumulative additions of extracellular Ca^{2+} (mM; *arrows*) in the presence of 80-mM K depolarization. Force is shown above the simultaneously collected Ca^{2+} values

These observations might call into question the relevance of Ca^{2+} to the contraction of aortic smooth muscle, but in fact clear evidence exists that Ca^{2+} is a component of force production in this muscle. For example, if the muscle is first depolarized with high K, in the absence of extracellular Ca^{2+}, and then Ca^{2+} is readmitted and increased step by step (Fig. 5), then a clear correspondence is observed between the stepwise rise in the fura-2-reported intracellular Ca^{2+} and the staircase of vasoconstrictive force.

Similarly, if intracellular Ca^{2+} is pushed up by ionophores, such as ionomycin, the increase in Ca^{2+} is constantly associated with vasoconstriction (Fig. 4; see also [3]). However these contractions take place when intracellular Ca^{2+} is raised to micromolar levels, well above the typical physiological range [3].

Finally, one can understand – and summarize – all of these results by admitting that vasoconstriction depends on intracellular Ca^{2+} in every case, but that the relation linking Ca^{2+} and contraction is variable [3]. Agonists such as norepinephrine and, in our hands, also high-K depolarization shift the force-to-Ca^{2+} curve toward lower values of Ca^{2+}, a phenomenon of "calcium sensitization." Evidence of calcium sensitization was produced also with aequorin in the intact coronary artery [4] and with α-toxin permeabilization in the rabbit mesenteric artery [5]. Sensitization occurs also with phorbol esters, putative activators of protein kinase C (PKC) [6]. With these compounds, force can be developed without a detectable increase in intracellular Ca^{2+}. There is an open question about the participation of PKC in vascular smooth

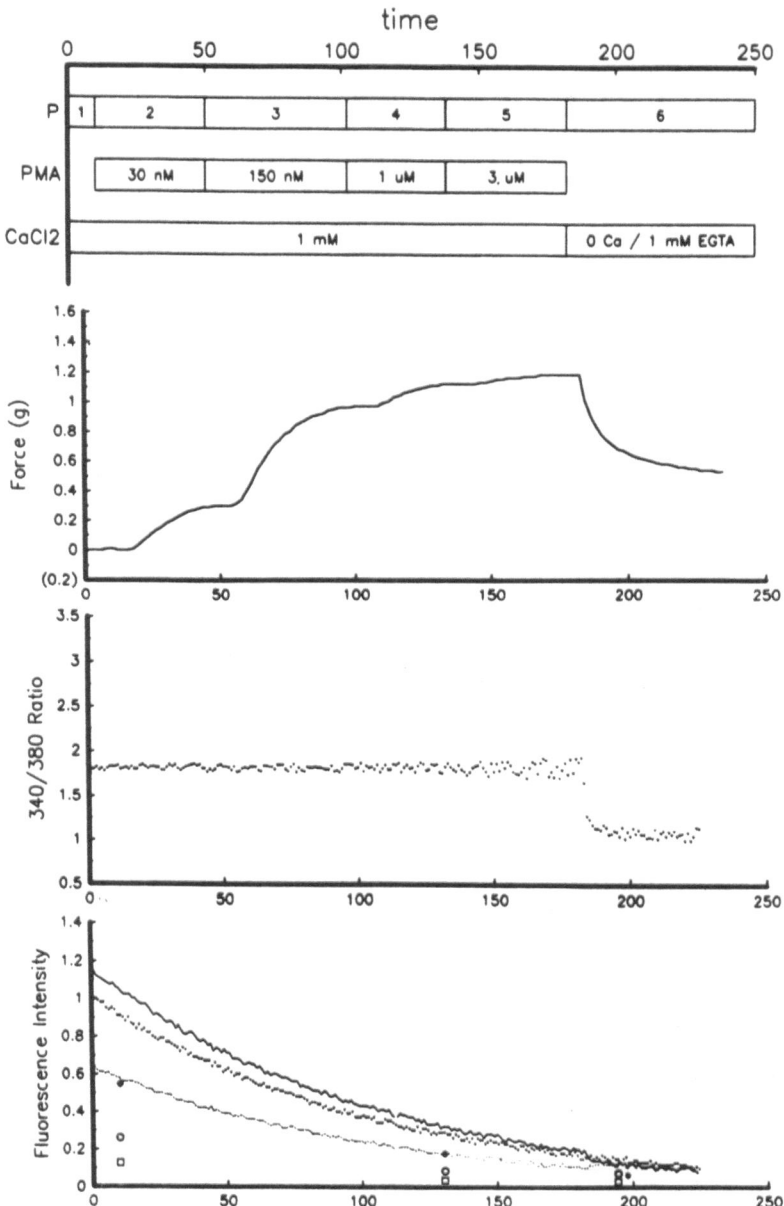

Fig. 6. Response of rat aortic smooth muscle to increasing concentrations of phorbol myristate acetate (PMA). The *top* tracing is the force record. The *middle* tracing represents the 340/380 fluorescence ratio of fura-2. The *bottom* tracing shows the fluorescence intensity at single wavelengths. The *thick solid line* represents fluorescence excited at 340 nm; *the faint line*, 380-nm excitation; the *heavy dotted line*, 359-nm excitation. Emission was read at 510 nm in every case. Note the time scale on the abscissa and the delay between PMA additions and the resulting force response. The last period represents the passage to a Ca²⁺-free, EGTA-containing medium. Abbreviations as in Fig. 2

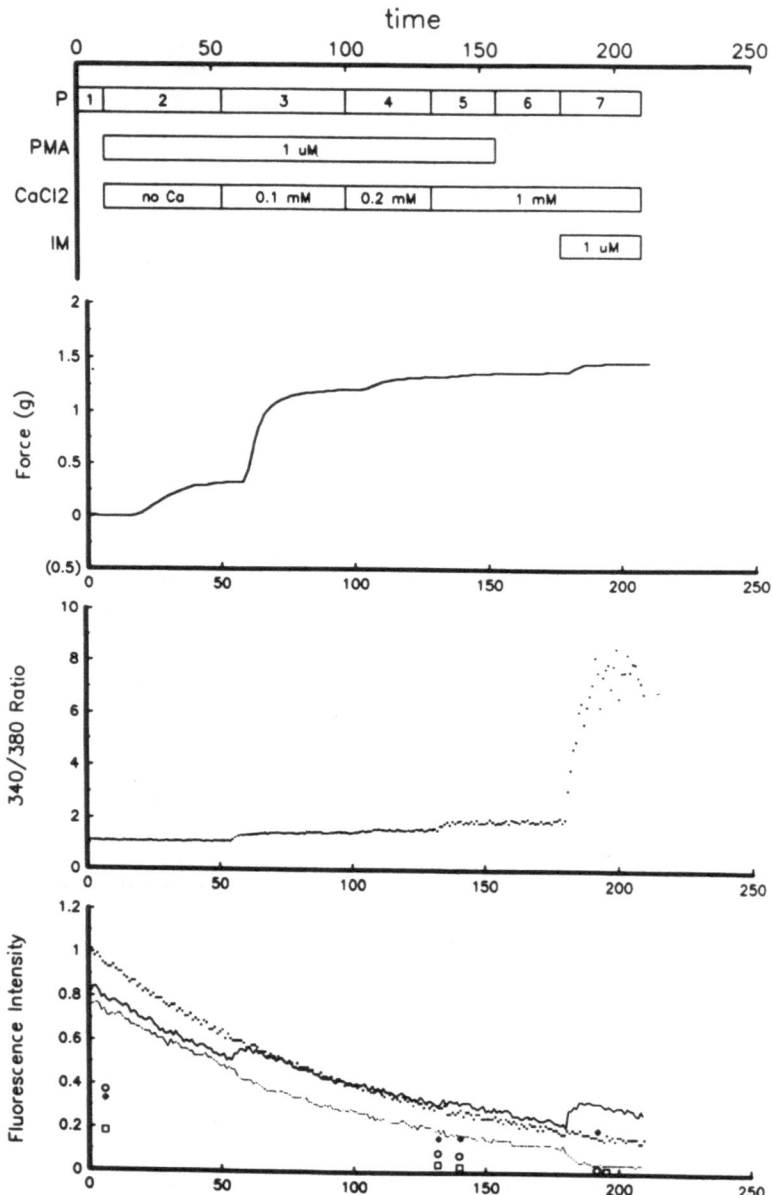

Fig. 7. Force response to changes in extracellular Ca^{2+} in the presence of $1\,\mu M$ PMA. The changes of intracellular Ca^{2+} are represented by the 340/380 fluorescence ratio in the *middle panel*. The fluorescence intensity at 340, 359 and 380 nm excitation is plotted in the *bottom panel*. Abbreviations and symbols as in Fig. 6; *IM*, ionomycin

muscle physiology, on the basis of at least two arguments: (1) phorbol esters do act as vasoconstrictors, (2) a number of agonists induce phosphoinositide hydrolysis, resulting in production of diacylglycerols, the natural activators of PKC [7]. PKC activation, then, might account, to some extent, for calcium sensitization.

During exposure of rat aortic smooth muscle to phorbol ester, such as phorbol myristate acetate (PMA, Fig. 6), the pattern reported by fura-2 is similar to that reported by aequorin in coronary artery [6]. The Ca^{2+} signal of fura-2 is practically unchanged. One point to notice is that contractions induced by phorbol esters (PMA in Fig. 6) are very slow and are delayed relative to the time of PMA addition. There is a silent period of about 10–15 min before tension starts to rise. It should also be noted that when extracellular Ca^{2+} is removed and EGTA is added to the medium, making intracellular Ca^{2+} fall to very low levels, the force is approximately halved, but not zeroed. The question then arises whether contractions induced by phorbol esters are actually Ca^{2+}-dependent or independent.

Figure 7 shows again a Ca^{2+} staircase experiment, with extracellular. Ca^{2+} raised step by step up to $1 mM$ in the presence of PMA. Intracellular Ca^{2+} rises accordingly, and a slow force response follows each Ca^{2+} addition. This suggests that phorbol ester-induced force development is Ca^{2+}-dependent, although the tensile capacity is saturated at very low levels of intracellular Ca^{2+}.

To summarize, phorbol esters are calcium-sensitizing agents, in that they evoke vasoconstriction at very low levels of intracellular Ca^{2+}. However, some kinetic aspects of phorbol ester-induced contractions are difficult to reconcile with physiologic vasoconstriction: phorbol ester-induced contractions are an order of magnitude slower than those induced by normal agonists; they are not readily reversed by washout of these agents; and finally, Ca^{2+} changes are rather slowly coupled to contraction changes in the presence of phorbol esters (Fig. 7). The relevance of these observations to the physiologic role of PKC is uncertain; stimulation by phorbol esters is not necessarily equivalent to natural activation by diacylglycerols.

In conclusion, we have observed that vascular tone can be modulated by changes in intracellular Ca^{2+}, but many vascular agents also modulate the sensitivity of contractile – regulatory proteins to Ca^{2+}. This implies that the force-to-calcium relationship is a variable quantity. The mechanism of calcium sensitization appears to be at least as important as the changes in intracellular Ca^{2+}. This mechanism of calcium sensitization remains to be understood.

References

1. Morgan JP, Morgan KG (1984) Stimulus-specific patterns of intracellular calcium levels in smooth muscle of ferret portal vein. J Physiol (Lond) 351:155–167
2. Grynkiewicz G, Poenie M, Tsien RY (1985) A new generation of Ca^{2+} indicators with greatly improved fluorescence properties. J Biol Chem 260:3440–3450
3. Bruschi G, Bruschi ME, Regolisti G Borghetti A (1988) Myoplasmic Ca^{2+}-force relationship studied with fura-2 during stimulation of rat aortic smooth muscle Am J Physiol 254:H840–H850
4. Bradley AB Morgan KG (1987) Alterations in cytoplasmic calcium sensitivity during porcine coronary artery contractions as detected by aequorin J Physiol (Lond) 385: 437–448
5. Nishimura J, Kolber M, Van Breemen C (1988) Norepinephrine and GTP-γ-S increase myofilament Ca^{2+} sensitivity in α-toxin permeabilized arterial smooth muscle. Biochem Biophys Res Commun 157:677–683
6. Jiang MJ Morgan KG (1987) Intracellular calcium levels in phorbol ester-induced contractions of vascular muscle. Am J Physiol 253:H1365–H1371
7. Volpe P, Di Virgilio F, Bruschi G, Regolisti G Pozzan T (1989) Phosphoinositide metabolism and excitation – contraction coupling in smooth, cardiac and skeletal muscles. In: Mitchell RH, Drumond AH, Downes CP (eds) Inositol lipids in cell signalling. Academic Press London

Superficial Sarcoplasmic Reticulum Regulates Activated and Steady-State Cytosolic Ca^{2+} Concentrations in Vascular Smooth Muscle

J. Barakeh[1], M.B. Cannell[1], R. Khalil[1], O. Thastrup[2], and C. van Breemen[1]

[1] Department of Pharmacology (R189), University of Miami, PO Box 016189, Miami, FL 33101, USA
[2] Department of Clinical Chemistry, University Hospital, Rigshopitalet, Blegdamsvej 9, DK-2100 Copenhagen, Denmark

Introduction

Force development of depolarized smooth muscle has been found to be more strongly correlated with the rate of Ca^{2+} entry than with net Ca^{2+} uptake [27]. These observations led to the superficial buffer barrier hypothesis, which states that a fraction of the Ca^{2+} entering vascular smooth muscle cells during depolarization is accumulated by sarcoplasmic reticulum located near the cell surface (superficial SR) before reaching the myofilaments. Subsequently, Loutzenhiser and van Breemen [11] were able to dissociate (at least transiently) stimulated Ca^{2+} influx from contraction by a prior depletion of SR Ca^{2+} using norepinephrine (NE) in a Ca^{2+}-free solution. Tension did not develop until SR Ca^{2+} levels were restored to near resting concentrations, which suggests that the agonist-releasable Ca^{2+} pool can be refilled by voltage-stimulated Ca^{2+} entry, and supports the superficial buffer barrier hypothesis. In addition, the response of this tissue to U-44069 (a prostaglandin H_2 analogue which causes the release of a fraction of the NE-releasable intracellular Ca^{2+} pool) was tested in the presence of La^{3+} and D600. Both these agents reduced the contractile response to U-44069 by approximately 50%. However, repetitive contractions could be elicited in the presence of D600, while only a single response was obtainable in the presence of La^{3+}, which suggested that the agonist-releasable Ca^{2+} pool can be refilled by passive Ca^{2+} entry. A later study by Hwang and van Breemen [5] demonstrated that caffeine releases the Ca^{2+} taken up by the superficial SR during BAY K8644-stimulated Ca^{2+} influx. From these findings it may be concluded that tension development from activation of voltage-gated calcium channels is dependent upon the Ca^{2+} content of the SR, which can remove a fraction of Ca^{2+} entering the cell, and that this stored Ca^{2+} is at least a part of the agonist-releasable Ca^{2+} pool.

The superficial buffer barrier hypothesis predicts that microscopic Ca^{2+} gradients will exist near the inner surface of the cell membrane. Rembold [18] has recently provided evidence for such intracellular Ca^{2+} gradients in

G. Bruschi A. Borghetti (Eds.)
Cellular Aspects of Hypertension
© Springer-Verlag Berlin · Heidelberg 1991

arterial smooth muscle. He found that the rate of Ca^{2+} entry as measured by aequorin could be uncoupled from the rate of force development and myosin phosphorylation by depleting the SR with histamine. A subsarcolemmal Ca^{2+} gradient was inferred from the observation that the muscle was completely relaxed at a time when the average Ca^{2+} concentration, as indicated by aequorin luminescence, was well in excess of the threshold for myofilament activation.

With the advent of fluorescent imaging techniques, it has been possible to obtain direct visualization of intracellular Ca^{2+} ($[Ca^{2+}]_i$) gradients. Rose and Lowenstein [19] first provided evidence that living cells, in this case the salivary gland of *Chironomus*, could support local internal Ca^{2+} gradients. Video imaging of fura-2 fluorescence in bovine adrenal chromaffin cells has shown that the rise in $[Ca^{2+}]_i$ in response to high K^+ occurs in three distinct phases: (1) a rise in $[Ca^{2+}]_i$ highly localized to the subplasmalemmal space, (2) a general, homogeneous rise in $[Ca^{2+}]_i$ occurring $1-2\,s$ after the first phase, and (3) a large increase in $[Ca^{2+}]_i$ observable in central regions of the cell [6]. Recent electrophysiological studies have shown that Ca^{2+}-sensitive K^+ channels in smooth muscle plasmalemma exhibit spontaneous activity which is not matched to force production, suggesting that there can be increases in Ca^{2+} near the plasmalemma that are functionally distinct from those detected by the contractile machinery [1]. These findings are consistent with a model of Ca^{2+} homeostasis in smooth muscle in which peripherally located SR will sequester Ca^{2+} entering the cell, and release Ca^{2+} preferentially towards the plasmalemma to be ultimately extruded to the extracellular space.

We propose that the SR-mediated Ca^{2+} uptake determines to some extent the cytosolic force – $[Ca^{2+}]_i$ relationship of high K^+-induced contractions. We therefore examined the role of SR Ca^{2+} storage and uptake in this relationship by simultaneous measurement of cytosolic Ca^{2+} using fura-2 and isometric contraction [21]. The relationship of cytosolic Ca^{2+} to force is compared before and after depletion of SR Ca^{2+} with NE. To further evaluate the functional integration of the SR and plasmalemma (a necessary implication of the superficial buffer barrier hypothesis), we employed the plant alkaloid ryanodine and the sesquiterpene lactone thapsigargin. Both compounds selectively interfere with sequestration of Ca^{2+} by the SR. The use of such agents is a rational attempt to examine the effects of perturbation of SR Ca^{2+} buffering on steady-state Ca^{2+} homeostasis more selectively than with agonists. Finally, a model for steady-state Ca^{2+} regulation by the SR is presented.

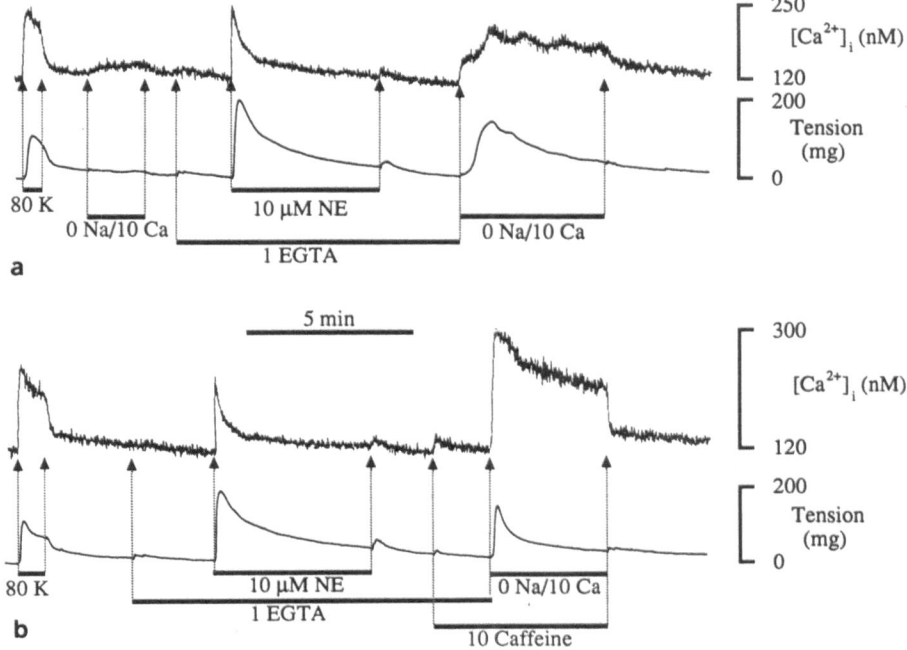

Fig. 1a,b. Effect of NE with Ca^{2+}-free/1 mM EGTA (1G) on [Ca^{2+}]$_i$ and tension response to 130 mM Li$^+$/10 mM Ca^{2+} in rabbit inferior vena cava in the absence (**a**) and presence (**b**) of 10 mM caffeine. Rabbit inferior vena cava loaded with fura-2 was equilibrated in PSS under 500 mg tension for 30 min. The tissue was stimulated with 80 mM KCl, washed with PSS, and stimulated with 130 Li$^+$/10 mM Ca^{2+}. PSS was then replaced with 1 mM EGTA, PSS, and the tissue exposed to NE for 8 min. Replacement of 1 mM EGTA, PSS with 130 Li$^+$/10 mM Ca^{2+} was then performed (**a**). In (**b**), 10 mM caffeine was included in the 1 mM EGTA, PSS following NE stimulation and during subsequent replacement with 130 Li$^+$/10 mM Ca^{2+}

Materials and Methods

The inferior vena cava from male albino New Zealand rabbits weighing 2–3 kg was isolated and opened longitudinally. Endothelium was removed by gently rubbing the intimal surface with filter paper moistened with physiological saline solution (PSS) [3]. The tissue was then cut into rectangular sheets (7 mm width and 15 mm length), and tied to stainless steel tissue holders [7]. The mounted preparation was placed in a 3.5-ml polystyrene cuvette containing PSS, and positioned in the temperature-controlled chamber of the spectrofluorimeter (Spex Fluorolog, Spex Industries, New Jersey) to measure the autofluorescence. The luminal side of the tissue was illuminated with alternating wavelengths of 340 and 380 nm controlled by an electronic chopper, and 505 nm emitted light from the strip was collected in the front face mode via a photomultiplier. Excitation at either 340 or 380 nm was synchronized with emission by software designed by Spex Industries. The mounted muscle strips were then loaded with fura-2 by incubation in the dark in a solution containing $5 \mu M$ fura-2/AM, 0.5% cremophorel, and 3 mg/ml bovine albumin [17] for 2.5 h at room temperature. The tissue was then rinsed with PSS for 20 min.

Experiments were performed using the fluorimeter described above. The muscle strip was held vertically in the cuvette with one end attached to a strain gauge transducer to measure contraction. After the individual auto-fluorescences at 340 nm and 380 nm were subtracted, the ratio of fluorescence due to excitation at 340 nm (F_{340}) to that at 380 nm (F_{380}) was calculated and converted to $[Ca^{2+}]_i$ levels using the expression [4]

$$[Ca^{2+}]_i = K_d(R - R_{min})/(R_{max} - R)S_{f2}/S_{b2}$$

in which K_d, the dissociation constant of the Ca^{2+}–fura-2 complex, was assumed to be 224 nM [28], R_{max} was the measured ratio $R_{340/380}$ in the presence of $10 \mu M$ ionomycin in normal PSS, and R_{min} the ratio in the presence of $10 \mu M$ ionomycin int Ca^{2+}-free (10 mM EGTA) PSS. S_{f2} and S_{b2} are the signals recorded due to 380 nm excitation in $10 \mu M$ ionomycin/Ca^{2+}-free and $10 \mu M$ ionomycin/normal PSS solutions, respectively.

PSS contained (in millimolar): NaCl 140, KCl 5, CaCl₂ 1.5, MgCl₂ 1, glucose 10, HEPES 5, pH 7.2. For 80-mM K⁺ solutions equimolar concentrations of KCl were substituted for NaCl, and $10 \mu M$ phentolamine was included. NaCl was completely replaced with LiCl in 130 mM Li⁺/10 mM Ca^{2+} solutions. In Ca^{2+}-free solutions, CaCl₂ was omitted and 1 mM EGTA was included. Drugs were obtained from the following sources: norepinephrine and caffeine (Sigma), phentolamine (Ciba), ionomycin (Calbiochem), ryanodine (a gift from Dr. M.T. Nelson), thapsigargin (supplied by O. Thastrup). All other chemicals used were of analytical grade.

Results

The objective of the experiment the results of which are shown in Fig. 1 is to stimulate a moderate Ca^{2+} influx through the leak and Na⁺/Ca^{2+} exchange pathways and then compare its effects on tension and $[Ca^{2+}]_i$ when the SR is taking up Ca^{2+} to those when Ca^{2+} accumulation is blocked by caffeine. The vena cava was initially stimulated with 80 mM K⁺ to elicit a control contraction. Tension increased in a parallel fashion with $[Ca^{2+}]_i$ with a delay of less than 1 s and could be reversed by washout. Addition of 130 mM Li⁺/10 mM Ca^{2+} had little effect on either $[Ca^{2+}]_i$ or force, suggesting that Na⁺/Ca^{2+} exchange exerts a relatively minor influence on vascular tone. Depletion of the SR with $10 \mu M$ NE in Ca^{2+}-free solution caused a transient increase in $[Ca^{2+}]_i$ similar in magnitude to that evoked by 80 mM KCl, but a much greater level of tension was developed. Replacement of Ca^{2+}-free solution

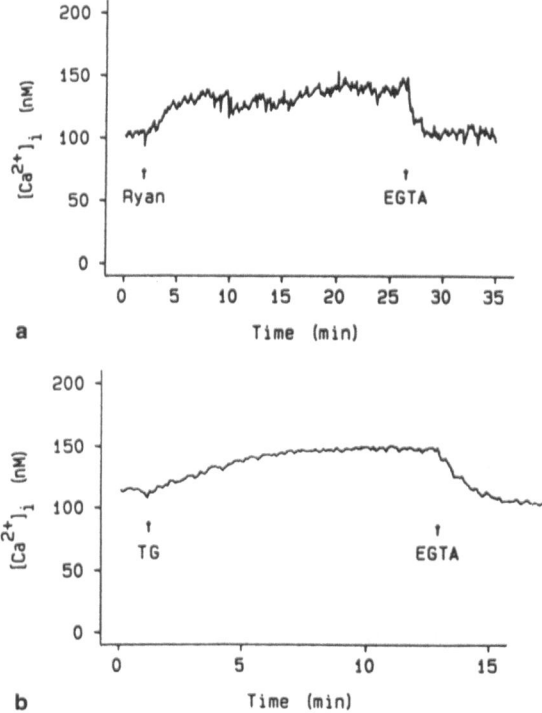

Fig. 2a,b. Effect of ryanodine (**a**) and thapsigargin (**b**) on [Ca^{2+}]$_i$ in rabbit inferior vena cava. Rabbit inferior vena cava loaded with fura-2 was equilibrated in PSS under 500 mg tension for 30 min. Ryanodine (30 μM) or thapsigargin (2 μM) was added and the changes in [Ca^{2+}]$_i$ were recorded. Ca^{2+}-free PSS containing 1 mM EGTA was added and the changes in ryanodine or thapsigargin response were observed

with 130 mM Li$^+$/10 mM Ca^{2+} produced a biphasic rise in [Ca^{2+}]$_i$ and an increase in tension in SR-depleted muscle strips (Fig. 1a). The first phase of the rise in [Ca^{2+}]$_i$ occurred immediately following addition of 130 mM Li$^+$/ 10 mM Ca^{2+}. However, force was partially uncoupled from [Ca^{2+}]$_i$, with the onset of developed tension delayed by approximately 30 s. When caffeine is used to block the accumulation of Ca^{2+} by the SR, both the uncoupling of tension and biphasic rise in [Ca^{2+}]$_i$ are abolished (Fig. 1b). The presence of caffeine also resulted in a lower contractile response to the tonic phase of [Ca^{2+}]$_i$ compared to Li$^+$ alone (Fig. 1b).

In order to study the possible contribution of the SR to steady state [Ca^{2+}]$_i$, ryanodine, a more specific inhibitor of SR Ca^{2+} accumulation, was applied as shown in Fig. 2a. Although release of Ca^{2+} from the SR or inhibition of SR Ca^{2+} uptake would be expected to only transiently increase [Ca^{2+}]$_i$, the increase in [Ca^{2+}]$_i$ is maintained throughout the period of ryanodine exposure. This increase is dependent on Ca^{2+} influx through the

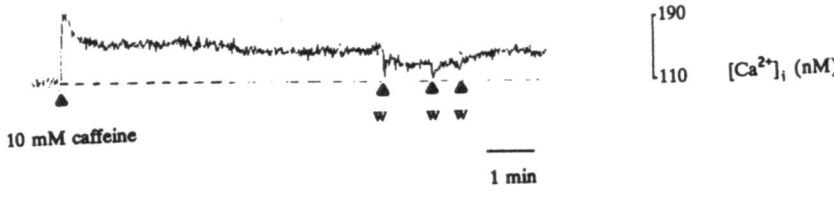

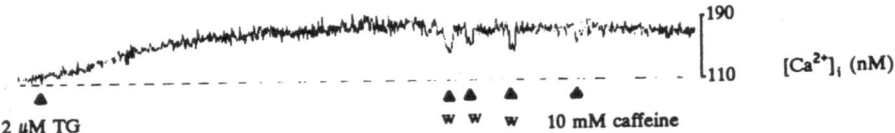

Fig. 3. Effect of thapsigargin (*TG*) on caffeine-mediated changes in $[Ca^{2+}]_i$ in rabbit inferior vena cava. Rabbit inferior vena cava loaded with fura-2 was equilibrated in PSS under 500 mg tension for 30 min. Caffeine (10 m*M*) was added and washed out. Thapsigargin (2 μ*M*) was added, changes in $[Ca^{2+}]_i$ were recorded, and the tissue was thoroughly washed out. Caffeine (10 m*M*) was again added after the washout period

leak, since removal of extracellular Ca^{2+} leads to an immediate reversal of the ryanodine effect. In Fig. 2b, inhibition of Ca^{2+} uptake into the SR by 2 μ*M* thapsigargin also produces a maintained increase in $[Ca^{2+}]_i$ which is reversed by removal of extracellular Ca^{2+}.

A previous study [5] showed that ryanodone caused the release of the caffeine-sensitive SR Ca^{2+} store in smooth muscle. To test whether thapsigargin also affects this caffeine-sensitive store, we compared the effects of caffeine on $[Ca^{2+}]_i$ before and after application of 2 μ*M* thapsigargin. Caffeine (10 m*M*) produced an immediate increase in $[Ca^{2+}]_i$ which declined to a plateau after 2–3 min (Fig. 3). This phasic increase in $[Ca^{2+}]_i$ was accompanied by a transient increase in tension (data not shown). Application of 2 μ*M* thapsigargin resulted in an increase in $[Ca^{2+}]_i$ which, unlike the caffeine-induced increase had neither a phasic component nor a concomitant change in tension. Following washout of thapsigargin, $[Ca^{2+}]_i$ remained elevated and reapplication of 10 m*M* caffeine produced no change in $[Ca^{2+}]_i$. Thus, the effect of thapsigargin on $[Ca^{2+}]_i$ is similar to that of ryanodine despite their different modes of action (see below). However, both of these compounds reduce SR Ca^{2+} accumulation.

Discussion

Depletion of SR Ca^{2+} with NE in Ca^{2+}-free solution resulted in a transient dissociation between increased $[Ca^{2+}]_i$ and force in the rabbit inferior vena cava. Since neither high K^+ nor high Li^+ have been shown to change myofilament sensitivity, the delay in the onset of force, given the time course of $[Ca^{2+}]_i$ elevation, may be due to the entry of Ca^{2+} into a region where it

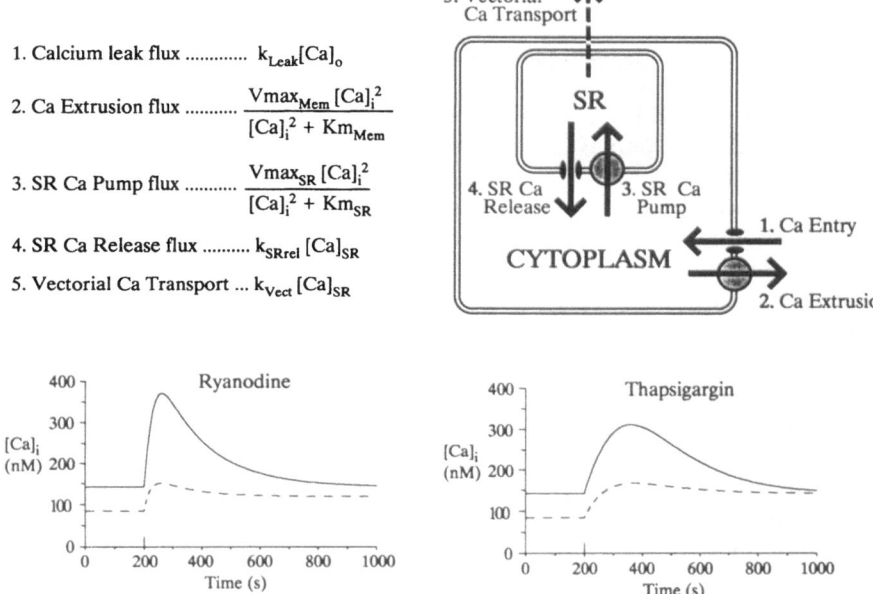

1. Calcium leak flux .... $k_{Leak}[Ca]_o$

2. Ca Extrusion flux .... $\dfrac{Vmax_{Mem}\,[Ca]_i^2}{[Ca]_i^2 + Km_{Mem}}$

3. SR Ca Pump flux .... $\dfrac{Vmax_{SR}\,[Ca]_i^2}{[Ca]_i^2 + Km_{SR}}$

4. SR Ca Release flux .... $k_{SRrel}\,[Ca]_{SR}$

5. Vectorial Ca Transport ... $k_{Vect}\,[Ca]_{SR}$

Fig. 4. A model to describe the factors affecting Ca²⁺ homeostasis in vascular smooth muscle. The following processes are assumed to be operating. *1* Passive leakage of Ca²⁺ across the plasmalemma, the rate of which depends upon the rate constant K_{Leak} and external [Ca²⁺]ᵢ. *2* Active extrusion of [Ca²⁺]ᵢ via a plasmalemmal Ca²⁺-ATPase. *3* Active uptake into the SR via another Ca²⁺-ATPase, inhibitable with thapsigargin. *4* Passive release of Ca²⁺ from the SR, the rate of which depends upon the rate constant K_{SRrel}, supposedly increased by ryanodine, and [Ca²⁺]ᵢ in the SR. *5* Preferential release of Ca²⁺ from the SR towards the plasmalemma leading to extrusion of Ca²⁺, indicated in the diagram with a *broken arrow*. Graphs depict basal [Ca²⁺]ᵢ when parameter 5 is excluded (*solid lines*) or included (*dotted lines*). The graph labeled "Ryanodine" depicts theoretical [Ca²⁺]ᵢ when K_{SRrel} of parameter 4 is increased five-fold, while the plot labeled "Thapsigargin" shows the theoretical consequence of setting $V_{SR} = 0$

can be taken up into the SR rather than activating tension directly. With this in mind, we expect that prior SR depletion would increase the delay between influx and tension production. Once the SR has refilled, the rate of Ca²⁺ accumulation by the SR will decrease and this will result in an increase in $d[Ca^{2+}]_i/dt$, as well as an increase in force production. This idea is consistent with our observations that the first phase of Ca²⁺ influx is not accompanied by force production, and that caffeine, which prevents SR Ca²⁺ accumulation, abolished both the plateau between the two phases of [Ca²⁺]ᵢ increase and the delay in force development. It is also notable that, while producing an equivalent increase in [Ca²⁺]ᵢ, NE gives a greater level of tension production than does high K⁺, most probably by increasing the sensitivity of myofilaments of Ca²⁺ [2,15]. In relation to this point, we also observe that caffeine, a phosphodiesterase inhibitor (which increases cellular concentrations of cAMP), leads to decreased myofilament sensitivity, as has been previously shown for α-toxin permeabilized arteries [14].

The superficial buffer barrier may be important in the maintenance of basal $[Ca^{2+}]_i$ as well as in buffering stimulated Ca^{2+} entry. The SR can affect steady-state $[Ca^{2+}]_i$ only if the SR Ca^{2+} pump contributes to Ca^{2+} extrusion, and this would require vectorial Ca^{2+} transport towards the surface membrane and exterior of the cell. The dynamics of this contribution of the superficial buffer barrier to basal $[Ca^{2+}]_i$ can be described with a three-compartment pump-leak model of Ca^{2+} homeostasis, illustrated in Fig. 4. The three compartments are cytosol, SR, and the extracellular space. The broken arrow indicating vectorial Ca^{2+} transport represents preferential Ca^{2+} extrusion from the SR to the extracellular space, as suggested by the superficial buffer barrier hypothesis, via a narrow junctional area where both membranes are closely apposed [22]. Dissociation and rate constants assigned to parameters in equations 1–5 listed in Fig. 4 have been chosen in accordance with ^{45}Ca flux experiments conducted in our laboratory. These equations were numerically integrated with a microcomputer and the results of the integration plotted as pictured.

The effect on steady-state $[Ca^{2+}]_i$ of substances which can alter SR Ca^{2+} handling, such as ryanodine and thapsigargin, can be predicted using the above model. In previous studies, it was found that ryanodine does not affect ^{45}Ca influx nor does it influence the rates of turnover of the Ca^{2+}-ATPase and Na^+/Ca^{2+} exchanger [8,9,23]. The specific action of ryanodine on the SR may be to lock SR Ca^{2+} channels in a subconductance open state, leading to a release of Ca^{2+} [20]. Thapsigargin will induce release of Ca^{2+} from an intracellular pool in a number of cell types [25]. In rat parotid acinar cells, thapsigargin was shown to release Ca^{2+} from an I_3-sensitive intracellular Ca^{2+} pool [24]. This action of thapsigargin does not appear to be mediated by production of inositol phosphates [6,24], nor by any activation of a G protein, phospholipase C, or a plasmalemmal receptor [26]. Thapsigargin has been shown to inhibit the 100 kD Ca^{2+}-ATPase of endoplasmic reticulum (ER) in rat hepatocytes, with no effect on the plasmalemmal Ca^{2+}-ATPase [26]. Also, ATP depletion in these cells closely paralleled the Ca^{2+}-releasing action of thapsigargin, suggesting that thapsigargin may release Ca^{2+} from ER (and possibly also SR) solely by inhibition of Ca^{2+}-ATPase-mediated uptake. Also, we have demonstrated that both thapsigargin (Fig. 3) and ryanodine [5], deplete the affeine-releasable SR Ca^{2+} store in smooth muscle. In the present study both thapsigargin and ryanodine produced a mild but maintained increase in $[Ca^{2+}]_i$, despite the fact that the SR Ca^{2+} supply is transient. These results can be simulated using the pump-leak model in Fig. 4 if vectorial SR Ca^{2+} extrusion to the extracellular space (equation 5) is added to the conventional model described by equations 1–4. In the absence of vectorial Ca^{2+} transport (solid lines), increasing the rate constant of SR Ca^{2+} release (K_{SRrel}) five-fold to simulate the action of ryanodine induces a transient increase in $[Ca^{2+}]_i$. A similar result is produced when application of thapsigargin is simulated by setting the rate of SR Ca^{2+} uptake (V_{SR}) equal to 0. However, if we include vectorial Ca^{2+} transport in the model (Fig. 4, dotted lines), simulating the actions of either ryanodine or thapsigargin produces a maintained increase in $[Ca^{2+}]_i$ as seen in the experimental results shown in Fig. 2. This simplified model of the superficial buffer barrier shows that inhibition of SR Ca^{2+}-ATPase or an increase in the SR Ca^{2+} permeability can give rise to an increase in the steady-state $[Ca^{2+}]_i$ as was seen in the experimental applications of thapsigargin and ryanodine, and is consistent with the original superficial buffer barrier hypothesis [27].

In addition to contributing to Ca^{2+} extrusion, the superficial buffer barrier may serve to generate and maintain localized Ca^{2+} gradients within the cell and, in particular, near the plasmalemma. Such Ca^{2+} gradients have been observed in several other cell types, with various attendant functions. Intracellular Ca^{2+} gradients can be linked to cell division and directional growth in neurons. In frog sympathetic neurons, depolarization results in a rise in $[Ca^{2+}]_i$ in the growth cones, which are responsible for neurotransmitter release, as well as in the soma [10]. Application of caffeine produces a rise in $[Ca^{2+}]_i$ in the soma only. Exocytosis from bovine adrenal chromaffin cells has been shown to be triggered by a subplasmalemmal rise in $[Ca^{2+}]_i$, but release of internally stored Ca^{2+} does not produce secretion [16]. A selective sub-plasmalemmal rise in $[Ca^{2+}]_i$ may also be important in epinephrine-mediated relaxation in guinea-pig taenia caecum [12]. Apamin, a toxin extracted from bee venom known to selectively antagonize Ca^{2+}-activated K^+ channel activity, converted the hyperpolarization normally caused by ATP or epine-phrine in this tissue to a transient depolarization. These results could be explained if the superficial SR effectively divides the cytoplasm into super-ficial and deep zones. In this same tissue, Nelemans and den Hertog [13] noted that α_1-adrenoceptor stimulation led to a significant increase in quin-2 fluorescence while tension decreased. Thus, the hyperpolarization and con-comitant relaxation of the taenia caecum appeared to be due to an increase in Ca^{2+} concentration in an area restricted near the inner plasmalemmal sur-face. This model would also explain the recent electrophysiological findings of Benham and Bolton [1] mentioned earlier, in which Ca^{2+}-activated K^+ channel activity was observed in smooth muscle cells without contraction. The functions of the superficial buffer barrier may therefore include separate Ca^{2+} regulation of membrane channels and cytoplasmic enzymes as well as protecting cells with large surface-to-volume ratios from activation by a relatively large Ca^{2+} leak.

References

1. Benham CD, Bolton TB (1986) Spontaneous outward currents in single visceral and vascular smooth muscle cells of the rabbit. J Physiol (Lond) 381:385–406
2. DeFeo TT, Morgan KG (1985) Calcium-force relationships as detected with aequorin in two different vascular smooth muscles of the ferret. J Physiol (Lond) 369:269–282
3. Furchgott RF, Zawadzki JV (1980) The obligatory role of endothelial cells in the relaxation of arterial smooth muscle by acetylcholine. Nature 288:373–376
4. Grynkiewicz G, Poenie M, Tsien RY (1985) A new generation of Ca^{2+} indicators with greatly improved fluorescence properties. J Biol Chem 260:3440–3450
5. Hwang KS, van Breemen C (1985) Effects of the Ca agonist Bay K8644 on ^{45}Ca influx and net Ca uptake into rabbit aortic smooth muscle. Eur J Pharmacol 116:299–305
6. Jackson TR, Patterson SI, Thastrup O, Hanley MR (1988) A novel tumour promoter, thapsigargin, transiently increases cytoplasmic free Ca^{2+} without generation of inositol phosphates in NG115–401L neuronal cells. Biochem J 253:81–86
7. Khalil R (1989) A study of the calcium regulatory mechanisms involved in maintained agonist-induced vascular smooth muscle tone. Doctoral dissertation University of Micmi, pp 33
8. Lai FA, Anderson K, Rousseau E, Liu Q, Meissner G (1988) Evidence for a Ca^{2+} channel within the ryanodine receptor complex from cardiac sarcoplasmic reticulum. Biochem Biophys Res Commun 151:441 449
9. Lai FA, Erickson HP, Rousseau E, Liu Q, Meissner G (1988) Purification and reconstitution of the calcium release channel from skeletal muscle. Nature 331:315–319
10. Lipscombe D, Madison DV, Poenie M, Reuter H, Tsien RL, Tsien RW (1988) Spatial distribution of calcium channels and cytosolic calcium transients in growth cones and cell bodies of sympathetic neurons. Proc Natl Acad Sci USA 85:2398–2407
11. Loutzenhiser R, van Breemen C (1983) Mechanisms of stimulated Ca^{2+} influx and consequences of Ca^{2+} influx inhibition. In: Merril GF, Weiss HR, Scriabine A (eds) Symposium on Ca^{2+} entry blockers, adenosine and neurohumors. Urban and Schwarzenberg, Munich, pp 73–91
12. Mass AJJ, den Hertog J, Ras R, den Akker J (1980) The action of apamin on guinea pig taenia caeci. Eur J Pharmacol 67:265–274
13. Nelemans A, den Hertog A (1987) Calcium translocation during activation of α_1-adrenoceptor and voltage-operated channels in smooth muscle cells. Eur J Pharmacol 140:39–46
14. Nishimura J, van Breemen C (1989) Direct regulation of smooth muscle contractile elements by second messengers. Biochem Biophys Res Commun 163:929–935
15. Nishimura J Kolber M, van Breemen C (1988) Norepinephrine and GTP-γ-S increase myofilament Ca^{2+} sensitivity in α-toxin permeabilized arterial smooth muscle. Biochem Biophys Res Commun 157:677–683
16. O'Sullivan AJ, Cheek TR, Moreton RB, Berridge MJ, Burgoyne RD (1989) Localization and heterogeneity of agonist-induced changes in cytosolic calcium concentration in single bovine adrenal chromaffin from video imaging of fura-2. EMBO J 8:401–411
17. Poenie M, Alderton J, Steinhardt R, Tsien RY (1986) Calcium rises abruptly and briefly throughout the cell at the onset of anaphase. Science 233:886–889
18. Rembold C (1989) Desensitization of swine arterial smooth muscle to transplasmalemmal Ca^{2+} influx. J Physiol (Lond) 416:273–290
19. Rose B, Lowenstein WR (1975) Calcium ion distribution in cytoplasm visualized by aequorin: diffusion in cytosol restricted by energized sequestering. Science 190:1204–1206

20. Rousseau E, Smith JS, Meissner G (1987) Ryanodine modifies conductance and gating behavior of single Ca^{2+} release channel. Am J Physiol 253:C364–368
21. Sato K, Ozaki H, Karaki H (1988) Changes in cytosolic calcium level in vascular smooth muscle strip measured simultaneously with contraction using fluorescent calcium indicator fura 2. J Pharmacol Exp Ther 246:294–300
22. Somlyo AV, Franzini-Armstrong C (1985) New views of smooth muscle structure using freezing, deep-etching and rotary shadowing. Experientia 41:841–856
23. Sutko JL, Ito K, Kenyon JL (1985) Ryanodine: a modifier of sarcoplasmic reticulum calcium release in striated muscle. Fed Proc 44:2984–2988
24. Takemura H, Hughes AR, Thastrup O, Putney JW Jr (1989) Activation of calcium entry by the tumor promoter thapsigargin in parotid acinar cells. J Biol Chem 264:12266–12271
25. Thastrup O (1990) Role of Ca^{2+}-ATPases in regulation of cellular Ca^{2+} signalling, as studies with the selective microsomal Ca^{2+}-ATPase inhibitor, thapsigargin. Agents Actions 29:8–15
26. Thastrup O, Dawson AP, Scharff O, Foder B, Cullen PJ, Drobak BK, Bjerrum PJ, Christensen SB, Hanley MR (1989) Thapsigargin, a novel molecular probe for studing intracellular calcium release and storage. Agents Actions 27:17–23
27. van Breemen C (1977) Calcium requirement for activation of intact aortic smooth muscle. J Physiol (Lond) 272:317–329
28. Williams DA, Fogarty KE, Tsien RY, Fay FS (1985) Calcium gradients in single smooth muscle cells revealed by the digital imaging microscope using fura-2. Nature 318:558–561

Subsarcolemmal Increase in Intracellular Ca^{2+} in Vascular Muscle Cells from Spontaneously Hypertensive Rats*

K. Hermsmeyer and P. Erne

Cardiovascular Research Laboratory, Earle A. Chiles Research Institute, Providence Medical Center, 4805 NE Glisan Street, Portland, OR 97213, USA

Introduction

Ca^{2+} channel function appears to be altered in vascular muscle from the venous side of even newborn spontaneously hypertensive rats (SHR), suggesting that genetically determined alterations in vascular muscle cells occur which are likely to contribute to development of increased peripheral resistance [7]. Specifically, there is an increase in the probability for opening of the L-type Ca^{2+} channels (thought to be important for the filling of the Ca^{2+} stores necessary for contraction) compared to that seen in vascular muscle cells from normotensive rats [16]. Recent reports from our laboratory have indicated that this alteration in Ca^{2+} channel function does not result from inherent conductance properties of the channel, but can be identified with Ca^{2+} inactivation of the L-type Ca^{2+} channel [10]. As in other kinds of cells, the high-threshold, L-type Ca^{2+} channel normally inactivates as a function not only of depolarization, but also of the accumulation of Ca^{2+} in the region immediately inside the cell surface membrane, called the subsarcolemmal space [14,17]. These experiments were carried out to determine whether there was a change in the free Ca^{2+} ion concentration in the subsarcolemmal space that might explain the altered Ca^{2+} channel function.

Earlier studies by our laboratory had suggested that there is an increase occurring specifically in the periphery of the vascular muscle cells which show abnormally increased L-channel function [5]. Using the fluorescent Ca^{2+} indicator fura-2, we have determined that while average Ca^{2+} concentration within vascular muscle cells is not abnormal in cells from hypertensives, the free Ca^{2+} concentration is elevated even at rest in localized regions ("hot spots") of vascular muscle cells from SHR. Our earlier study suffered from the limitation which characterizes all fluorescence measurements made from

* This study was supported by NIH grants HL38537 and HL38645, by the Institute for Nutrition and Cardiovascular Research of the National Dairy Board, and by the Schweizerische Stiftung für Medizinische-Biologische Stipendien.

G. Bruschi A. Borghetti (Eds.)
Cellular Aspects of Hypertension
© Springer-Verlag Berlin · Heidelberg 1991

living cells by conventional microscope techniques, which is that the depth of focus is several times the thickness of a cell, and thus would fail to detect some of the more vertically localized areas of high Ca^{2+} activity. In fact, given this limitation, it is remarkable that local areas of high Ca^{2+} activity were observable with the lower resolution fluorescence quantitation of conventional microscopes.

In these experiments, we used laser scan microscopes of confocal design to increase the resolution in the z-axis for studies of Ca^{2+}-sensitive fluorescence intensity in vascular muscle cells from hypertensive and normotensive rats. The increased resolution of the confocal laser scan microscope allows structures less than $0.5 \mu m$ in diameter to be truly resolved in three dimensions. Although the whole-image temporal resolution is not as good as can be achieved with a non-confocal microscope, time frames of $0.5 s$ can be resolved, which is sufficient to detect hot spots of Ca^{2+} activity which had been inferred previously [5], and to allow this separate examination of the subcellular regulation theory advanced earlier [5].

Methods

Cell Isolations from Azygos Veins

Primary cultures of vascular muscle cells were prepared from azygos veins of 3-day-old SHR and genetically matched Wistar-Kyoto rats (WKY), as described in detail elsewhere [8]. The cells were dissociated and plated at low density (50–100000 cells) onto polylysine-coated or uncoated glass coverslips.

Calcium Quantitation from Isolated Single Cells

Two days later, the coverslips were placed in a laminar flow chamber [9] and a single, spontaneously contracting vascular muscle cell was identified and observed with a 50×1.00 NA water immersion Leitz objective on an inverted microscope. In order to load cells with fluo3, $20 \mu l$ fluo3 acetoxymethylester (AM) stock solution ($10 \mu M$ fluo3-AM in dimethyl sulfoxide containing 0.05% pluronic 127) was added to the 300-μl chamber. After 5 min, extraneous fluo3 was washed away with ionic solution for mammals (ISM) consisting of (mM): 130 NaCl, 16 NaHCO$_3$, 4.7 KCl, 1.8 CaCl$_2$, 0.4 MgCl$_2$, 0.4 MgSO$_4$, 0.5 NaH$_2$PO$_4$ and 17 HEPES, with 5.5 dextrose and 0.03 CaNa$_2$EDTA. Temperature during the entire procedure was kept constant at 37°C.

High resolution fluorescence intensity quantification was carried out using a BioRad MRC 600 confocal laser scan microscope, using an argon ion laser at 488 nm. Incident light fluorescence excitation was used along with a DK 510 dichroic mirror and an LP 520 long-pass filter. The sensitivity of the photomultiplier was optimized for each cell being studied and then kept constant throughout that series of measurements. An electronic noise filter was optimized and kept constant as well, to produce the highest level of signal-to-noise ratio. Images were acquired by complete scanning of the entire field, with the time resolution limited by the speed of the wobbler deflection mirror responsible for the scanning. These images were acquired at a rate of 2/s.

The advantage of the laser scan confocal microscope is the very great improvement in z-axis resolution, with out-of-focus fluorescence suppressed from focal planes more than $0.7\mu m$ from the plane of focus. These high resolution images allow optical sectioning of fluorescence to better reveal localized regions of high intensity.

Determination of Ca^{2+} is based on analysis of the Ca^{2+}-dependent fluorescence ratio on a pixel/pixel basis. For analysis of intracellular Ca^{2+}, digital frame memories at standardized time points were interrogated at each pixel for fluorescence intensity in images. Regions of high or low fluo3 concentration were found to be nearly constant throughout the experiment, in contrast to Ca^{2+} activity, which was changing with contraction and relaxation. We do not stress the absolute values of Ca^{2+} activity, only the relative values, as the fluo3 is a single wavelength Ca^{2+} indicator and might also alter Ca^{2+} activity in the cytoplasm. Statistical analysis was by unpaired t tests. The $p = 0.05$ level of significance was defined as the criterion for acceptance of differences.

The drug used in these experiments was norepinephrine. Nonfading fluorescent calibration beads were provided by GTE Sylvania, Towanda, Pa. Fluo3 was purchased from Molecular Probes Inc., Eugene, Oregon and other chemicals were purchased from Sigma, St. Louis, Missouri or Fisher Scientific.

Results

Studies of the isolated, single cells from azygos veins from newborn rats showed localized areas of high Ca^{2+} intensity referred to as hot spots. Figure 1 presents a series of images collected at intervals of 0.5 s to give a total elapsed time of 6 s, arranged in horizontal rows. Peaks of contractions can be seen by areas of increased fluorescence. Calling the frames A–R sequentially, A, G, K, and N show contractions, while the lower Ca^{2+} activity occurring just before or after peak contraction is evident in the other frames. In all frames, there are notable areas of high Ca^{2+} intensity near the periphery of the cell, which must, in these laser scan confocal microscope images, result from small, three-dimensional structures we estimate to be $0.5-1.0\mu m$ in diameter. During the entire contraction and relaxation phase, there is higher-than-average Ca^{2+} activity in the subsarcolemmal (just inside the surface membrane) space.

Fig. 1. The sequential series of images represents one cell laser scanned at 2 images/s with the images arranged in horizontal rows to show contraction and relaxation cycles occurring spontaneously in this single vascular muscle cell. The cell was loaded with $10\,\mu M$ fluo3 for 5 min and each image represents a single scan. Images are referred to as *A-F* in the top row, *G-L* in the second row, and *M-R* in the third row. Images *A*, *G*, and *N* show the peak of contractions occurring in the cell, which was contracting spontaneously. The largest increase in Ca^{2+} is in *A*, where the high concentration of Ca^{2+} occurring just adjacent to the cell membrane is evident. Areas of high Ca^{2+} very near the membrane can be seen in each of the images, even extending out to the periphery of the cell. This is remarkable because the fluo3 indicator does not allow correction for thin regions of the cell where the submicron-diameter process of the cell contains such a small light path that the fluorescence intensity is much decreased. However, even in these regions, significant fluorescence is apparent in every image. Furthermore, the fluorescence intensity is rather granular in these regions. The circular region to the upper right of each of the cell images represents a separate cell that was in the field and took up a small amount of fluo3, but was not contracting.

In the contracting cell, the distribution of intracellular Ca^{2+} is consistent with small organelles such as sarcoplasmic reticulum and other internal membrane system, but does not show the nucleus, which would appear to have Ca^{2+} increases no different from other regions of the central part of the cell. This cell contracted continuously during laser observation for over 15 min. The resolution through the 1.00 water immersion objective is calculated to be $0.5\,\mu m$. These increases in intracellular Ca^{2+} during contraction are consistent with subcellular control of intracellular Ca^{2+}, based on ion regulatory proteins known to be found in intracellular membranes.

In all of the cells we studied, Ca^{2+} activity was notably heterogeneously distributed, consistent with release from intracellular organelles and Ca^{2+} cycling in the subsarcolemmal space. Ca^{2+} activity was higher, even at rest, in peripheral regions of cells from SHR, suggesting that there should have been increased inactivation of the L-type Ca^{2+} channels if the inactivation mechanism, probably dependent on a Ca^{2+}-binding protein at the internal surface of the Ca^{2+} channel, were functioning properly.

Discussion

In this study we used confocal laser scan microscopy to test the hypothesis we had previously suggested [5], that hot spots of Ca^{2+} occur in vascular muscle cells in the subsarcolemmal regions and that these hot spots can be found, under certain conditions, to be increased in vascular muscle cells of hypertensive rats as compared with normotensive rats. It is clear that in vascular muscle, released Ca^{2+} is avidly taken up by the sarcoplasmic reticulum, and there is evidence for significant numbers of high affinity Ca^{2+} pumps in both the sarcoplasmic reticulum and the sarcolemma [2,3]. Our experiments have shown that there are localized regions of high intensity within a vascular muscle cell. This corroborates the earlier studies we carried out with fura2 which suggested heterogeneity of intracellular Ca^{2+} and vascular muscle [4,5]. As we suggested earlier, physical barriers other than internal membrane systems are not necessary for occurrence of different concentrations of free Ca^{2+} within vascular muscle cells. The occurrence of such localized regions near the surface membrane is consistent with the hypothesis that this localized region could be very important for modulation of ion channels.

Alterations in control of intracellular Ca^{2+} have been one of the most consistent of the observations from hypertensive animals and humans [6]. In a recent review of vascular muscle and blood platelets, we have suggested that Ca^{2+} alterations occurring in both animal models and human hypertension could be regarded as a clue for a fundamental mechanism in cell membranes (occurring as a genetically determined factor) causing increased blood pressure [4]. Similar suggestions have also been made by others [1,11,13]. Indeed, there is evidence that abnormally low Ca^{2+} may contribute to hypertension in at least some humans [15]. The evidence accumulating (for example, the strong correlation of platelet Ca^{2+} with blood pressure [4]) provides a compelling challenge for further work on mechanisms of Ca^{2+} control in hypertension. Indeed, Kwan et al. [12] have concluded that the abnormal biochemistry of vascular muscle membranes derives largely from changes in Ca^{2+} that increasingly appears to be related to Ca^{2+} transport and Ca^{2+} binding systems. This very early report of the application of the newly developed confocal laser scan microscope to the resolution of intracellular ion dynamics suggests that there is great value in further exploration of subcellular Ca^{2+} localization related to normal functions and disease mechanisms.

References

1. Bohr DG, Webb RC (1988) Vascular smooth muscle membrane in hypertension. Ann Rev Pharmacol Toxicol 28:389–409
2. Carafoli E, Zurini M (1982) The Ca^{2+}-pumping ATPase of plasma membranes: purification, reconstruction, and properties. Biochim Biophys Acta 683:279–301
3. Daniel EE, Grover AK, Kwan CY (1982) Isolation and properties of plasma membrane from smooth muscle. Fed Proc 41:19–28

4. Erne P, Hermsmeyer K (1988) Desensitization to norepinephrine includes refractoriness of calcium release in myocardial cells. Biochem Biophys Res Commun 151:333–338
5. Erne P, Hermsmeyer K (1989) Intracellular vascular muscle calcium modulation in genetic hypertension. Hypertension 14:145–151
6. Erne P, Bolli P, Burgisser E, Buhler FR (1984) Correlation of platelet calcium with blood pressure. N Engl J Med 310:1084–1088
7. Hermsmeyer K, Erne P (1989) Cellular calcium regulation in hypertension. Am J Hypertens 2:655–658
8. Hermsmeyer K, Mason R (1982) Norepinephrine sensitivity and desensitization of cultured single vascular muscle cells. Circ Res 50:627–632
9. Hermsmeyer K, Robinson R (1977) High sensitivity of cultured cardiac muscle to autonomic agents. Am J Physiol 233:C172–C179
10. Hermsmeyer K, Rusch NJ (1989) Calcium channel alteration in genetic hypertension. Hypertension 14:(4)453
11. Kwan CY (1985) Dysfunction of calcium handling by smooth muscle in hypertension. Can J Physiol Pharmacol 63:366–374
12. Kwan CY, Belbeck L, Daniel EE (1979) Abnormal biochemistry of vascular smooth muscle plasma membrane as an important factor in the initiation and maintenance of hypertension in rats. Blood Vessels 16:259–268
13. Lau K, Eby B (1985) The role of calcium in genetic hypertension. Hypertension 7:657–667
14. Loriand G, Pacaud P, Mironneau C, Mironneau J (1986) Evidence for two distinct calcium channels in rat vascular smooth muscle cells in short-term primary culture. Pflugers Arch 407:566–568
15. McCarron DA (1982) Low serum concentrations of ionized calcium in patients with hypertension. N Engl J Med 307:226–228
16. Rusch NJ, Hermsmeyer K (1988) Calcium currents are altered in the vascular muscle cell membranes of spontaneously hypertensive rats. Circ Res 63:997–1002
17. Sturek M, Hermsmeyer K (1986) Calcium and sodium channels in spontaneously contracting vascular muscle cells. Science 233:475–478

Pathogenesis of Essential Hypertension: The Sodium Pump Inhibitor (Natriuretic Hormone) – Sodium/Calcium Exchange – Hypertension Hypothesis*

M.P. Blaustein

Department of Physiology, University of Maryland School of Medicine,
Baltimore, MD 21201, USA

Introduction

The importance of sodium and the key role of the kidneys in the pathogenesis of hypertension is widely recognized. Excessive dietary salt may not play an etiologic role in all individuals with essential hypertension, but the majority of patients with this disease, especially those with low-renin essential hypertension, respond to reduced dietary sodium and/or natriuretic drugs. Despite these well-established observations, the relationship between sodium and hypertension is still vigorously debated because the precise pathogenic mechanisms are not known.

The sodium/calcium exchanger in the plasma membrane of vascular smooth muscle cells serves as a unique link between sodium metabolism and vascular contractility [4,24]. Therefore, it seems logical to consider the possibility that this exchanger plays a central role in the elevation of blood pressure under conditions of altered sodium metabolism [3]. Until recently, however, a critical link in the chain that apparently leads from salt retention to elevated blood pressure, namely, an endogenous digitalis-like substance (EDLS), was missing. Now that this substance has been purified and characterized [18,22], it seems appropriate to re-examine the hypothesis that the EDLS and sodium/calcium exchange are intimately involved in the pathogenesis of at least some forms of essential hypertension [3]. This article describes the hypothesis and the evidence that has accrued to support it.

*Supported by a research grant from the National Institutes of Health (AR-32276).

G. Bruschi A. Borghetti (Eds.)
Cellular Aspects of Hypertension
© Springer-Verlag Berlin · Heidelberg 1991

The Role of the Kidneys

Sodium retention, as a result of renal parenchymal disease and renal failure, if untreated and uncompensated, invariably leads to the development of (secondary) hypertension. In patients with "essential" (primary) hypertension, however, there is no evidence that nonvascular renal parenchymal disease is an initiating factor – although nephrosclerosis and renal failure may develop as a consequence of chronic blood pressure elevation. A number of observations, however, provide evidence that altered renal function predisposes to the development of essential hypertension.

Several groups of investigators have compared renal sodium handling, as manifested by lithium clearance, in the offspring of normotensive and hypertensive individuals. These investigators observed that the young, normotensive offspring of hypertensive subjects, who are themselves likely to develop hypertension later in life, tend to have a lower renal lithium clearance than do the offspring of normotensive parents [25,26]. The lower lithium clearance indicates that these individuals tend to retain a salt load for a longer period of time than the offspring of normotensive parents, due to increased renal tubular reabsorption of sodium. The precise mechanisms responsible for this enhanced reabsorption of sodium (and lithium) are not known.

Additional evidence that the kidneys play a critical role in the etiology of essential hypertension comes from two retrospective studies of renal transplant patients. One detailed study [8] of six black hypertensive patients revealed that initially these patients exhibited essential hypertension with no evidence of overt renal disease. As a result of uncontrolled hypertension, however, these patients eventually developed nephrosclerosis and renal failure which required nephrectomy and renal transplantation. The fact that these individuals were subsequently able to maintain normal blood pressures without antihypertensive medication for many years implies that their hypertension was directly related to their kidneys, and was probably the consequence of a functional defect.

In another type of study [13], posttransplant hypertension in normotensive kidney transplant recipients was correlated with the incidence of hypertension in the families of the donors. Thus, in humans, as in genetic animal models [13], hypertension "follows the kidneys."

The aforementioned evidence indicates that the kidneys play a fundamental role in the pathogenesis of essential hypertension [14]. The primary disturbance seems to be a relatively reduced ability of these kidneys to excrete a salt load. This is, of course, a "permissive factor' because the development of hypertension also depends upon the dietary salt intake – the "environmental factor."

The Role of Salt

Much has been written about the importance of salt in the etiology of hypertension, even though its precise mechanism has not yet been resolved. The following are some of the key observations:

1. There is a well-documented correlation between dietary salt intake and the incidence of hypertension in various populations [2]. In those populations in which dietary sodium is less than about 25 mEq/day (600 mg/day), hypertension is virtually nonexistent, and blood pressure does not rise with age. Conversely, in populations with a very high salt intake, the incidence of hypertension and stroke is very high.
2. Low salt diets and natriuretic agents reduce blood pressure in the majority of patients with essential hypertension [23], and can also help to reduce blood pressure in patients with renal parenchymal disease. The effectiveness of the natriuretic agents can be circumvented by increasing dietary salt.
3. Sodium retention induced by exogenous administration or excessive secretion of mineralocorticoids leads to elevation of blood pressure that can be prevented by dietary sodium restriction.

The retention of sodium and subsequent development of hypertension depends upon the presence of an appropriate accompanying anion, namely, chloride. Indeed, if dietary chloride is severely restricted, positive sodium balance cannot be induced, and hypertension will not develop despite a large excess of dietary sodium [16,20]. The reason is that the main locus of the retained sodium is the blood plasma, where most of the accompanying anion must be chloride because the concentrations of the other prevalent plasma anions such as bicarbonate and phosphate are controlled by a variety of other metabolic pathways [16].

Retention of sodium (and chloride) must be accompanied by sufficient water to maintain normal plasma osmolarity. Thus, the net result is an expansion of plasma volume. Indeed, volume expansion, rather than the sodium ion, itself, appears to be the prerequisite to the elevation of blood pressure [16,20,27]. The critical question, then, is: How does the (tendency to) expanded plasma volume lead to elevation of blood pressure?

The Interrelationship Between the Cardiovascular System and the Kidneys

Systemic blood pressure (BP) is the product of the cardiac output (CO) and the total peripheral resistance (TPR): i.e., BP = CO × TPR. Initially, with an expansion of plasma volume, cardiac output will increase and thereby cause blood pressure to rise. The pressure-induced natriuresis will then tend to restore the volume to normal and thereby feed back to remove the stimulus that initially led to the increased cardiac output. Even if there is a continuing tendency toward expansion of plasma volume, however, physiological regulatory readjustments will soon return cardiac output to normal. But, under these circumstances, with a chronic tendency to increase plasma volume, the elevated blood pressure will then be maintained as a result of a normal cardiac output and increased peripheral resistance – as is the case in chronic essential hypertension. In chronic essential hypertension, plasma volume is usually normal or low-normal, but there is a continuing *tendency* toward an increased plasma volume (as a result of the *tendency* to retain salt and water); it is this *tendency* to expand the plasma volume that maintains the increase in blood pressure [16]. The elevated blood pressure helps to restore (reduce) the plasma volume back to normal via a pressure natriuresis; this critical interrelationship between the cardiovascular system and the kidneys has been emphasized by Guyton and his colleagues [14]. As long as the heart is functioning normally, with kidneys that are unable to maintain normal salt balance (and plasma volume) because of a tendency toward excessive sodium reabsorption and retention, the body is poised to maintain a normal plasma volume – even at the expense of an elevated blood pressure! In other words, the cardiovascular reflexes are reset so that cardiac output is defended (i.e., maintained at the normal level) while total peripheral resistance rises so that plasma volume can be maintained at a normal level. Note that this is the compensated state; it is a mistake to think that the normal plasma volume implies that sodium retention was not an underlying factor in the development of the hypertension. Even in established mineralocorticoid hypertension, which is induced by sodium retention, plasma volume is usually normal [16].

The transition from an elevated blood pressure caused by a high cardiac output with a normal peripheral resistance to a high blood pressure with normal cardiac output and high peripheral resistance has bee termed "whole body *autoregulation*" [14]. New mechanisms have been elucidated, however, that may explain precisely how this "autoregulation" takes place. But before we discuss how the tendency toward an expanded plasma volume leads to an increase in peripheral resistance, we should consider a corollary to the foregoing statements, namely: What happens when the kidneys are normal and the heart is compromised (i.e., cardiac output is reduced at the normal venous pressure and plasma volume)? Cardiac output will again be defended, but in this case as a result of increased plasma volume and venous pressure which stretches the heart and increases cardiac contractility (as explained by the Frank – Starling "Law of the Heart"). Taken to its extreme, when the heart (cardiac output) can no longer keep up with demand because of cardiac disease, congestive heart failure ensues. This situation will also occur in patients with essential hypertension whose hearts are excessively compromised – for example, by severe left ventricular hypertrophy. When the heart is no longer able to maintain a normal cardiac output in the presence of an elevated total peripheral resistance and normal plasma volume, the blood pressure will fall as a consequence of the reduced cardiac output.

Moreover, with the onset of congestive heart failure, the plasma volume will rise and peripheral resistance may fall in an attempt to maintain cardiac output and tissue perfusion. Thus, congestive heart failure is precipitated by a condition in which, in terms of cardiac output, demand exceeds supply, whereas hypertension results when supply exceeds demand.

Since the control of plasma volume is so central to this interrelationship between the renal and cardiovascular systems, we need to understand the fundamental mechanisms that regulate plasma volume and peripheral resistance.

Restoration and Maintenance of Normal Plasma Volume: Role of Natriuretic Hormones

When plasma volume is acutely expanded, humoral mechanisms are recruited that tend to promote natriuresis and diuresis. These include (but are not limited to) a reduction in the secretion of aldosterone (and renin) and vasopressin (antidiuretic hormone, ADH), and an increase in the secretion of atrial natriuretic peptides and a sodium pump inhibitor (EDLS). The atrial natriuretic peptides are secreted primarily by the cardiac atria and induce both natriuresis and vasodilation; reduced vasopressin secretion has similar effects. The circulating inhibitor of sodium pumps, a hormone first suggested by the work of de Wardener and his colleagues [10] will, like digitalis, also induce a mild natriuresis – in this case, by (partially) inhibiting sodium reabsorption by the renal tubule cells. These initial compensatory mechanisms will attempt to restore plasma volume to normal without elevating blood pressure: even though the EDLS has a vasoconstrictor action (see below), this will be offset by the vasodilatory effects of the atrial natriuretic peptides and the reduced ADH level.

If the plasma volume is not adequately controlled by these initial measures, however, and there is a continuing tendency toward plasma volume expansion, secretion of the EDLS will continue at a relatively high rate, and its vasoconstrictor actions will begin to prevail. The idea of a natriuretic hormone with hypertensinogenic properties was first suggested by Dahl and his co-workers more than 20 years ago [9]: Their studies on salt-sensitive and salt-resistant hypertension in rats led them to conclude that the elevation of blood pressure in the salt-sensitive animals on a high-salt diet was due to the action of a salt-excreting (natriuretic) hormone with a hypertensinogenic action. Subsequently, plasma from patients with low-renin essential hypertension was observed to contain elevated levels of a sodium pump inhibitor [17]. This hormone, EDLS, which has been purified from human plasma [18], is a low molecular weight (<1000) steroid-like compound. It behaves very much like digoxin in its ability to bind with high affinity to the sodium pump and to inhibit, selectively, cellular uptake of potassium and extrusion of sodium by the digoxin-sensitive sodium pump

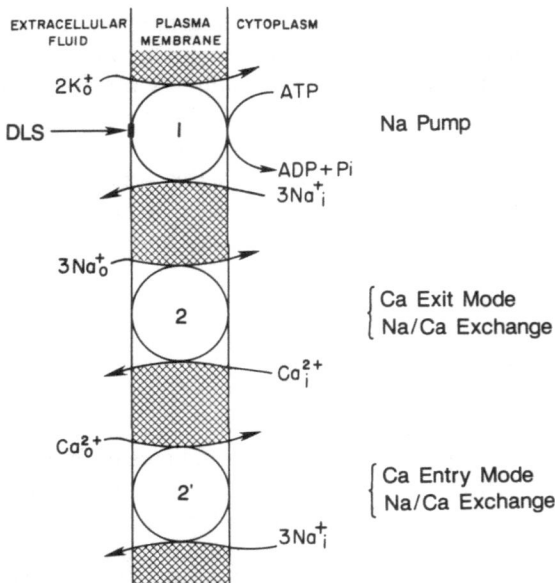

EXTRACELLULAR FLUID | PLASMA MEMBRANE | CYTOPLASM

Fig. 1. The parallel operation of *1* the sodium pump (Na$^+$, K$^+$-ATPase), the site of action of the endogenous digitalis-like substance (*DLS*), and 2, 2' the sodium/calcium exchanger in the vascular smooth muscle cell plasma membrane. The exchanger is shown operating in both the calcium exit mode (2) and the calcium entry mode (2'). The subscripts *i* and *o* refer to intracellular and extracellular ions, respectively

[18], and in its cardiotonic action. The plasma levels of this substance are elevated in humans and animals during volume expansion, and in animals treated with mineralocorticoids and salt [15].

The structure of EDLS, and its tissue of origin, are not yet known, although high levels have been found in adrenal glands [19]. Nevertheless, the fact that its action is indistinguishable from that of digoxin has enabled us to begin to elucidate the mechanisms by which it causes the blood pressure to rise. Indeed, chronic digoxin administration does not elevate blood pressure in normal humans or animals, but it does raise blood pressure in animals with compromised renal function [11] – thereby demonstrating that a digitalis-like substance can have a hypertensinogenic effect.

How a Digitalis-Like Substance Elevates Blood Pressure: Role of Sodium/Calcium Exchange

The increase in peripheral resistance that is the hallmark of chronic essential hypertension is the consequence of functional and/or structural vasoconstriction. The immediate trigger for vascular smooth muscle contraction

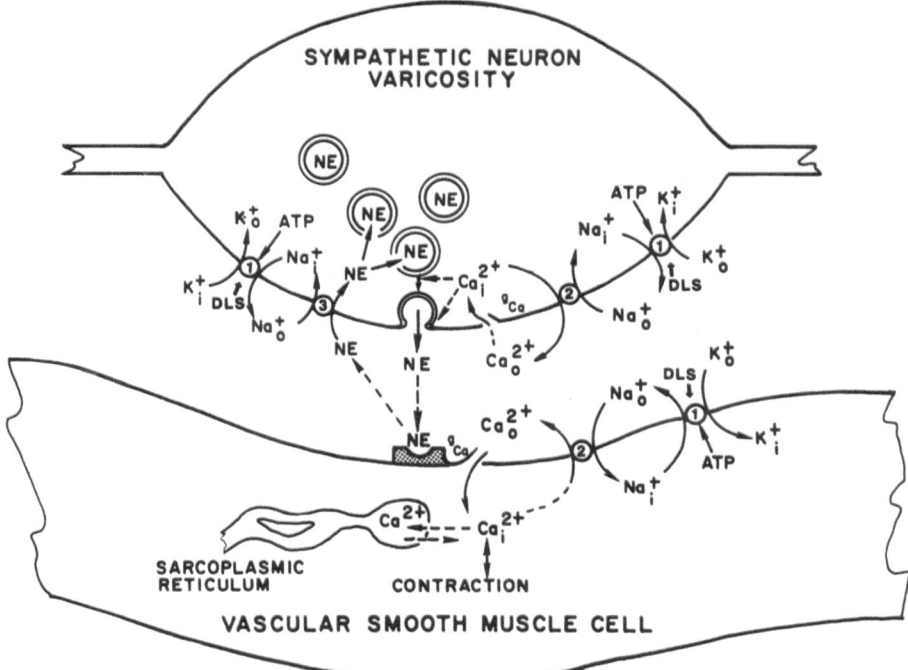

Fig. 2. Diagram of a sympathetic neuron varicosity and a vascular smooth muscle cell showing the various mechanisms described in this article that can influence the free calcium ion concentration in the vascular smooth muscle cell. *NE*, norepinephrine; *DLS*, digitalis-like substance; g_{Ca}, norepinephrine receptor-operated calcium conductance mechanisms (channels). *1*, sodium pump; *2*, sodium/calcium exchanger; *3*, sodium-norepinephrine co-transporter. See text and Blaustein and Hamlyn [5] for further details

(and thus, vasoconstriction) is a rise in the cytosolic free calcium concentration. Moreover, raised cytosolic calcium may also promote secretion (of the interstitial collagen matrix in the vascular wall, for example) and may induce cell hypertrophy and proliferation (hyperplasia), all of which will increase vascular wall thickness and narrow the vascular lumen. We must therefore determine how defective sodium metabolism and elevated plasma levels of a sodium pump inhibitor alter vascular wall calcium metabolism. A critical link in this chain is the sodium/calcium exchanger located in the plasma membrane of the vascular smooth muscle cells (as well as in most other types of cells). This exchanger helps to modulate the cytosolic free calcium concentration and contractility; moreover, as a result of its influence on cytosolic calcium, the exchanger also indirectly controls the amount of calcium that is stored in the vascular smooth muscle sarcoplasmic reticulum and, thus, is available for release into the cytosol whenever the cells are activated [1,4,6,28]. These effects may be synergistic with those

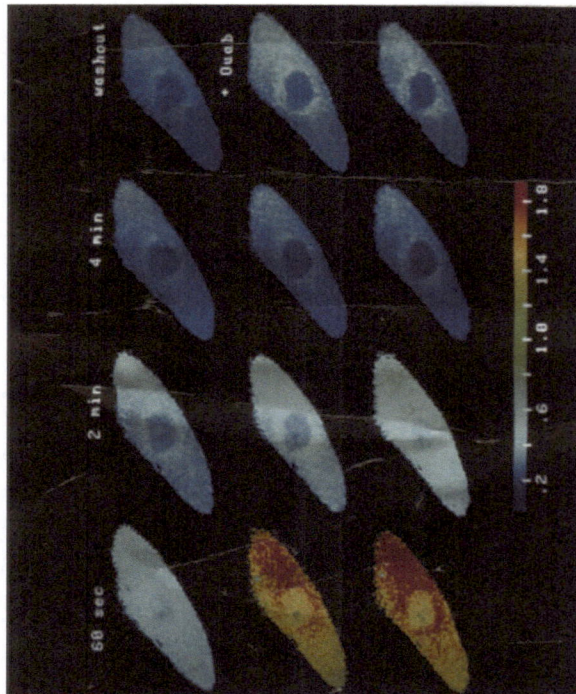

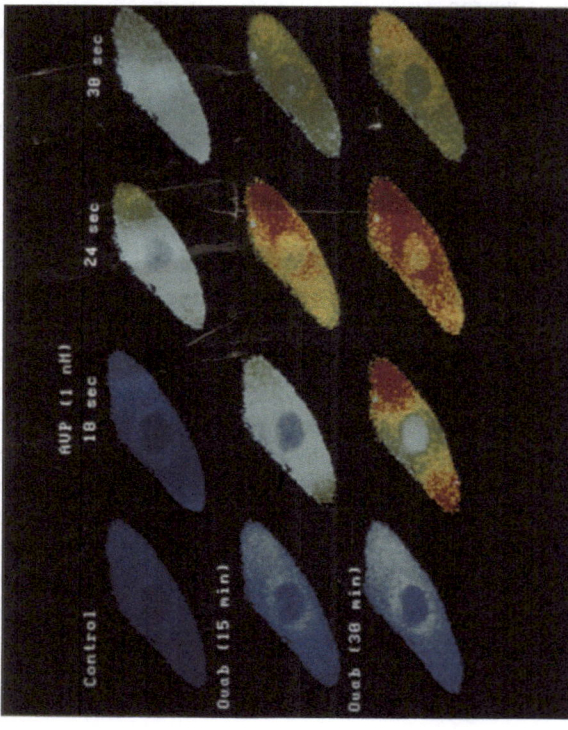

Fig. 3. Effects of ouabain on resting and arginine vasopressin- (*AVP-*) evoked apparent intracellular free calcium levels, [Ca²⁺]$_{App}$, in a cultured A7r5 cell (derived from fetal rat aorta). The cell was loaded with the membrane-permeable acetoxymethyl ester of the calcium-sensitive fluorochrome, fura-2; [Ca²⁺]$_{App}$ was determined by digital imaging methods. *Upper row* of images was obtained under "control" conditions: *upper left* image shows [Ca²⁺]$_{App}$ in the unstimulated cell, and subsequent images illustrate the time-course of changes in [Ca²⁺]$_{App}$ evoked by 1 n*M* AVP. After washout of AVP (*upper right hand image*) 3 m*M* ouabain was added to the medium. Images in *second* and *third rows* illustrate, respectively, the [Ca²⁺]$_{App}$ after 15 and 30 min of exposure to ouabain in the unstimulated cell (*lefthand* images) and at various times (as shown) after the introduction of 1 µ*M* AVP. Scale indicates [Ca²⁺] in µ*M*. Note that ouabain increased [Ca²⁺]$_{App}$ in the resting cell, especially in the perinuclear region, where most of the intracellular organelles, including the sarcoplasmic reticulum, are located. Ouabain also augmented the transient increase in [Ca²⁺]$_{App}$ evoked by AVP, and caused [Ca²⁺]$_{App}$ to oscillate. (Reprinted from Bova et al. [7], with permission)

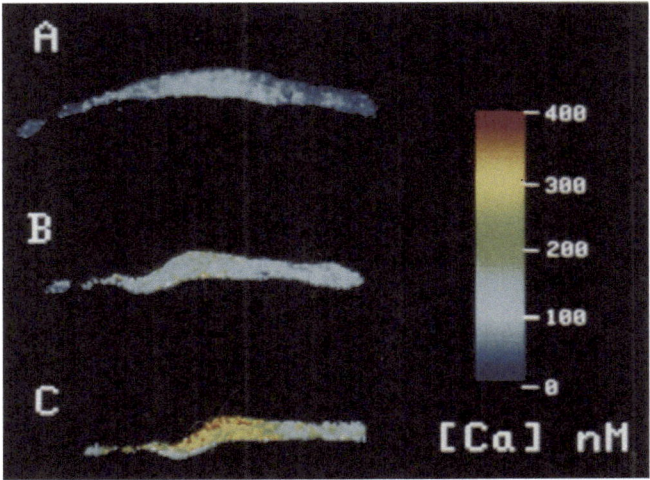

Fig. 4. Effects of potassium-free media and removal of external sodium on a fura-2-loaded bovine tail artery cell. Image *A*: $[Ca^{2+}]_{App}$ under control conditions Image *B*: When external potassium was removed, $[Ca^{2+}]_{App}$ rose, especially in the perinuclear region, and the cell shortened. Image *C*: Subsequent replacement of external sodium by *N*-methyl glucamine caused $[Ca^{2+}]_{App}$ to increase further and the cell to shorten more. Scale indicates $[Ca^{2+}]$ in n*M* (Data from Goldman and Blaustein [12], published with permission)

of sodium pump inhibition in vascular sympathetic neurons: EDLS can be expected to enhance norepinephrine secretion and inhibit re-uptake [5,21].

The sodium/calcium exchanger can move calcium either into or out of the cells, in exchange for sodium, which moves in the opposite direction (Fig. 1). The direction in which the exchanger moves net calcium depends upon the sodium and calcium concentration gradients and upon the electrical potential difference across the plasma membrane. Indeed, during a cell's activity cycle, the exchanger may move calcium into the cell when it is depolarized, and out of the cell when it is repolarized [4]. The plasma sodium and calcium concentrations are normally maintained relatively constant, and the cytosolic sodium concentration is controlled by the plasma membrane sodium pump (which operates in parallel with the sodium/calcium exchanger: Fig. 1). Now consider what happens when the cytosolic sodium concentration is elevated slightly and the inwardly-directed sodium gradient is therefore reduced as a result of partial inhibition of the sodium pump by the EDLS: This will tend to elevate the cytosolic free calcium concentration by inhibiting calcium extrusion and, simultaneously, promoting calcium entry via the sodium/calcium exchanger (Fig. 1 and 2). Because the exchanger's coupling ratio is 3 sodium ions:1 calcium ion [4], a relatively small increase in the cytosolic sodium concentration should have a relatively large effect on the cytosolic free calcium concentration: For example, a 10%

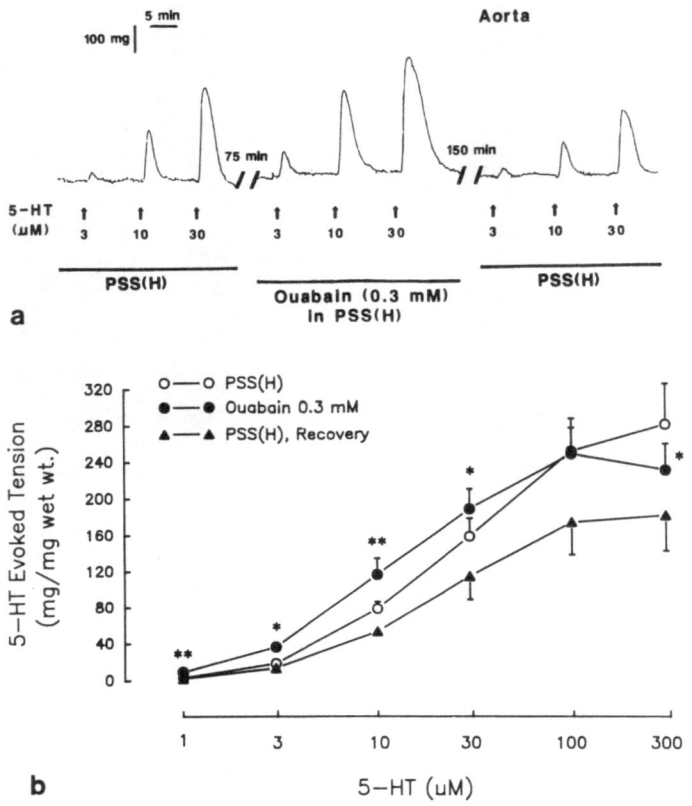

Fig. 5 a,b Effect of ouabain on serotonin- (5-HT-) evoked contractions of rings of rat aorta. **a** Isometric tension responses of an aortic ring to various concentrations of 5-HT (introduced very briefly into the superfusion fluid) under control condition (*left*), after addition of 0.3 m*M* ouabain (*middle*), and after washout of ouabain (*right*). **b** 5-HT dose-response curves before (*open circles*), during (*solid circles*), and after (*solid triangles*) treatment of rat aortic rings with 0.3 m*M* ouabain. *PSS(H)*, HEPES-buffered physiological salt solution. (Reprinted from Bova et al. [7], with permission)

increase in cytosolic sodium might be sufficient to increase cytosolic free calcium by 30%–40% or more. Perhaps even more important, the store of calcium in the intracellular compartment (the sarcoplasmic reticulum) will also be increased, so that more calcium will now be available for release into the cytosol whenever the cells are activated by vasoconstrictors such as norepinephrine, angiotensin II, and vasopressin.

Increased cytosolic free calcium levels and increased storage of calcium in intracellular organelles (including the sarcoplasmic reticulum) in response to sodium pump inhibition can be observed directly with digital imaging methods in vascular smooth muscle cells loaded with fura-2 (Figs. 3, 4; see also [7]). Vasoconstrictors evoke augmented calcium transients (transient increases in the cytosolic free calcium level) in these sodium pump-inhibited cells (Fig. 3). Such elevated cytosolic and sarcoplasmic reticulum calcium levels in resting cells, and augmented calcium transients in activated cells, probably account for the amplified contractile responses that are induced by vasoconstrictors in sodium pump-inhibited vascular smooth muscle (Figs 5, 6; see also [7]). Comparable augmented contractile responses are observed in arteries exposed to purified human EDLS [6].

The dominant influence on blood pressure is manifested in the smooth muscle cells of the small muscular arteries, in which cytosolic free calcium is constantly maintained above the contraction threshold (i.e., these vessels exhibit tonic vasoconstriction, or "tone", because the smooth muscle cells are always partially contracted). In these tonically activated cells, elevation of cytosolic sodium, and thus also calcium, will directly enhance vasoconstriction and thereby increase peripheral vascular resistance. In addition, the chronically elevated cytosolic free calcium may enhance collagen secretion and promote hypertrophy and hyperplasia, because these are also calcium-regulated phenomena.

Inhibition of the sodium pump in sympathetic neurons by the EDLS will contribute further to the increased vascular tone in two ways [5] (Fig. 2): (1) These neurons also have a sodium/calcium exchanger in their plasma membranes [5,21], and catecholamine secretion is triggered by a rise in cytosolic free calcium. Sodium pump inhibition will therefore enhance catecholamine secretion. (2) Termination of catecholamine action on the smooth muscle cells occurs primarily by re-uptake of the transmitter into the neurons, and re-uptake is mediated by a sodium gradient-dependent co-transport of catecholamine and sodium. Therefore, sodium pump inhibition will cause the catecholamine levels to remain elevated in the clefts between the sympathetic neuron varicosities and the vascular smooth muscle cells for a longer-than-normal period of time.

Clearly, all of the aforementioned mechanisms, mediated by EDLS-dependent inhibition of sodium pumps in the smooth muscle cells and sympathetic neurons, will have synergistic effects in promoting vasoconstriction and elevating peripheral resistance and blood pressure.

Summary: Putting the Components Together

All the pieces of the puzzle can now be fitted into place with the aid of the diagram in Figure 7. When more sodium is ingested than the kidneys are able to handle, there is a (transient) slight positive sodium balance; as a result, sodium, chloride, and water are retained, leading to an expansion of plasma volume. The initial physiological responses include (increased) secretion of atrial natriuretic peptides and the EDLS, and inhibition of vasopressin and aldosterone secretion. The net effect is directly enhanced natriuresis and diuresis, and a reduction in plasma volume, with no significant effect on blood pressure. However, if there is a continuing tendency to sodium retention and volume expansion, the capacity of the aforementioned

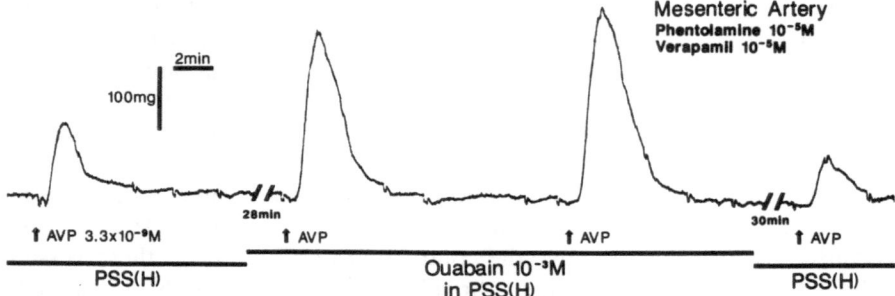

Fig. 6. Effect of 1 mM ouabain on 5-HT-evoked contractions in a ring of a second-order branch of rat mesenteric artery. In this experiment, 10 μM verapamil was present to block calcium channels and 10 μM phentolamine was present to block α-receptors. As in the aorta (Fig. 5), ouabain reversibly augmented the response to 5-HT. (Reprinted from Bova et al. [7], with permission)

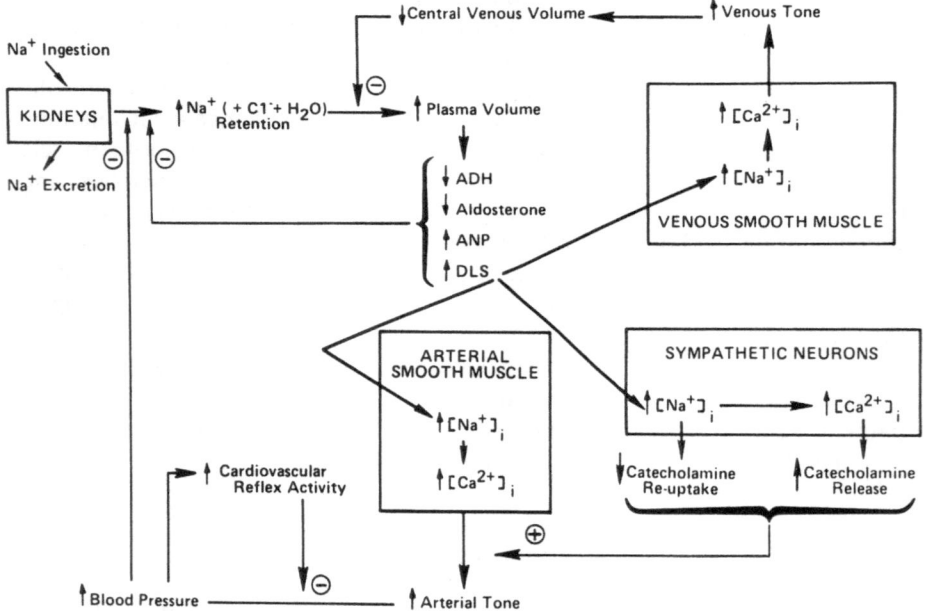

Fig. 7. Diagram showing the various feedback loops that help prevent plasma volume expansion when excessive sodium is ingested, relative to the kidneys' innate ability to excrete the sodium load. + and −: Positive and negative feedback loops, respectively. *ADH*, antidiuretic hormone; *ANP*, atrial natriuretic peptides; *DLS*, endogenous digitalis-like substance (sodium pump inhibitor); $[Na^+]_i$ and $[Ca^{2+}]_i$, intracellular sodium and calcium concentrations, respectively

mechanisms to control plasma volume will be exceeded; then the chronically elevated plasma level of the EDLS will inhibit the sodium pumps in the arterial and venous smooth muscle cells and in the sympathetic neurons. The increased venous tone will help to reduce plasma volume directly by reducing central venous volume. Arterial tone will be increased by direct action of the EDLS on the arterial smooth muscle and, indirectly, via the hormone's action on the sympathetic neurons. Initially, of course, blood pressure will be maintained in the normal range (but will be labile) because of the compensating cardiovascular reflexes. Once the capacity of these reflexes to control blood pressure is exceeded, however, the blood pressure will begin to rise; this will induce a pressure natriuresis to help restore plasma volume to normal. Therefore, in the established phase of the disease, we observe normal plasma volume and cardiac output in the presence of increased total peripheral resistance and blood pressure. This is a chronic state of "*virtual* hypervolemia": The *tendency* to retain salt and water provides the stimulus (via as yet unidentified pathways) to maintain the hypertension so that the renal arterial pressure can compensate for the excessive sodium intake by promoting adequate sodium excretion via the pressure natriuresis to maintain a normal plasma volume.

References

1. Ashida T, Blaustein MP (1987) Regulation of cell calcium and contractility in arterial smooth muscle: the role of Na/Ca exchange. J Physiol (Lond) 392:617–635
2. Blackburn H, Prineas R (1983) Diet and hypertension: anthropology, epidemiology and public health implications. Prog Biochem Pharmacol 19:31–79
3. Blaustein MP (1977) Sodium ions, calcium ions, blood pressure regulation, and hypertension: a reassessment and a hypothesis. Am J Physiol 232(1):C165–C173
4. Blaustein MP (1989) Sodium/calcium exchange in cardiac, smooth and skeletal muscle: key to the control of contractility. Curr Top Membr Transport 34:289–330
5. Blaustein MP, Hamlyn JM (1984) Sodium transport inhibition, cell calcium, and hypertension: the natriuretic hormone/Na-Ca exchange/hypertension hypothesis. Am J Med 77:45–59
6. Bova S, Blaustein MP, Harris DW, Ludens JH, DuCharme DW, Hamlyn JM (1990) Effects of an endogenous digitalis-like factor (EDLF) on heart and aorta (abstract). Hypertension 16:316–317
7. Bova S, Goldman WF, Yuan X -J, Blaustein MP (1990) Influence of the Na^+ gradient on Ca^{2+} transients and contraction in vascular smooth muscle. AM J Physiol 259: H409–H423
8. Curtis JJ, Luke RG, Dustan HP, Kashgarian M, Whelchel JD, Jones P, Diethelm AG (1983) Remission of essential hypertension after renal transplantation. N Engl J Med 309:1009–1015
9. Dahl LK, Knudsen KD, Iwai J (1969) Humoral transmission of hypertension: evidence from parabiosis. Circ Res 24 [Suppl I]:I-21–I-33
10. DeWardener HE, Mills IH, Clapham WF, Hayter CJ (1961) Studies on the efferent mechanism of the sodium diuresis which follows the administration of intravenous saline in the dog. Clin Sci 21:249–258

11. Fujimura A, Ebara A, Yaoka O (1984) The effect of long term digoxin administration on blood pressure of rat. Jpn J Hypertens 7(1):49
12. Goldman WF, Blaustein MP (1988) Stimulation-induced regional alteration of Ca^{2+} levels in single arterial smooth muscle cells. J Cardiovasc Pharmacol 12 [Suppl 5]: S13–S19
13. Guidi E, Bianchi G, Rivolta E, Ponticelli C, Quarto di Palo F, Minetti L, Polli E (1985) Hypertension in man with a kidney transplant: role of familial vs. other factors. Nephron 41:14–21
14. Guyton AC, Coleman TG, Cowley AW, Scheel KW, Manning RD, Norman RA (1977) Arterial pressure regulation: overriding dominance of the kidneys in long-term regulation and in hypertension. Am J Med 52:584–594
15. Hamlyn JM (1989) Increased levels of a humoral digitalis-Like factor in deoxycorticosterone acetate-induced hypertension in the pig. J Endocrinol 122:409–420
16. Hamlyn JM, Blaustein MP (1986) Sodium chloride, extracellular fluid volume, and blood pressure regulation. Am J Physiol 251:F563–F575
17. Hamlyn JM, Ringel R, Schaeffer J, Levinson PD, Hamilton BP, Kowarski AA, Blaustein MP (1982) A circulating inhibitor of (Na+K)-ATPase associated with essential hypertension. Nature 300:650–652
18. Hamlyn JM, Harris DW, Ludens JH (1989) Digitalis-like activity in human plasma. Purification, affinity, and mechanism. J Biol Chem 264:7395–7404
19. Hamlyn JM, Harris DW, Resau J, Ludens JH (1990) Digitalis-like activity in human and bovine adrenals. FASEB J 4:A295 (abstract)
20. Kurtz TW, Al-Bander HA, Morris RC Jr (1987) "Salt-sensitive" essential hypertension in men: is the sodium ion alone important? N Engl J Med 317:1043–1048
21. Magyar K, Nguyen TT, Török TL, Tóth PT (1987) [^{3}H]noradrenaline release from rabbit pulmonary artery: sodium-pump-dependent sodium-calcium exchange. J Physiol (Lond) 393:29–42
22. Matthews WR, Harris DW, DuCharme DW, Lutzke BS, Mandel F, Clark MA, Ludens JH, Hamlyn JM (1990) Endogenous digitalis-like factor (EDLF). III, Mass spectral characterization (abstract) Hypertension 16:316
23. Morgan T, Carney S, Myers J (1979) Sodium and hypertension. A review of the role of sodium in pathogenesis and the action of diuretic drugs. Pharmacol Ther 9:395–418
24. Reuter H, Blaustein MP, Haeusler G (1973) Na-Ca exchange and tension development in arterial smooth muscle. Philos Trans R Soc Lond [Biol] 265:87–94
25. Skrabal F, Herholz H, Neumayr M, Hamberger L, Ledochowski M, Sporer H, Hortnagl H, Schwarz S, Schonitzer D (1984) Salt sensitivity in humans is linked to enhance sympathetic responsiveness and to enhanced proximal tubular reabsortion. Hypertension 6:152–158
26. Weder AB (1986) Red cell lithium-sodium countertransport and renal lithium clearance in hypertension. N Engl J Med 314:198–201
27. Weinberger MH (1987) Sodium chloride and blood pressure. N Engl J Med 317: 1084–1086
28. Woolfson RG, Hilton PJ, Poston L (1990) The effects of ouabain and low sodium on the contractility of human subcutaneous resistance vessels. Hypertension 15: 583–590

Calcium Metabolism and Its Relationship to Blood Pressure in Humans

A. Hvarfner

Department of Internal Medicine, University Hospital, S-75185 Uppsala, Sweden

Introduction

Essential hypertension (EHT) is a condition carrying increased risk of vascular morbidity and death. Whether the blood pressure per se or hither to unidentified factors associated with EHT are the cause of this morbidity is not well understood, neither are the mechanisms behind the development of hypertension yet known. Let us imagine how this group of subjects at risk for vascular disease might have been identified if the invention of the calcium atomic absorption technique, the ionized calcium analyzer, and the parathyroid hormone assay had preceded the invention of the sphygmomanometer – and furthermore had been at least as easy to employ in the screening of populations.

In this paper all analyses have been performed with multivariate statistical techniques in order as far as possible to eliminate and analyze confounding differences between subjects in sex, age and body mass index. Figures are given as mean ± standard deviation.

Epidemiology

Ninety-seven apparently healthy, normotensive subjects aged between 16 and 82 years were recruited in an age- and sex-stratified manner from a general health survey [1]. In that population we found an inverse relationship between plasma ionized calcium concentrations and mean blood pressure (MBP; $r = -0.37$; $p = 0.0005$), the slope indicating that going from +2SD to −2SD in plasma ionized calcium (i.e., from 1.30 to 1.10 mmol/l) would correspond to a 10 mmHg increase in MBP. The total serum calcium concentration did not show any such relationship. We also observed a positive relationship to fasting renal calcium excretion and an inverse relationship to serum phosphate, but taking body mass index into account these metabolic indices proved to be more associated to obesity than to blood pressure.

G. Bruschi A. Borghetti (Eds.)
Cellular Aspects of Hypertension
© Springer-Verlag Berlin · Heidelberg 1991

In another study [2] 58 patients with EHT aged from 19 to 64 years were compared with the healthy population just described. They were receiving no medication, and were regard in to antihypertensive medication virgins. The patients diastolic blood pressure (DBP) ranged from 90 to 135 mmHg, and thier systolic blood pressure (SBP) from 130 to 235 mmHg; thus, most of them were mildly hypertensive. The EHT patients had a lower plasma ionized calcium concentration than the healthy normotensive subjects (1.19 $\pm$ 0.04 vs 1.21 $\pm$ 0.05 mmol/l, $p = 0.04$). There were no differences in serum total calcium, whether adjusted for albumin concentrations or not. Serum parathyroid hormone (PTH) concentrations were higher among the patients than among the normotensives (0.79 $\pm$ 0.13 vs 0.74 $\pm$ 0.14 arbU/l, $p = 0.02$). The fasting renal calcium excretion was more pronounced among the patients than among the reference population (29.5 $\pm$ 16.6 vs 22.7 $\pm$ 12.9 μmol/l, $p = 0.005$). As in the previous study, there was an inverse relationship between plasma ionized calcium and MBP, and there was also a positive relationship between MBP and fasting urinary calcium excretion.

To summarize these two studies: subjects with higher blood pressures had lower plasma concentrations of ionized calcium. Since the total serum calcium concentration did not demonstrate a corresponding pattern, this implies increased plasma calcium binding at high blood pressures. There was also increased renal calcium leak at high blood pressures, and more PTH secretion.

Free intracellular calcium concentrations were examined with the quin-2 technique in platelets from 86 subjects: 22 apparently healthy, normotensive volunteers from a general health survey and 64 patients investigated for possible EHT [3]. Of the patients 29 were receiving various antihypertensive drugs at the time of investigation, the other 35 had been free of medication for at least 6 months. The MBP of the untreated patients ranged from 108 to 152 mmHg. The basal concentration of free calcium in platelets among the healthy subjects and the treated patients was essentially the same (99.4 $\pm$ 26 vs 105 $\pm$ 52 nmol/l), but the untreated patients had higher levels of free intracellular calcium in thier platelets (122 $\pm$ 50 nmol/l, $p = 0.04$ vs reference subjects). There was a univariate relationship between cytosolic free calcium and MBP among the total untreated sample of normo- and hypertensive subjects ($r = +0.30$; $p = 0.02$), but taking into account the inverse relationship between serum PTH and cytosolic calcium seen in this study ($r = 0.26$; $p = 0.05$), the cytosolic free calcium corresponded only to 5% of the variability in blood pressure. Neither the total serum nor plasma ionized calcium concentrations nor renal calcium excretion showed any relationship to platelet cytosolic calcium levels.

Acute Experimental Hypercalcemia

In an experimental model we performed constant-rate intravenous calcium infusions for 120 min in 22 volunteer subjects with normotension and mild essential hypertension [4]. None of the volunteers had ever had any anti-hypertensive treatment. Their basal MBP ranged from 74 to 124 mmHg. Fourteen of the subjects were reexamined on a second 0 occasion, when verapamil was administered as an initial bolus and then in parallel with the intravenous calcium infusion in order to achieve a steady-state therapeutic level.

Plasma ionized clacium, which in the basal state in about 1.20 mmol/l, was increased by 0.40 mmol/l after 120 min calcium infusion. The MBP increased by an average of 12 mmHg. A round calculation of plasma volume and extracellular volume showed that at the end of the infusion about 13% of the administered calcium was still in the plasma, 8% in the free ionized form and 5% as bound calcium. Twenty-eight percent was distributed in the interstitial space as free ionized calcium, while 17% had been excreted in the urine, leaving 42% to be explained by extrarenal clearance. Verapamil did not interfere in any significant way with this distribution between the compartments. Thus, the dominant clearance pathway was the extrarenal.

The changes during calcium infusion were integrated and expressed as incremental area under the curve. The increase in DBP was directly related to the basal level of plasma ionized calcium ($r = +0.45$; $p = 0.04$) and inversely related to the basal serum PTH levels ($r = -0.45$; $p = 0.05$); i.e., those with the most pronounced blood pressure increase were those with the highest plasma ionized calcium and the lowest serum PTH in the basal state.

Then we turned the question the other way round: are calcium kinetics related to basal blood pressure? Initial blood pressure was inversely related to increase in plasma ionized calcium ($r = -0.45$; $p = 0.02$) – i.e., at higher basal blood pressure levels, plasma ionized calcium increased less during the infusion of calcium – but this was not reflected in the renal excretion of calcium. We therefore interpreted this relationship as an index of more pronounced extrarenal calcium clearance among the hypertensive subjects.

Calcitonin gene-related peptide (CGRP) concentrations in serum were also determined during the calcium infusions [5]. CGRP is the most potent endogenous vasodilator known so far. The CGRP concentrations decreased among those with high basal blood pressure, whereas the concentrations were increased among those with lower blood pressure, the change in CGRP describing an inverse relationship to basal blood pressure ($r = -0.72$; $p = 0.006$).

When calcium was administered in parallel with verapamil, the increase in plasma ionized calcium was not related to basal blood pressure. However, the suppression of serum PTH was less pronounced in subjects with higher than in those with lower basal blood pressure ($r = -0.68$; $p = 0.007$). There was also a positive relationship between basal blood pressure and calcium excretion during the calcium infusion in the presence of verapamil ($r = +0.63$; $p = 0.02$) – which was not the case in the absence of verapamil.

The two observations (a) that in the presence of verapamil there was no relationship between basal blood pressure and extrarenal calcium clearance, and (b) that in the presence of verapamil there was more pronounced renal calcium excretion among those with higher blood pressure, indicate that the above hypothesized increased tissue calcium uptake among the hypertensive subjects may be blocked by verapamil, resulting in increased renal excretion of calcium due to a greater filtration load in the presence of verapamil.

Were there any associations between the changes in metabolic indices and the changes in blood pressure? Yes, the increase in blood pressure during calcium infusion was inversely related to the increase in ionized calcium ($r = -0.45$; $p = 0.05$) and to the decrease in urinary cyclic adenosine monophosphate (AMP) excretion ($r = -0.50$; $p = 0.03$), which is a measure of parathyroid activity. These patterns were seen both in the absence and presence of verapamil. Thus, the plasma clearance of calcium, which was predominantly extrarenal, was directly related to the blood pressure increase. The extent of parathyroid suppression, determined as renal excretion of cyclic AMP, was also directly related to the pressor effect of acute hypercalcemia.

This study shows that hypertensive subjects have greater extrarenal calcium clearance, possibly due to increased tissue uptake of calcium, which may be abolished by verapamil – and that greater tissue calcium uptake and more pronounced parathyroid suppression are associated with a greater pressor response in acute hypercalcemia. Thus, both direct action of the calcium ion and indirect effects such as parathyroid suppression and CGRP modulations may induce the blood pressure response to acute hypercalcemia.

Metabolic Effects of Antihypertensive Treatment

We have seen how manipulations of the calcium metabolism may interfere with blood pressure. To turn the question the other way round, we entered 35 patients with previously untreated essential hypertension into an open randomized study for 6 months [6]. They were treated with three different types of antihypertensive drug: either propranolol or bendroflumethiazide or verapamil. The DBP generally decreased by 10 mmHg in each of the three groups. Testing whether this drop was related to changes in calcium metabolic indices within each of the groups or within the total sample showed that there were no such significant relationships. Nevertheless, there were apparent influences of the drugs on the calcium metabolic indices.

Propranolol induced a distinct rise in plasma ionized calcium ($+0.06 \pm 0.01$ mmol/l; $p = 0.001$), although there was no significant change in serum total calcium; i.e., propranolol induced a change in the plasma binding of calcium. The concentration of free fatty acids was concomitantly reduced by about 50% and in the basal state there was an inverse relationship between plasma ionized calcium and free fatty acids. A reduction in the binding of calcium to free fatty acids, and in turn to albumin, may explain the redistribution of calcium from the bound to the free fraction during propranolol treatment. The parathyroid hormone concentrations were somewhat decreased during treatment with propranolol, while there were no evidence of changes during the thiazide or verapamil treatments. Renal calcium excretion was not changed by propranolol, but increased during verapamil treatment and decreased in the classical manner during thiazide treatment.

Calcium and Vascular Retinopathy

In another study we examined 55 patients with untreated essential hypertension ophthalmoscopically [7]. They were classified into two groups: one of patients with no signs of vascular retinopathy, the other containing all patients with arteriolar narrowing, tortuosity, localized narrowing, or crossing phenomena. None of the patients had any signs of more severe vascular retinopathy. Their MBP was in the range of 109–144 mmHg, with an average of 120 mmHg.

In an analysis of covariance taking into account not only age, sex, and body mass index, but also differences in blood pressure between the subgroups, we compared calcium metabolic indices between the groups. This comparison demonstrated that those with signs of vascular retinopathy had lower levels of plasma ionized calcium (1.15 ± 0.01 vs 1.18 ± 0.01 mmol/l; $p = 0.01$) and higher 24-h renal excretion of cyclic AMP (39.6 ± 1.8 vs $34\,2 \pm 1.9$ mmol/l; $p = 0.05$). This observation suggests that ionized calcium may serve as a marker for vascular damage as well as for high blood pressure. Whether perturbations of the calcium metabolism make the vascular system more fragile, or whether these are phenomena secondary to vascular affection, cannot be decided from this study.

Discussion

Intracellular Calcium

The implications of increased basal cytosolic calcium levels are as yet poorly understood. The reports on platelet cytosolic calcium in relation to blood pressure are among the few concerned with measurements in the resting state. Most experience has accumulated in the experimental field with assessments of intracellular calcium responses to acute stimulation in vitro. In platelets stimulated by paltelet activating factor [8] or vasopressin [9] there was an immediate but transient increase in intracellular calcium on initiation of the shape change and secretion. Activation also occurred under conditions in which the intracellular calcium increase was small or suppressed, indicating that intracellular activators other than cytosolic calcium are involved. Consequently, myosin phosphorylation induced in platelets by thrombin requires less cytosolic calcium than that induced by the calcium ionophore ionomycin [10].

A humoral factor which differed in normotensive and hypertensive subjects has been shown to increase platelet cytosolic calcium [11]. In our investigation of basal cytosolic calcium a standardized buffer was used, so the effect of such a humoral factor would have been diminished. Thus, the observed positive relationship between cytosolic calcium and blood pressure should be regarded as a relationship between blood pressure and resting platelet calcium homeostasis per se. Whether the platelet cytosolic calcium reflects the calcium metabolism in vascular smooth muscle, as previously suggested [12,13], remains to be confirmed.

Calcium is able to bind to various molecular sites in the cell membrane, and bound calcium has important effects on membrane function. Defective calcium binding might be expected to increase the activity of the potential-operated channels, with a resulting increase in calcium entry [14–17]. However, since the plasma concentrations of calcium are $10\,000-1\,000\,000$ times higher than the basal levels of free cytosolic calcium, it seems unlikely that the minute differences in plasma ionized calcium concentration would alter the membrane permeability or cause significant changes in the calcium concentration within the cell.

Acute hypercalcemia is associated with increased peripheral vascular resistance [18,19]. The rate of extrarenal calcium clearance during the calcium infusions was highest in the subjects with higher basal blood pressure and in those with the greatest pressor response. These findings together suggest that extrarenal calcium clearance may be a measure of cellular calcium uptake in contractile and/or secretory cells, and corroborate the hypothesis of greater cellular calcium avidity in essential hypertension.

Parathyroid Hormone

Present knowledge about the effects of PTH upon blood pressure is somewhat equivocal. It has been proposed that in essential hypertension PTH may promote an increase in the cytosolic calcium in smooth muscle cells, thereby increasing the peripheral vascular resistance [20]. Chronic elevation of serum PTH in primary hyperparathyroidism is associated with hypertension [21], and 12-day PTH infusion in normal subjects [22] raised the blood pressure. Increased secretion of PTH may increase the intracellular calcium concentration [23,24] and could serve as an ionophore in this respect; for example, in cultured renal proximal tubular cells, PTH induced a sustained increase in cytosolic calcium and cyclic AMP [25], but in freshly prepared proximal tubular cells in another study PTH caused a decrease in cytosolic calcium [26].

Parathyroid hormone has an acute dilatory effect of vascular smooth muscle in vitro [27]. This effect appears to be mediated by a PTH-specific receptor which exerts its effect by an increase in intracellular cyclic AMP via adenylate cyclase. PTH has also been found to induce vasodilation in acute experiments in vivo [28]. In our study of intracellular calcium in platelets, an inverse relationship between serum PTH and intracellular calcium was seen, mimicking the well-established feedback relationship seen in parathyroid cells, in which parathyroid secretion is lower at higher intracellular calcium levels [29,30]. During the calcium infusions the pressor response was more pronounced in subjects with lower preinfusion serum PTH levels, and also in those with the greatest suppression of parathyroid activity during the infusions. Thus, higher basal parathyroid activity may provide protection against calcium-induced pressor effects, and the suppression of parathyroid activity during calcium infusion may eliminate a physiological vasodilatory effect exerted by PTH. This suggestion is supported by an earlier observation that the pressor effect of intravenously administered calcium was absent in totally parathyroidectomized patients [31], and by another model in which the hypotension induced by acute hypocalcemia proved to be dependent on parathyroid activation [32].

Calcitonin Gene-Related Peptide

CGRP is a 37-amino-acid neuropeptide that is present in the central and peripheral nervous systems, with binding sites located in the heart and peripheral arteries [33]. Its physiological role is not fully established. On a molar basis CGRP is the most potent endogenous vasodilator known, and general or local alterations in CGRP may influence peripheral vascular tone [33,34]. Infusions of CGRP have also been shown to lower the blood pressure in humans [33,34]. The close relationship between CGRP response to calcium infusion and basal blood pressure indicates that CGRP may be involved in the complex interpaly between calcium metabolism and blood pressure regulation.

Conclusion: A Common Denominator?

There is an indication that the serum concentration of free fatty acid is increased in essential hypertension [35]. Free fatty acids may themselves bind calcium and enhance calcium binding to albumin [36–38]. We observed an inverse relationship between plasma ionized calcium and serum free fatty acids in patients with essential hypertension. In the same study propranolol induced a redistribution of plasma calcium from the bound to the ionized fractions, and in parallel to this reductions in serum free fatty acid and PTH concentrations occurred. The ability of propranolol to decrease free fatty acid concentrations has been demonstrated preveiously [39,40]. If EHT is a state of β-adrenergic hyperactivity, this implies that such a state might induce the low concentrations of ionized calcium seen in EHT.

Human parathyroid glands have adrenergic innervation [41], and β-adrenergic agonists have been found to stimulate parathyroid secretion in vitro [42,43] and in vivo [44,45]. Conversely, PTH secretion has been inhibited by propranolol in some patients with secondary hyperparathyoridism or renal failure [46–48], whereas in studies of normal subjects and hyperparathyroid patients [49–51] no consistent effect of this drug on parathyroid secretion was observed. The effect of propranolol on the distribution of serum calcium fractions probably depends on the pretreatment free fatty acid turnover, which will differ in different patient groups, and this might explain the seemingly contradictory effect of propranolol upon PTH secretion in different studies.

It seems reasonable to propose that raised serum free fatty acid levels at higher blood pressures, probably a reflection of increased catecholaminergic stimulation, might result in enhanced binding of serum calcium, thereby reducing the free calcium concentration, which might secondarily stimulate PTH secretion. Increased complex binding of serum calcium, increased serum PTH concentrations, and enhanced urinary calcium excretion, in parallel with higher serum free fatty acids, have also been observed in morbid obesity [52].

There are reports on hyperinsulinemia in essential hypertension [53,54]. Insulin stimulates urinary calcium excretion with concomitant sodium retention [55]. Dietary sodium may modify urinary calcium excretion [56]. Sodium retention, due to excessive intake or some renal limitation of sodium excretion, may induce secretion of natriuretic hormone [57], thus inhibiting sodium pumps in renal tubules and thereby inducing natriuresis. Increased intracellular sodium will, in turn, decrease the sodium–calcium countertransport, thereby increasing the intracellular calcium concentration [58].

There are a number of possible mechanisms underlying the relationship between calcium metabolism and blood pressure, and no single explanation can be expected to clarify this association. Calcium metabolism is involved in normal blood pressure regulation; it is therefore reasonable to assume that the above relationship indicate that calcium metabolism plays a role in hypertension. However, there is also evidence of other factors that may cause changes in both blood pressure and calcium metabolism in parallel. The association of ionized calcium with vascular retinopathy, independently of blood pressure, is an indication that calcium endocrinology may have a role in the vascular disease and not only in the hypertension.

References

1. Hvarfner A, Ljunghall S, Mörlin C, Wide L, Bergström R (1986) Indices of mineral metabolism in relation to blood pressure in a sample of a healthy population. Acta Med Scand 219:461–468
2. Hvarfner A, Bergström R, Mörlin C, Wide L, Ljunghall S (1987) Relationships between calcium metabolic indices and blood pressure in patients with essential hypertension as compared with a healthy population. J Hypertens 5:451–456
3. Hvarfner A, Larsson R, Mörlin C, Rastad J, Wide L, Akerström G, Ljunghall S (1988) Cytosolic free calcium in platelets: relationship to blood pressure and indices of systemic calcium metabolism. J Hypertens 6:71–77
4. Hvarfner A, Mörlin C, Wide L, Ljunghall S (1989) Interactions between indices of calcium metabolism and blood pressure during an intravenous calcium infusion in humans. J Hum Hypertens 3:211–220
5. Hvarfner A, Bergström R, Mörlin C, Theodorsson E, Ljunghall S (1988) The calcitonin gene related peptide (CGRP) response to intravenous calcium is related to blood pressure. Acta Physiol Scand 132:439–440
6. Hvarfner A, Bergström R, Lithell H, Mörlin C, Wide L, Ljunghall S (1988) Changes in calcium metabolic indices during long-term treatment of patients with essential hypertension. Clin Sci 75:543–549
7. Hvarfner A, Mörlin C, Präntare H, Wide L, Ljunghall S (1990) Calcium metabolic indices, vascular retinopathy and plasma renin activity in essential hypertension. Am J Hypertens 3(12) (in press)
8. Hallam TJ, Sanchez A, Rink TJ (1984) Stimulus-response coupling in human platelets. Biochem J 218:819–827
9. Hallam TJ, Thompson NT, Scrutton MC, Rink TJ (1984) The role of cytoplasmic free calcium in the response of quin2-loaded human platelets of vasopressin. Biochem J 221:897–901
10. Hallam TJ, Daniel JL, Kendrick-Jones J, Rink TJ (1985) Relationship between cytoplasmic free calcium and myosin light chain phosphorylation in intact platelets. Biochem J 232:373–377
11. Lindner A, Keuny M, Meacham AJ (1987) Effects of a circulating factor in patients with essential hypertension on intracellular free calcium in normal platelets. N Engl J Med 316:509–513
12. Niedermann R, Pollard TD (1975) Human platelet myosin. II In vitro assembly and structure of myosin filaments. J Cell Biol 67:72–79
13. Erne P, Bolli P, Bürgiesser E, Bühler FR (1984) Correclation of platelet calcium with blood pressure effect of antihypertensive therapy. N Engl J Med 310:1084–1088
14. Webb RC, Bohr DF (1978) Mechanism of membrane stabilization by calcium in vascular smooth muscle. Am J Physiol 235:227–232
15. Orlov SN, Postnov YV (1982) Ca^{+2} binding and membrane fluidity in essential and renal hypertension. Clin Sci 63:281–284
16. Robinson BF (1984) Altered calcium handing as a cause of primary hypertension. J Hypertens 2:543–460
17. Sprenger KGB (1985) Alteration of cellular calcium metabolism as primary cause of hypertension. Clin Physiol Biochem 3:208–220
18. Marone C, Beretta-Piccolo C, Weidmann P (1980) Acute hypercalcemic hypertension in man: role of hemodynamics, catecholamines and renin. Kidney Int 20: 92–96
19. Berl T, Levi M, Ellis M, Chaimovitz C (1985) Mechanism of acute hypercalcemic hypertension in the conscious rat. Hypertension 7:923–930
20. Lau K, Eby B (1985) The role of calcium in genetic hypertension. Hypertension 7:657–667

21. Rosenthal FD, Roy S (1972) Hypertension and hyperparathyroidism. Br Med J 4:396–397
22. Hulter HN, Melby JC, Peterson JC, Cooke CR (1986) Chronic continuous PTH infusion results in hypertension in normal subjects. J Clin Hypertens 4:360–370
23. Borle AB, Uchikawa T (1978) Effects of parathyroid hormone on the distribution and transport of calcium of cultured kidney cells. Endocrinology 102:1725–1732
24. Marcus C, Orner FB (1980) Parathyroid hormone as a calcium ionophore in bone cells: tests of specificity. Calcif Tissue Int 5:159
25. Hruska KA, Goligorsky M, Scolbe J, Tsutsumi M, Westbrook S, Moskowitz D (1986) Effects of parathyroid hormone on cytosolic calcium in renal proximal tubular primary cultures. Am J Physiol 251:F188–F198
26. Dolson GM, Hise MK, Weinman ED (1985) Relationship among parathyroid hormone, cAMP and calcium on proximal tubule sodium transport. Am J Physiol 249: F409–F416
27. Nickols GA, Netz MA, Cline WH (1986) Endothelium-independent linkage of parathyroid hormone receptors of rat vascular tissue with increased adenosine 3'-5'- monophosphate and relaxation of vascular smooth muscle. Endocrinology 119: 349–356
28. Dowe JP, Joshua IG (1987) In vivo arteriolar dilation in response to parathyroid hormone fragments. Peptides 8:443–448
29. Shoback DM, Thatcher J, Leombruno R, Brown EM (1984) Relationship between parathyroid hormone secretion and cytosolic calcium concentration in dispersed bovine parathyroid cells. Proc Natl Acad Sci USA 81:3113–3117
30. Larsson R, Wallfelt C, Abrahamsson H, Gylfe E, Ljunghall S, Rastad J, Rorsman P, Wide L, Akerström G (1984) Defective regulation of the cytosolic Ca^{2+} activity in parathyroid cells from patients with hyperparathyroidism. Biosci Rep 4:909–915
31. Gennari C, Nami R, Bianchini C, Aversa AM (1985) Blood pressure effects of acute hypercalcemia in normal subjects and thyroparathyroidectomized patients. Miner Electrolyte Metab 11:369–373
32. Zawada ET Jr, Johansson M, McClung D, TreWee J, MacKenzie T (1987) Systemic and renal hemodynamic consequences of manipulation of serum calcium and/or parathyroid hormone in the intact conscious mongrel dog. J Am Coll Nutr 6:131–138
33. Franco-Cereceda A, Gennari C, Nami R, Agnusdei D, Pernow J, Lundberg JM, Fischer AJ (1987) Cardiovascular effects of calcitonin gene-related peptides I and II in man. Circ Res 60:393–397
34. Struthers AD, Brown MJ, MacDonald DWR, Beacham JL, Stevenson JC, Morris HR, MacIntyre I (1986) Human calcitonin gene-related pepide: a potent endogenous vasodilator in man. Clin Sci 70:389–393
35. Singer P, Gödicke W, Voigt S, Hajdu I, Weiss M (1985) Postprandial hyperinsulinemia in patients with mild essential hhpertension. Hypertension 7:182–186
36. Ladenson JH, Shyong JC (1977) Influence of fatty acids on the binding of calcium to human serum albumin. Clin Chim Acta 75:293–302
37. Wortsmann J, Traycoff RB (1980) Biological activity of protein-bound calcium in serum. Am J Physiol 238(1):E104–E107
38. Zaloga GP, Willey S, Tomasic P, Chernow B (1987) Free fatty acids alter calcium binding: a cause for misinterpretation of serum calcium values and hypocalcemia in critical illness. J Clin Endocrinol Metab 64:1010–1014
39. Deacon SP (1978) The effects of atenolol and propranolol upon lipolysis. Br J Clin Pharmacol 5:123–125
40. Day JL, Metcalfe J, Simpson CN (1982) Adrenergic mechanisms in control of plasma lipid concentrations. Br Med J 284:1145–1148
41. Norberg K-A, Persson B, Granberg P-O (1975) Adrenergic innervation of the human parathyroid gland. Acta Chir Scand 141:319–322

42. Brown EM, Hurwitz S, Aurbach GD (1977) Beta-adrenergic stimulation of cyclic AMP content and parathyroid hormone release from isolated parathyroid cells. Endocrinology 100:1696–1702

43. Hanley DA, Taktsuki K, Birnbaumer ME, Schneider AB, Sherwood LM (1980) In vitro perfusion for the study of parathyroid hormone secretion: effects of extracellular calcium concentration and β-adrenergic regulation on bovine parathyroid hormone secretion in vitro. Calcif Tissue Int 32:19–27

44. Fischer JA, Blum JW, Binswanger U (1973) Acute parathyroid hormone response to epinephrine in vivo. J Clin Invest 52:2434–2440

45. Mayer GP, Hust JG, Barto JA, Keaton JA, Moore MP (1979) Effect of epinephrine on parathyroid hormone secretion in calves. Endocrinology 104:1181–1187

46. Kukreja SC, Williams GA, Vora NM, Hargis GK (1981) Normal responsiveness of serum parathyroid hormone to β-adrenergic blockade in patients with secondary hyperparathyroidism. Horm Metab Res 13:233–235

47. Farrington K, Hamzeh J, Varghese Z, Moorhead JF (1980) Effects of oral propranolol on parathyroid hormone secretion in chronic renal failure. Br Med J 281:1320

48. Coevoet B, Desplan C, Sebert JL, Makdassi R, Andrejak M, Gheerbrant JD, Tolani M, Calmette C, Moukthar MS, Fournier A (1980) Effect of propranolol and metoprolol on parathyroid hormone and calcitonin secretion in uraemic patients. Br Med J 280:1344–1346

49. McCarron DA, Muther RS, Plant SB, Krutzik S (1982) Ionized calcium and the in vivo response of normal and hyperplastic parathyroid glands to β-adrenergic agents. Nephron 32:149–154

50. Ljunghall S, Rudberg C, Åkerström G, Wide L (1982) Effects of β-adrenergic blockade on serum parathyroid hormone in normal subjects and patients with primary hyperparathyroidism. Acta Med Scand 211:27–30

51. Epstein S, Heath III H, Bell NH (1983) Lack of influence of isoproterenol, propranolol, and dopamine on immunoreactive parathyroid hormone and calcitonin in normal man. Calci Tissue Int 35:32–36

52. Andersen T, McNair P, Fogh-Andersen N, Nielsen TT, Hyldstrup L, Transböl I (1986) Increased parathyroid hormone as a consequence of changed complex binding of plasma calcium in morbid obesity. Metabolism 35:147–151

53. Modan M, Halkin H, Almog S, Lusky A, Eshkol A, Shefi M, Shitrit A, Fuchs Z (1985) Hyperinsulinemia – a link between hypertension, obesity and glucose intolerance. J Clin Invest 75:809–817

54. Pollare T, Lithell H, Berne C (1990) Insulin resistance is a characteristic feature of primary hypertension independently of obesity. Metabolism 39:167–174

55. DeFronzo RA, Cooke CR, Andres R, Faloona GR, Davis PJ (1975) The effect on insulin on renal handling of sodium, potassium, calcium and phosphate in man. J Clin Invest 55:845–855

56. Muldowney FP, Freaney R, Moloney MF (1982) Importance of dietary sodium in the hypercalciuria syndrome. Kidney Int 22:292–296

57. deWardener HE (1978) The control of sodium excretion. Am J Physiol 235: F163–F173

58. Blaustein MP, Hamlyn JM (1984) Sodium transport inhibition, cell calcium and hypertension. Am J Med 77:45–49

Parathyroids, Hypertension, and Vascular Reactivity

A. Gairard, R. Schleiffer, F. Pernot, C. Bergmann,
and B. van Overloop

[1] Laboratoire de Pharmacologie Cellulaire et Moléculaire CNRS URA600,
Faculté de Pharmacie, Université Louis Pasteur, Strasbourg, 74 Route du Rhin, BP 24,
F-67400 Illkirch-Graffenstaden, France

Bone, kidney, and intestine are the main targets for parathyroid hormone (PTH) [22]. Nevertheless, cardiovascular effects with vasodilatation and hypotension have been described, initially with parathyroid extracts, then confirmed with the synthetic hormone (bovine PTH1-84) and fragments (PTH1-34, rat, bovine, human sequences). Only very recently, high affinity binding sites for PTH fragments have been described on rabbit renal microvessels [52].

These are acute effects, mainly observed with pharmacological doses. On the other hand, in humans primary hyperparathyroidism is associated with hypertension [23,65], and parathyroidectomy performed in young of the spontaneosly rats hypertensive (SHR) and Lyon hypertensive (LH) strains delays the development and attenuates the level of hypertension [60,69].

This review presents and discusses observations of both acute and chronic effects of calcium-regulating hormones, chiefly PTH, calcitonin, and vitamin D. Since calcium is involved in not only cardiac and vascular muscle contraction [54] but also in hormonal and nervous activity, multifactorial mechanisms are involved and probably competing at different levels.

Cardiovascular Effects of PTH

Experiments on vascular and cardiac preparations clearly establish that PTH induces vascular relaxation and cardiac stimulation. From in vivo studies it appears that the cardiac, renal, adrenal, celiac, and hepatic vascular beds are especially sensitive to its dilator effect [56]. Isolated perfused rat mesenteric vascular bed with periarterial nerve stimulation [51] and isolated perfused rat kidney under nonfiltering conditions [48] are very sensitive preparations with CE_{50} in the nanomolar range. Moreover, the effect of PTH on contractile responses from isolated arteries has been investigated and relaxation of precontracted vessels and inhibition of norepinephrine- or

G. Bruschi A. Borghetti (Eds.)
Cellular Aspects of Hypertension
© Springer-Verlag Berlin · Heidelberg 1991

vasopressin-evoked contraction obtained [5,57,73]. EC_{50} of PTH is approximately $10\,nM$ in isolated rat caudal artery.

The mechanisms of PTH's vasoactive effects are regarded as specific, since adrenergic and histaminergic antagonists do not modify the hypotensive effect [56,67]. However, indomethacin has been shown to abolish the vasodilator effect in some but not all studies, suggesting a minor role for vasodilator prostaglandins.

Among the proposed cellular mechanisms of action are stimulation of cAMP production and inhibition of ^{45}Ca influx. For example, microvessels isolated from rabbit kidney cortex display a PTH-sensitive adenylate cyclase [24], as do microvessels from the cerebral bed [28]. Recently, studies have confirmed that PTH enhances cAMP production in rat aortic myocytes in culture and also during relaxation of rat aorta [4,50]. Structure-activity relationships with analogs and antagonists, (3–34) and (7–34)-bPTH, reveal differences in the adenylate cyclase response from renal microvascular and tubular tissues [25]. Specific vascular binding sites for rat PTH1–34 and human PTH1–34-related protein, which is released by tumoral tissues, have been very recently characterized in vascular and endothelial cells [27] and in renal microvessel from rabbit kidney [52]. The dissociation constant Kd remains in the nanomolar range, as EC_{50} for vasodilatation in the same vascular bed; these two values differ nearly by a factor of 100 from circulating levels of PTH. Nevertheless, the same difference appears with angiotensin II, another vasoactive peptide.

Calcium turnover in cells is generally increased by PTH. By contrast, both PTH extract and bPTH1–34 reduce ^{45}Ca influx in isolated rat aorta and tail artery without any effect on efflux [58,68,73]. Moreover, a recent report indicates that bPTH1–34 inhibits the L calcium channel, as shown by whole-cell patch-clamp techniques in cultured vascular smooth muscle cells from rat tail artery [59]. Two different approaches have shown that PTH appears to act on vascular smooth muscle cell like a calcium channel blocking agent, thus inducing vasodilatation. To date, the relative contribution of the effects of PTH on cAMP and Ca^{2+} fluxes has not been precisely delineated. More recently, nevertheless, a transient rise in intracellular free calcium has been described for hPTH, measured with fura-2 in cultured fetal rat aortic cell line [38]. The observed affinity is low (EC_{50} about $220\,nM$) compared to those obtained during vasodilatation and biochemical studies on cAMP production, and may be due to cell line characteristics.

To conclude with the vascular actions of PTH: the kinetics of cAMP production and ^{45}Ca influx are both compatible with the vasodilating effects in coronary, mesenteric, and renal beds. These data probably have physiological relevance.

Acute cardiac effects are also obtained with PTH1–34. Bolus injections of PTH decrease mean arterial pressure and enhance heart rate [56], but positive inotropic effects have been reported in isolated papillary muscle at $10^{-11}\,M$ [36]. PTH also induces an increase in the beating rate of rat atria

(EC_{50} $10^{-8} M$) [74]. PTH enhances in a dose-dependent fashion the beating rate [41] and cAMP production of cardiac myocytes cultured from neonatal rats, an effect which is prevented by verapamil [10,11]. In addition, PTH action in the heart is additive to phenylephrine and isoproterenol. In guinea pig papillary muscles [53], bPTH enhances the slow action potential during the inotropic action and increases the slow inward current as measured using voltage clamp methods [39]. On the other hand, chronic in vivo administration in the rat shows that both bPTH1–84 and bPTH1–34 fragments increase total myocardial calcium content, impair mitochondrial energy production, and decrease blood pressure and cardiac index [3]. In agreement with this, clinical observations have demonstrated an association between enhanced myocardial calcium content, left ventricular hypertrophy, and parathyroid hyperactivity during myocardial dysfunction (decrease in the ejection fraction) in chronic renal failure in rats [33] and patients undergoing dialysis [26,66]. Parathyroidectomy has been shown to enhance left ventricular function in hemodialysis patients [18].

Cardiovascular effects of Calcitonin, Calcitonin-Gene-Related Peptide, and 1,25 Dihydroxyvitamin D_3

Although PTH has been shown to have vasoactive properties, less information is available for calcitonin (CT). Intravenous injection of CT does not induce any modification of blood pressure in normotensive rats [70]. However, enhanced hepatic and reduced coronary circulation occur after CT injection in anesthetized dog [47]. Flushing, a rapid and transient facial and thoracic vasodilatation, is frequently reported as a side effect of CT administration. More recently, calcitonin-gene-related peptide (CGRP), a neuropeptide encoded by the calcitonin gene, has been found in the nervous system, and also in heart and vascular tissues. This is a potent vasodilating substance [12] which acts on heart, blood vessels, and kidney via direct and indirect mechanisms [20]. Another major calcium-regulating hormone is vitamin D (1,25 $(OH)_2D_3$) which is involved in cardiovascular function [77,78]. Recently, specific binding sites for 1,25 $(OH)_2D_3$ were described in heart [76] and vascular smooth muscle cells [32,37]. In normotensive Sprague-Dawley rats 1,25 $(OH)_2D_3$ stimulates ^{45}Ca uptake by cultured aortic rat vascular smooth muscle cells in culture but suppresses aortic cell growth and antagonizes the mitogenic effect of epidermal growth factor [43]. By contrast, in SHR but not in WKY rats, 1,25 $(OH)_2D_3H$ stimulates the growth of myocytes cultured from the superior mesenteric artery and increases the sensitivity of isolated mesenteric resistance arteries to norepinephrine [15,16].

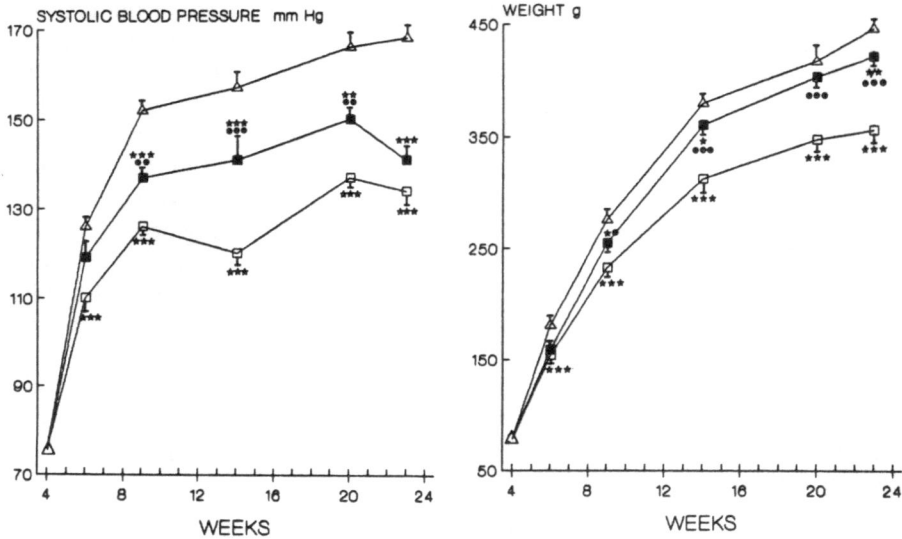

Fig. 1. Parathyroidectomy (PTX) and evolution of systolic blood pressure (*left*) and body weight (*right*) in 20 Lyon hypertensive rats (LH △), 20 PTX LH rats (□) and 9 PTX LH rats fed a calcium-enriched diet (■). PTX was performed at 4 weeks of age, just after the first measurement, and the calcium-enriched diet (calcium content 1.2%) was given from 5 weeks of age. Data are mean ± sem. ★ $P < 0.05$; ★/★$P < 0.02$; ★★$P < 0.01$; ★★★$P < 0.001$ vs LH; •$P < 0.05$; ••$P < 0.01$; •••$P < 0.001$ vs LH PTX (From [61])

Hypertension, Parathyroid Gland, and Calcium

Numerous reports indicating an involvement of parathyroid function in blood pressure regulation have appeared during the last decade [44]. Clinical observations show that primary hyperparathyroidism is frequently associated with enhanced serum calcium and systolic blood pressure, [17] and, similarly, young mild by hypertensive persons have raised serum PTH levels [21]. Moreover, chronic PTH infusion results in hypertension in normal subjects [29]. Finally, relationships between dietary calcium and sodium [63,64], calcium metabolic indices [30,31,81], and blood pressure are apparent in patients with essential hypertension. Dietary sodium and calcium are linked to changes in blood pressure in hypertensive black adults [82].

Experimentally the importance of the parathyroid gland for the overall development of hypertension was first described in mineralocorticoid hypertensive rats [8] and in the genetically hypertensive SHR [69]. Another genetically hypertensive rat strain, the Lyon hypertensive rat, is also sensitive to bilateral parathyroidectomy performed in youth (5 weeks of age) [60]. These data have been confirmed and expanded in the SHR by several groups [2,19,46,55]. Further, blood pressure is elevated in conscious male offspring in the sixth generation of rats parathyroidectomized on day 5 of pregnancy [49].

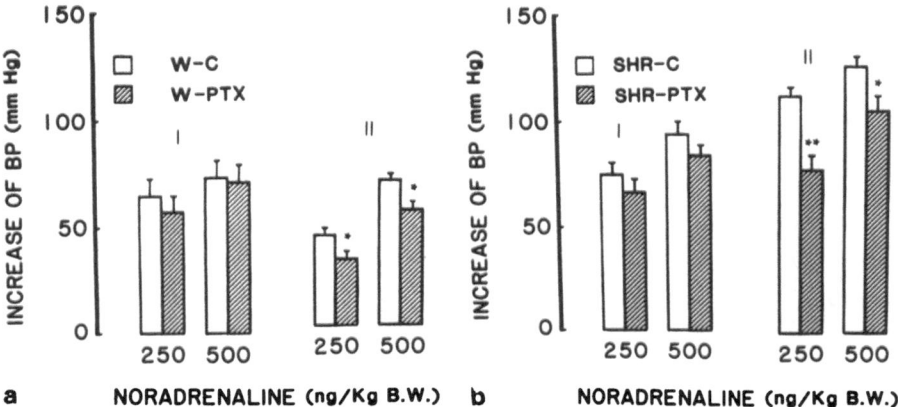

Fig. 2a Noradrenaline-evoked increases in blood pressure (*BP*) in spontaneously hyper-tensive control (□) and parathyroidectomized (▨) rats. Surgery was performed at 6 weeks of age. Values are means ± SEM in eight to nine rats. (*I*) 8-week-old rats; (*II*) 28-week-old rats. Statistical significance: ★$P < 0.05$; ★★$P < 0.01$ (Student's t test). **b** Noradrenaline-evoked increases in blood pressure (*BP*) in normotensive (▨) rats. Surgery was performed at 6 weeks of age. Values are means ± SEM in four to six rats. (*I*) 8-week-old rats; (*II*) 28-week-old-rats. Statistical significance: ★$P < 0.05$ (Student's t test). (from [71])

Very recently a novel cell type was described in the parathyroid gland of the SHR [35], and transplantation of SHR parathyroid gland into pre-viously parathyroidectomized (PTX) Sprague-Dawley rats results in a chronic rise in blood pressure [55]. Moreover, in the SHR a circulating hypertensive factor which enhances blood pressure and tail artery calcium uptake in normotensive rat [42] disappears after parathyroidectomy [55]. Thus clearly shown in the SHR for the moment, parathyroids produce or activate a factor (parathyroid hypertensive factor, PHF) which is involved in the develop-ment of hypertension and obviously differs from PTH. Results obtained in humans after parathyroidectomy are contradictory: parathyroidectomy cor-rects hypertension only in some studies with primary hyperparathyroidism [13], but not in others [40]. The secretion of the parathyroid gland is modu-lated by calcium-enriched or deprived diets and numerous results, obtained in rats but also in dogs, show that blood pressure is modified in normotensive or hypertensive animals. Calcium-enriched diets (two-to four-fold, mainly with $CaCO_3$) lowers blood pressure level in experimental hypertension: mineralocorticoid in rats [9], renal in rats [34], spontaneous in SHR [1,14], Dahl rats [62], and LH rats [61]. As expected, a low-calcium diet increases blood pressure in normotensive rats [70,75] and in LH rats [61] (Fig. 1).

In order to elucidate the mechanisms by which parathyroidectomy delays the development and attenuates the level of hypertension, cardiovascular reactivity was studied. Blood pressure response to bolus noradrenaline administration was measured in anesthetized parathyroidectomized (PTX)

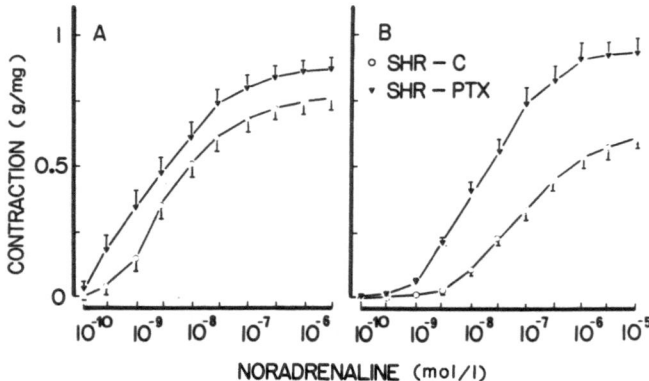

Fig. 3a,b. Cumulative concentration-effect curves elicited by noradenaline (10^{-10} to 10^{-5} mol/l) in aorta from control (SHR-C) and parathyroidectomized (SHR-PTX) spontaneously hypertensive rats. Parathyroidectomy was performed at 6 weeks of age. Values are mean ± SEM. **a** 8-week-old rats; **b** 28-week-old rats (n = 8 − 13) (from [72])

rats whose blood pressure regulation was suppressed by vagotomy and atropine treatment. In vivo cardiovascular reactivity to norepinephrine in PTX SHR was reduced (Fig. 2), like calcium content in aortic and heart fragments. However, these results are not specific for hypertensive animals since similar data are also obtained on normotensive PTX Wistar rats [71]. A change in the hind limb vascular reactivity to norepinephrine and vasopressin is not observed in PTX SHR and WKY rats [79]. The plasma level of norepinephrine and angiotensin II remains unchanged in PTX stroke-prone SHR (SHR-SP) and PTX WKY [46]. Nevertheless, although no, major effect was seen in blood pressure in the following studies, parathyroidectomy ameliorates vascular lesions induced by deoxycorticosterone in heart and kidney [80] and prevents stroke in SHR-SP (Stasch et al., this volume). In vitro vascular reactivity can be measured on conduit vessels such as aorta or resistance vessels such as mesenteric arteries [14]. With calcium-enriched diets which decrease PTH secretion, no difference in mesenteric artery sensitivity to KCl or norepinephrine is obtained [16]. However, with PTX SHR and normotensive PTX WKY rats (Fig. 3) unexpected results relating to aortic contraction are seen: an enhanced response to norepinephrine in which extracellular calcium concentrations play a key role [72]. With PTX LH rats, whose serum calcium is restablished at normal values with a calcium-fortified diet, vascular reactivity to norepinephrine is also increased in the presence but not in the absence of the endothelium (F. Pernot, personal communication). This new result is currently under investigation, since endothelial cells are known to release both contracting and relaxing factors.

In conclusion: calciotropic hormones can modulate vascular contractility and cellular calcium turnover. Vascular and cardiac muscle are to be considered as target organs for parathyroid secretions, PTH and PHF, and these factors may play a role in normal and pathophysiologic cardiovascular control. Their contribution to growth, multiplication, and pathological modifications of vascular and cardiac cells during hypertension and atherosclerosis remains to be further clarified.

References

1. Ayachi S (1979) Increased dietary calcium lowers blood pressure in the spontaneously hypertensive rat. Metabolism 28:1234–1238
2. Baksi SN (1988) Hypotensive action of parathyroid hormone in hypoparathyroid and hyperparathyroid rats. Hypertension 11:509–513
3. Baczynski R, Massry SG, Kohan R, Magott M, Saglikes Y, Brautbar N (1985) Effect of parathyroid hormone on myocardial energy metabolism in the rat. Kidney Int 27:718–725
4. Bergmann C, Schoeffter P, Stoclet JC, Gairard A (1987) Effect of parathyroid hormone and antagonist on aortic cAMP levels. Can J Physiol Pharmacol 65:2349–2353
5. Berthelot A, Gairard A (1975) Effet de la parathormone sur la pression artérielle et la contraction de l'aorte isolée de rat. Experientia 31:457–458
6. Berthelot A, Gairard A (1978) Effect of parathyroidectomy on cardiovascular reactivity in rats with mineralocorticoid-induced hypertension. Br J Pharmacol 62:199–205
7. Berthelot A, Gairard A (1980) Parathyroid hormone and DOCA induced hypertension in the rat. Clin Sci 58:365–371
8. Berthelot A, Schleiffer R, Gairard A (1979) Parathyroëde et hypertension arteriëlle à l'acétate de désoxycorticostérone chez le rat. Can J Physiol Pharmacol 57:157–162
9. Berthelot A, Gairard A, Goyault M, Pernot F (1980) Relation between calcium and cardiovascular reactivity in mineralocorticoid induced hypertension in the rat. Br J Pharmacol 70:301–306
10. Bogin E, Harary I (1987) The relationship of calcium and parathyroid hormone in their effect on heart cells. Mol Cell Biochem 77:29–36
11. Bogin E, Massry SG, Harary I (1981) Effect of parathyroid hormone on rat heart cells. Clin Invest 67:1215–1227
12. Brain SD, Williams TJ, Tippins JR (1985) Calcitonin-gene-related peptide is a potent vasodilator. Nature 313:54–56
13. Broulik PD, Horky K, Pacovsky V (1985) Blood pressure in patients with primary hyperparathyroidism before and after parathyroidectomy. Exp Clin Endocrinol 86:346–352
14. Bukoski RD, McCarron DA (1986) Altered aortic reactivity and lowered blood pressure associated with high calcium intake. Am J Physiol 251:H976–983
15. Bukoski RD, Xue H, McCarron DA (1987) Effect of 1,25(OH)$_2$ vitamin D3 and ionized Ca^{2+} on ^{45}Ca uptake by primary cultures of aortic myocytes of spontaneously hypertensive and Wistar Kyoto normotensive rats. Biochem Biophys Res Commun 146:1330–1335
16. Bukoski RD, DeWan P, McCarron DA (1989) 1,25(OH)$_2$, vitamin D3 modifies growth and contractile function of vascular smooth muscle or spontaneously hypertensive rats. Am J Hypertens 2:553–556
17. Christensson T, Helstrom K, Wengle B (1977) Blood pressure in subjects with hypercalcaema and primary hyperparathyroidism detected in a health screening programme. Eur J Clin Invest 7:109–113

130   A. Gairard et al.

18. Drüeke T, Fleury J, Toure Y, De Vernejoul P, Frauchet M, Lesourd P, Le Pailleur C, Crosnier J (1980) Effect of parathyroidectomy on left-ventricular function in haemodialysis patients. Lancet i:i112–i114
19. Geiger H, Bahner U, Palkovits M, Seewaldt B, Heidland A (1988) Is the effect of calcium diet or parathyroidectomy on the development of hypertension in spontaneously hypertensive rats mediated by atrial natriuretic peptide? Kidney Int 34 [Suppl 25]:S93–S97
20. Gennari C, Nami R, Agnusdei D, Bianchini C, Pavese G (1989) Acute cardiovascular and renal effects of human calcitonin gene related peptide. Am J Hypertens 2:45S–49S
21. Grobbee DE, Hackeng WHL, Berkenhagen JC, Hofman A (1988) Raised plasma intact parathyroid hormone concentrations in young people with mildly raised blood pressure. Br Med J 296:8145–8146
22. Habener JF, Rosenblatt M, Potts J (1984) Parathyroid hormone: biochemical aspects of biosynthesis, secretion, action and metabolism. Physiol Rev 64:985–1053
23. Hellstrom J, Birk G, Edvall CA (1958) Hypertension in hyperparathyroidism. Br J Urol 30:13–24
24. Helwig JJ, Schleiffer R, Judes C, Gairard A (1984) Distribution of parathyroid hormone-sensitive adenylate cyclase in isolated rabbit renal cortex microvessels and glomeruli. Life Sci 35:2649–2657
25. Helwig JJ, Yang MCM, Bollack C, Judes C, Pang PKT (1987) Structure-activity relationship of PTH: relative sensitivity of rabbit renal microvessel and tubule adenylate cyclases to oxidized PTH and PTH inhibitors. Eur J Pharmacol 140:247–257
26. Himelman RB, Helm CA, Schiller NB (1989) Is parathormone a cardiac toxin in uremia? Int J Cardiac Imaging 3:209–215
27. Hirata Y, Takata S, Takaji Y, Yoshimi H, Fujita T (1987) Parathyroid hormone receptor in vascular smooth muscle and endothelial cells. 1st international conference on new actions of PTH, Kobe, 1987
28. Huang M, Hanley DA, Rorstad OP (1983) Parathyroid hormone stimulates adenylate cyclase in rat cerebral microvessels. Life Sci 32:1009–1014
29. Hulter HN, Melby JC, Peterson JC, Cooke CR (1986) Chronic continous PTH infusion results in hypertension in normal subjects. J Clin Hypertens 4:360–370
30. Hvarfner A, Bergström R, Mörlin C, Wide L, Ljunghall S (1987) Relationships between calcium metabolic indices and blood pressure in patients with essential hypertension as compared with a healthy population. J Hypertens 5:451–456
31. Hvarfner A, Larsson R, Mörlin C, Rastad J, Wide L, Akerström G, Ljunghall S (1988) Cytosolic free calcium in platelets: relationships to blood pressure and indices of systemic calcium metabolism. J Hypertens 6:71–77
32. Inoue T, Kawashima H (1988) 1,25-dihydroxyvitamin D3 stimulates ^{45}Ca uptake by cultured vascular smooth muscle cells derived from rat aorta. Biochem Biophys Res Commun 152:1388–1394
33. Iseki K, Massry SG, Campese V (1986) Effects of hypercalcemia and parathyroid hormone on blood pressure in normal and renal-failure rats. Am J Physiol 250:F924–F929
34. Kageyama Y, Suzuki H, Arima K et al. (1987) Oral calcium tratment lowers blood pressure in renovascular hypertensive rats by suppressing the renin angiotensin system. Hypertension 10:375–382
35. Kaneko T, Ohtani R, Lewanczuk RZ, Pang PKT (1989) A novel cell type in the parathyroid glands of spontaneously hypertensive rats. Am J Hypertens 2:549–552
36. Katoh Y, Klein KL, Kaplan RA, Sanborn WG, Kurokawa K (1981) Parathyroid hormone has a positive inotropic action in the rat. Endocrinology 109:2252–2254

37. Kawashima H (1987) 1,25 Dihydroxyvitamin D3 receptors identified in rat heart. Biochem Biophys Res Commun 128–305
38. Kawashima H (1990) Parathyroid hormone causes a transient rise in intracellular ionized calcium in vascular smooth muscle cells. Biochem Biophys Res Commun 166:709–714
39. Kondo N, Shibata S, Tenner TE, Pang PKT (1988) Electromechanical effects of bPTH-(1–34) on rabbit sinus node cells and guinea pig papillary muscles. J Cardiovasc Pharmacol 11:619–625
40. Lafferty FW (1981) Primary hyperparathyroidism: changing clinical spectrum, prevalence of hypertension and discriminant analysis of laboratory test. Arch Intern Med 141:1761–1766
41. Larno S, Lhoste F, Auclair MC, Lechat P (1980) Interaction between parathyroid hormone and the beta-adrenoceptor system in cultured rat myocardial cells. J Mol Cell Cardiol 12:955–964
42. Lewanczuk RZ, Wang J, Zhang Z, Pang PKT (1989) Effects of spontaneously hypertensive rat plasma on blood pressure and tail artery calcium uptake in normotensive rats. Am J Hypertens 2:26–31
43. MacCarthy EP, Yamashita W, Hsu A, Ooi BS (1989) 1,25-dihydroxyvitamin D3 and rat vascular smooth muscle cell growth. Hypertension 13:954–959
44. McCarron DA (1987) Ca^{2+}, vitamin D, parathyroid hormone, and blood pressure: why should it matter? J Lab Clin Med 110:663–664
45. McCarron DA, Yung NH, Ugoretz BA, Krutzik S (1981) Disturbances in calcium metabolism in the spontaneously hypertensive rat. Hypertension 3 [Suppl I]:I161–I167
46. Mann JEF, Wiecek A, Bommer J, Ganten U, Ritz E (1987) Effects of parathyroidectomy on blood pressure in spontaneously hypertensive rats. Nephron 45:46–52
47. Moore PL, Strickland ML, Crass MF (1986) Vasoactive characteristics of calcitonin in coronary and hepatic circulations. J Hypertens 4 [Suppl 5]:S186–S188
48. Musso MJ, Barthelmebs M, Imbs JL, Plante M, Bollack C, Helwig JJ (1989) The vasodilator action of parathyroid hormone fragments on isolated perfused rat kidney. Arch Pharmacol 340:246–251
49. Nakanishi H, Fujii T (1989) Blood pressure response to norepinephrine and angiotensin II in the offspring of parathyroidectomized mother rats. Clin Exp Pharmacol Physiol 16:383–386
50. Nickols GA, Cline WH (1987) Parathyroid hormone-induced changes in cyclic nucleotide levels during relaxation on the rat aorta. Life Sci 40:2351–2359
51. Nickols GA, Metz MA, Cline WH (1986) Vasodilation of the rat mesenteric vasculature by parathyroid hormone. J Pharmacol Exp Ther 236:419–423
52. Nickols GA, Nickols MA, Helwig JJ (1990) Binding of parathyroid hormone and parathyroid hormone-related protein to vascular smooth muscle of rabbit renal microvessels. Endocrinology 126:721–727
53. Noriaki K, Shoji S, Tenner E, Pang PKT (1988) Electromechanical effects of bPTH-(1–34) on rabbit sinus node cells and guinea pig papillary muscles. J Cardiovas Pharmacol 11:619–625
54. Orlov SN, Postnov YV (1981) Calcium accumulation and calcium binding by the cell membranes of cardiomyocytes and smooth muscle of aorta in spontaneously hypertensive rats. Clin Sci 59 [Suppl 6]:207s–209s
55. Pang PKT, Lewanczuk R (1989) Parathyroid origin of a new circulating hypertensive factor in spontaneously hypertensive rats. Am J Hypertens 2:898–902
56. Pang PKT, Tenner TE, Yee YA, Yang M, Janssen HF (1980) Hypotensive action of parathyroid hormone preparation on rats and dogs. Proc Natl Acad Sci USA 77:675–678
57. Pang PKT, Yang MCM, Tenner TE, Kenny AD, Cooper CW (1986) Cyclic AMP and the vascular action of parathyroid hormone. Can J Physiol Pharmacol 64:1543–1547

58. Pang PKT, Yang MCM, Sham JSK (1988) Parathyroid hormone and calcium entry blockade in a vascular tissue. Life Sci 42:1395–1400
59. Pang PKT, Wang R, Shan J, Karpinski E, Benishin CG (1990) Specific inhibition of long lasting L-type calcium channel by synthetic parathyroid hormone. Proc Natl Acad Sci USA 87:623–627
60. Pernot F, Schleiffer R, Berthelot A, Vincent M, Sassard J, Gairard A (1986) Parathyroidectomy and development of genetic hypertension in the Lyon rat strain. Clin Exp Hypertens [A] 8:133–134 (abstract)
61. Pernot F, Schleiffer R, Bergmann C, Vincent M, Sassard J, Gairard A (1990) Dietary calcium, vascular reactivity and genetic hypertension in the Lyon rat strain. Am J Hypertens 3:846–853
62. Peuler JD, Morgan DA, Mark AL (1987) High calcium diet reduces blood pressure in Dahl salt-sensitive rats by neural mechanisms. Hypertension 9 [Suppl III]: III159–III165
63. Resnick LM, Laragh JH, Sealey JE, Alderman MH (1983) Divalent cations in essential hypertension: relations between serum ionized calcium, magnesium, and plasma renin activity. N Engl J Med 309:888–891
64. Resnick LM, Di Fabio B, Marion RM, James GD, Laragh JH (1986) Dietary calcium modifies the pressor effects of dietary salt intake in essential hypertension. J Hypertens 4 [Suppl 6]:S679–S681
65. Rosenthal FD, Roy S (1972) Hypertension in hyperparathyroidism. Br Med J 4: 396–397
66. Rostand SG, Sanders C, Kirk KA, Rutsky EA, Fraser RG (1988) Myocardial calcification and cardiac dysfunction in chronic renal failure. Am J Med 85:651–657
67. Saglikes Y, Massry SG, Iseki K, Nadler JL, Campese VM (1965) Effect of PTH on blood pressure and response to vasoconstrictor agonists. Am J Physiol 248: F674–F681
68. Schleiffer R, Berthelot A, Gairard A (1979) Action of parathyroid extract on arterial blood pressure and on contraction and ^{45}Ca exchange in isolated aorta of the rat. Eur J Pharmacol 58:163–167
69. Schleiffer R, Berthelot A, Pernot F, Gairard A (1981) Parathyroids, thyroids and development of hypertension in SHR. Jpn Circ J 45:1272–1279
70. Schleiffer R, Helwig JJ, Pernot F, Gairard A (1986) Vascular effects of calcitonin and parathyroid hormone. In: Doepfner W (ed) Calcitonin 1984. Excerpta Medica. Amsterdam. pp 15–24 (Current clinical practice series 2)
71. Schleiffer R, Pernot F, Gairard A (1986) Parathyroidectomy, cardiovascular reactivity and calcium distribution in aorta and heart of spontaneously hypertensive rats. Clin Sci 71:505–511
72. Schleiffer R, Gairard A (1988) Influence of parathyroidectomy on aortic responsiveness to noradrenaline in spontaneously hypertensive rats. Arch Int Pharmacodyn Ther 292:189–202
73. Schleiffer R, Bergmann C, Pernot F, Gairard A (1989) Parathyroid hormone acute vascular effect is mediated by decreased Ca^{2+} uptake and enhanced cAMP level. Mol Cell Endocrinol 67:63–71
74. Tenner TE, Ramanadham S, Yang MCM, Pang PKT (1983) Chronotropic actions of bPTH 1–34 in the right atrium of the rat. Can J Physiol Pharmacol 61:1162–1167
75. Togari A, Arai M, Shamoto T, Matsumoto S, Nagatsu T (1989) Elevation of blood pressure in young rats fed a low calcium diet. Biochem Pharmacol 38:889–893
76. Walters MR, Wicker DC, Riggle PC (1986) 1,25 Dihydroxyvitamin D3H receptors identified in rat heart. J Mol Cell Cardiol 18:67
77. Weishaar RE, Simpson RU (1987) Vitamin D3 and cardiovascular function in rats. J Clin Invest 79:1706–1712

78. Weishaar RE, Kim S, Saunders D, Simpson RU (1990) Involvement of vitamin D3 with cardiovascular function. III, Effects on physical and morphological properties. Am J Physiol 258 (21):E134–E142

79. Wiecek A, Kuczera M, Ganten U, Ritz E, Mann JFE (1989) Influence of parathyroidectomy on blood pressure and vascular reactivity in spontaneously hypertensive rats. Clin Exp Hypertens [A]11:1525–1530

80. Yang F, Nickerson PA (1988) Effect of parathyroidectomy on arterial hypertrophy, vascular lesions, and calcium content in deoxycorticosterone-induced hypertension. Res Exp Med 188:289–297

81. Zachariah P, Schwartz GL, Strong CG, Ritter S (1988) Parathyroid hormone and calcium: a relationship in hypertension. Am J Hypertens 1:79S–82S

82. Zemel MB, Geradoni SM, Walsh MF, Komanicky P, Standley P, Johnson D, Fitter W, Sowers JR (1986) Effects of sodium and calcium on calcium metabolism and blood pressure regulation in hypertensive black adults. J Hypertens 4 [Suppl 5]: S364–S366

Reviews

Campese VM (1989) Calcium, parathyroid hormone and blood pressure. Am J Hypertens 2:34S–44S

McCarron DA (1989) Calcium metabolism and hypertension. Kidney Int 35:717–736

Sowers JR, Zemel MB, Standley PR, Zemel PC (1989) Calcium and hypertension. J Lab Clin Med 114:338–348

Young EW, Bukoski RD, McCarron DA (1988) Calcium metabolism in experimental hypertension. Proc Exp Biol Med 187:123–141

Protection Against Hypertensive Cardiovascular Damage by Dihydropyridine Calcium Antagonists

J.P. Stasch, S. Kazda, G. Luckhaus, A. Knorr, B. Garthoff, and C. Hirth-Dietrich

Bayer AG, Pharma Reserch Centre, PO Box 10 17 09, W-5600 Wuppertal 1, FRG

Introduction

It is commonly acknowledged that calcium antagonists inhibit transmembrane calcium influx in arteriolar smooth muscle cells and thus reduce total peripheral resistance. This mechanism of action is supposed to be the prime basis of the antihypertensive action of these agents. However, despite their action as arteriolar vasodilators, therapeutically, calcium antagonists differ from vasodilators with other mechanisms of action (e.g., hydralazine or minoxidil) in having a marked natriuretic and diuretic action [6,8,17] and have been shown to preserve structure and function of heart, kidney, and mesenteric arterioles of hypertensive rats [2–4,10,13,18,24–26]. Therefore, tissue-protective effects of dihydropyridine calcium antagonists which might be unrelated to their hemodynamic actions, might add significantly to their long-term therapeutic efficacy in hypertension. We investigated the antihypertensive and tissue-protective effects of dihydropyridine calcium antagonists in long-term experiments: (1) nisoldipine in salt-loaded Dahl S and R rats in a preventive and therapeutic experiment, (2) nitrendipine in spontaneously hypertensive rats and Wistar-Kyoto rats in comparison to a classical vasodilator, and (3) nimodipine in stroke-prone spontaneously hypertensive rats in comparison to parathyroidectomy.

Salt-Induced Hypertension and Tissue Damage in Dahl Rats

Dahl rats are a suitable model for studying malignant hypertension and vascular damage. Fulminant hypertension in the salt-loaded S rat results in necrotizing arteriopathy with preferential localization in the kidney [7]. Our histological studies in Dahl rats revealed all the typical features such as fibrinoid intimal degeneration, medial hyperplasia, and periarteritis. Hypertensive changes of the afferent glomerular arteries of the hypertensive S rats on a high-salt diet led to collapse and degeneration of the glomeruli

G. Bruschi A. Borghetti (Eds.)
Cellular Aspects of Hypertension
© Springer-Verlag Berlin · Heidelberg 1991

Table 1. Effect of treatment for 5 weeks with the calcium antagonist nisoldipine in Dahl S and Dahl R rats: preventive experiment

	S rats + 8% NaCl		R rats + 8% NaCl	
	Controls	Nisoldipine	Controls	Nisoldipine
Systolic blood pressure (mmHg)	236 ± 3.6 (20)	168 ± 4.7 (22)***	130 ± 1.1 (19)	128 ± 0.8 (16)
Body weight (g)	262 ± 8.6 (20)	296 ± 5.9 (22)**	260 ± 7.6 (19)	266 ± 6.6 (18)
Plasma ANP (pg/ml)	470 ± 39 (19)	255 ± 30 (22)***	312 ± 15 (19)	279 ± 22 (15)
Relative heart weight (mg/100 g body weight)	396 ± 12 (20)	289 ± 4.0 (22)***	277 ± 3.2 (19)	274 ± 3.7 (18)
Hematocrit (%)	32 ± 1.7 (6)	48 ± 0.4 (10)***	46 ± 0.6 (8)	45 ± 0.6 (6)
Plasma renin activity (ng angiotensin I/ml h)	2.5 ± 0.4 (20)	0.12 ± 0.06 (22)***	0.6 ± 0.1 (16)	1.4 ± 0.3 (13)*

Values are means ± SEM (n); statistical significance was tested by Student's t test.
* $p < 0.02$; ** $p < 0.005$; *** $p < 0.001$, compared with values in controls.

[18]. Electron microscopic studies performed in our experimental series by J. Staubesand, Freiburg, revealed huge calcium deposits in the wall of the renal arteries of hypertensive Dahl rats [11]. No pathological findings were detected in S rats that remained normotensive on a low-sodium diet. Salt-loaded R rats also remained morphologically intact.

Even in the 1st week on high-salt diet, systolic blood pressure increased in the S control group from 136 to 158 mmHg, rising gradually to 236 mmHg in the 5th week. A 5-week nisoldipine treatment of S rats on a high-salt diet from the age of 5 weeks onwards largely prevented the development of hypertension throughout the treatment period. In control and treated R rats, blood pressure remained unchanged during the 5 weeks of high-salt regimen (Table 1; [5]).

The relative heart weights of S control rats on a high-sodium diet were considerably higher than those of the other groups. S nisoldipine-treated rats had significantly lower heart weights than S controls. In R rats there were no significant differences between controls and nisoldipine-treated animals with respect to heart weights (Table 1).

The plasma levels of ANP (atrial natriuretic peptides) were dramatically increased in the control rats by comparison to those in the nisoldipine-treated rats. In the treated and untreated salt-loaded S rats the level of ANP in plasma correlated positively with heart weight. Plasma ANP in treated control S rats was nearly in the same range as in treated and untreated R rats on a high-salt diet (Table 1). Similar results were achieved in a therapeutic trial in S rats on dietary salt loading for 10 weeks. Administration of nisoldipine was started after 5 weeks on this diet. Treatment with nisoldipine for 5 weeks not only reduced blood pressure and mortality (15 of 16 animals survived after addition of nisoldipine to the high-salt diet, in contrast to the control S rats, of which only 10 of 16 survived), but also produced a regression in cardiac hypertrophy and a regression in elevated plasma ANP levels by comparison to those in the untreated salt-loaded S controls (Table 2).

Since an increase in atrial filling pressure is known to be a strong stimulus for ANP release, these elevated ANP levels in S rats on a high-sodium diet can be regarded as a sign of cardiac overload due to volume expansion [1,9]. Additionally, it may be assumed that the low hematocrit values reflect increased plasma volume in salt-loaded S rats [12]. In parallel, plasma ANP levels in salt-loaded S rats are higher than in the corresponding R rats. Hence, the nisoldipine-induced normalization of plasma ANP levels in salt-loaded S rats must be regarded as the consequence of a protective effect against hypertension and cardiac volume overload (Table 2).

To study the influence of nifedipine on vascular wall structure, biopsies of mesenteric arterioles were obtained from Dahl S rats that, after 6 weeks on an 8% sodium diet, had established hypertension (238 mmHg). After 6 more weeks on an 8% sodium diet containing nifedipine 300 ppm, appropriate samples were taken from the same individuals for comparison.

Table 2. Effect of dietary salt loading (for 10 weeks) and nisoldipine therapy (introduced after 5 weeks) in Dahl S and Dahl R rats: therapeutic experiment

	Observation week	S rats + 8% NaCl		R rats + 8% NaCl	
		Controls	Nisoldipine	Controls	Nisoldipine
Systolic blood pressure (mmHg)	0	140 ± 2 (16)	138 ± 2 (16)	120 ± 2 (9)	120 ± 3 (10)
	5	252 ± 6 (16)	252 ± 6 (16)	136 ± 2 (9)	133 ± 3 (10)
	10	258 ± 4 (10)	183 ± 5 (15)***	130 ± 4 (9)	126 ± 2 (10)
Body weight (g)	0	162 ± 3 (16)	160 ± 3 (16)	144 ± 4 (9)	140 ± 4 (10)
	5	276 ± 8 (16)	259 ± 7 (16)	281 ± 4 (9)	273 ± 4 (10)
	10	279 ± 17 (10)	325 ± 11 (15)**	325 ± 7 (9)	317 ± 10 (10)
Plasma ANP (pg/ml)		488 ± 95 (9)	226 ± 38 (15)**	226 ± 36 (9)	152 ± 18 (8)
Relative heart weight (mg/100 g body weight)		473 ± 24 (10)	335 ± 15 (15)*	276 ± 3 (9)	273 ± 5 (10)
Hematocrit (%)		38 ± 2.4 (10)	48 ± 2.4 (15)**	45 ± 0.6 (9)	43 ± 1.0 (10)
Plasma renin activity (ng angiotensin I/ml h)		1.5 ± 0.2 (10)	0.2 ± 0.0 (15)***	0.1 ± 0.0 (9)	0.9 ± 0.4 (9)

Values are means ± SEM (n); statistical significance was tested by Student's t test.
* $p < 0.05$; ** $p < 0.02$; *** $p < 0.001$, compared with values in controls.

Table 3. Preventive and therapeutic experiment: the effect of nitrendipine treatment on systolic blood pressure, atrial natriuretic peptide (ANP) in plasma, relative heart weight, body weight, and plasma renin activity in SHR

	Controls	Nitrendipine		
		Weeks 62–71	Weeks 1–61	Weeks 1–71
Systolic blood pressure (mmHg)	210 ± 10	174 ± 8***	193 ± 4	172 ± 5**
Plasma ANP (pg/ml)	403 ± 43	264 ± 55*	223 ± 28	175 ± 21****
Relative heart weight (mg/100 g body weight)	434 ± 51	347 ± 4*	352 ± 16	313 ± 2***
Body weight (g)	376 ± 4	367 ± 6	370 ± 9	382 ± 6
Plasma renin activity (ng angiotensin I/ml h)	5.8 ± 0.8	4.2 ± 0.4	3.3 ± 0.3	2.9 ± 0.3****
n	5	7	10	10

Values are means $\pm$ SEM; statistical significance was tested by Student's t test.
* $p < 0.05$; ** $p < 0.02$; *** $p < 0.01$; **** $p < 0.001$, compared with values in untreated controls.

Examination showed that despite the continuous sodium load, nifedipine lowered systolic blood pressure to 164 mmHg, whereas it was 223 mmHg in the control group. In hypertensive rats before treatment, arterial endothelial damage was noted, with distension of the subendothelial space due to fibrin insudates, occasional rupture of the internal elastic lamina, myointimal proliferation, and mononuclear periarteritis; after nifedipine, obvious amelioration was seen, with intimal fibrin resorption. On the luminal side of the original internal elastic lamina, a new internal elastic lamina had formed [18,19].

Studies in SHR and WKY Rats

Similar results to those with nisoldipine in Dahl rats were obtained with nitrendipine in spontaneously hypertensive rats (SHR) and Wistar-Kyoto rats (WKY) (Table 3; [23]). This therapeutic experiment was performed for 61 weeks with two control groups and two nitrendipine-treated groups in parallel. After 61 weeks of observation treatment in one of the nitrendipine groups was stopped while the same therapeutic treatment was started in one of the untreated groups. The regimen in the other two groups was not changed. All groups were then observed for another 10 weeks. From the age of 8 weeks, the control SHR progressively manifested sustained hypertension. After 10 weeks of observation (18 weeks of age), systolic blood pressure exceeded 220 mmHg and then slowly increased during the following 40 weeks. After 50 weeks systolic blood pressure dropped to about 200 mmHg.

A 72-week nitrendipine treatment of SHR from the age of 8 weeks on effectively prevented the development of hypertension (Table 3; [23]).

The 10-week nitrendipine treatment in the 69-week old SHR with end-stage hypertension resulted in a reduction of systolic blood pressure; on the other hand, when treatment was stopped in SHR after 61 weeks, blood pressure increased almost to the level of the untreated controls within 10 weeks. As seen here, cardiac hypertrophy was also prevented in the long-term treated animals and reduced in old SHR after treatment with nitrendipine. End-stage hypertension in SHR is associated with cardiac failure, as can be concluded from the reduction in systolic blood pressure, called "decapitated hypertension", despite cardiac hypertrophy in old untreated SHR. In parallel, plasma levels of ANP are severalfold higher in 68-week-old SHR than in age-matched WKY. Measurement of ANP in plasma provides a sensitive marker for the severity of hypertension and cardiac volume overload (Table 3; [23]). Long-term treatment with nitrendipine in SHR not only prevented the development of hypertension and the associated cardiac hypertrophy, but also the rise in plasma ANP. Even in old SHR with pronounced cardiac hypertrophy and high plasma ANP levels, regression of these symptoms could be achieved by therapeutic nitrendipine treatment (Table 3; [23]). Long-term treatment with dihydropyridine calcium antagonists led to a reduction of plasma renin activity in old SHR in parallel with a drop in blood pressure. This is especially interesting in view of the fact that salt and water retention did not occur because of the natriuretic activity of the calcium antagonists [16]. Obviously, high renin levels in untreated controls were due to renal ischemia-stimulated renin release (Table 3; [23]). Nitrendipine may also have a protective effect against renal ischemia due to long-standing hypertension or old age.

In order to compare the influence of the dihydropyridine calcium antagonist nitrendipine on cardiac hypertrophy and ANP with that of the non-calcium antagonist vasodilator minoxidil, male SHR rats aged 22 weeks were treated with nitrendipine (1000 ppm in the solid feed) or minoxidil (80 mg/l in the drinking water) for 11 weeks. Effective control of systolic blood pressure was achieved by both types of therapeutic treatment (Table 4). In contrast to the calcium antagonist nitrendipine, which improves renal sodium excretion, vasodilators such as minoxidil or hydralazine are known to induce water and salt retention. Probably due to fluid retention, SHR receiving minoxidil became higher in body weight than the two other groups throughout the treatment period (Table 4; [24]).

These contrasting properties of calcium antagonistic and non-calcium antagonistic peripheral vasodilators are reflected also in their influence on plasma ANP, heart weight, and plasma renin activity, which were either normalized (ANP) or significantly ameliorated by long-term nitrendipine treatment. Minoxidil, however, caused further increases in ANP, heart weight, and plasma renin activity, aggravating the pathological increases in these variables in SHR (Table 4). This experiment shows that plasma

Table 4. Different effects of treatment (11 weeks) with nitrendipine or minoxidil on systolic blood pressure, atrial natriuretic peptide (ANP) in plasma, relative heart weight, body weight, and plasma renin activity in 33-week-old SHR and Wistar-Kyoto rats

	SHR			Wistar-Kyoto		
	Controls	Nitrendipine	Minoxidil	Controls	Nitrendipine	Minoxidil
Systolic blood pressure (mmHg)	247 ± 4	193 ± 3****	178 ± 3****	154 ± 4	145 ± 4	147 ± 4
Plasma ANP (pg/ml)	309 ± 38	135 ± 15****	401 ± 18*	115 ± 11	91 ± 12	263 ± 32****
Relative heart weight (mg/100 g body weight)	322 ± 3	304 ± 2****	360 ± 4****	261 ± 3	250 ± 3**	288 ± 2****
Body weight (g)	345 ± 5	340 ± 7	377 ± 4	369 ± 4	352 ± 4	396 ± 5
Body weight increase (%)	12	15	23	13	10	20
Plasma renin activity (ng angiotensin I/ml h)	4.6 ± 0.3	3.5 ± 0.3***	10.6 ± 2.0***	5.8 ± 0.6	5.4 ± 0.7	8.4 ± 0.7***
n	19	14	14	16	14	14

Values are means $\pm$ SEM; statistical significance was tested by Student's t test.
* $p < 0.05$; ** $p < 0.02$; *** $p < 0.01$; **** $p < 0.001$, compared with values in untreated controls.

volume load probably has a greater impact on cardiac hypertrophy than blood pressure. In addition, it may be concluded from these therapeutic experiments that volume status rather than blood pressure is the important factor for changes in ANP levels in plasma [24,25].

Tissue Damage and Calcium Overload in Stroke-Prone SHR Rats: Prevention by Nimodipine and by Parathyroidectomy

Stroke-prone SHR (SHRSP) were originally derived from the Okamoto strain of SHR [22]. At the age of 10–12 months they spontaneously develop cerebrovascular lesions and brain infarction. Dietary salt load in young SHRSP has been shown to intensify both hypertension and cerebro- and renovascular lesions [20]. In our experiments we used male adult SHRSP (aged 5 months) with established hypertension. The dietary salt load did not produce any additional increase in blood pressure in these rats, though it did drastically increase mortality and produce severe cerebro- and renovascular lesions [12,15].

Simultaneous treatment with the calcium antagonist nimodipine resulted in a dramatic reduction in mortality and vascular lesions in salt-loaded adult SHRSP [12,15]. In this experiment all control rats died within 9 weeks as a result of the high-salt diet. Histological investigations revealed severe cerebro- and renovascular lesions. More than half the nimodipine-treated rats survived for 24 weeks.

Nimodipine is a calcium antagonist derivative which has only a weak peripheral vasodilator effect. In this experiment the high blood pressure was not decreased by nimodipine, on the contrary, it was in very ill controls that blood pressure dropped before death, and on average it was lower than in the nimodipine-treated animals [12,15]. These experiments on SHRSP show that, rather than preventing high blood pressure, nimodipine primarily prevented the deterioration of hypertensive disease, the vascular damage, and the mortality. These findings support our assumption that the vascular damage in hypertension-prone rats is not entirely dependent on the systemic blood pressure. In addition to high blood pressure, other factors are presumably involved in the vascular damage induced by excessive transmembrane calcium influx [11,15].

There is increasing evidence that the activity of the parathyroid gland may be involved in human and experimental hypertension. In deoxycorticosterone acetate- (DOCA-)salt hypertension, renal and cardiac lesions have been reported to be largely reduced by surgical parathyroidectomy without influencing the high blood pressure [21]. In DOCA-salt hypertensive rats parathyroidectomy had a similar, blood-pressure-independent protective effect to that of nimodipine in our experiments with SHRSP. This similarity prompted us to compare the effect of nimodipine with that of parathyroidectomy in salt-loaded SHRSP. Parathyroidectomy was performed in one of three groups of 12-week-old male SHRSP; two groups, each of 18 rats, were sham-operated. After 17 days all three groups received a diet containing 8% NaCl, and one group of sham-operated animals was treated with nimodipine added to the high-salt diet [14].

The introduction of the high-salt diet resulted in an additional increase in blood pressure (more than 250 mmHg during the first 4 weeks) in all groups of these young SHRSP. This was in contrast with our previous studies with adult SHRSP, in which blood pressure increased only slightly after the same amount of salt was added to the diet [12,14]. All control animals died in the course of 4 weeks on a high-salt diet. After parathyroidectomy all rats developed tetany after the introduction of the high-salt diet, and more than half died in the early weeks. Six animals recovered quite quickly and survived the observation period of 17 weeks, but had extremely high blood pressure [12,14,16].

In the nimodipine-treated rats, too, blood pressure progressively increased on the high-salt diet. However, all rats developed well until the end of the observation period (17 weeks). Total serum calcium was normal in the two sham-operated groups (2.56 ± 0.06 and 2.53 ± 0.04 mmol/l, respectively) but reduced in the parathyroidectomy group (1.92 ± 0.07 mmol/l) [14]. Morphological lesions typical of hypertensive vascular disease in these rats [12] were severe in all control rats but completely absent in the survivors in the parathyroidectomy group and all nimodipine-treated rats. An involvement of renal damage can be also suspected in our experiment with salt-loaded SHRSP. Plasma renin activity reached high values in control rats despite the continuous high salt intake. Similar paradoxical increases in plasma renin activity after salt loading in SHRSP has already been described as an expression of progressive renal ischemia [14,15,20].

In general, there were some similarities in the effects of parathyroidectomy and nimodipine in this experiment. In contrast to controls, both nimodipine-treated rats and some of the parathyroidectomy group survived for a long period, but had extremely high blood pressure. Obviously, high blood pressure alone is not the cause of stroke and high mortality in salt-loaded SHRSP. It is conceivable that a high salt intake produces fatal tissue damage by inducing a harmful calcium overload. The lifesaving effect of nimodipine, mimicked by parathyroidectomy, may be due to inhibition of an excessive increase of cellular calcium concentration [11]. To answer this question we decided to investigate the influence of parathyroidectomy and of nimodipine on the natural development of stroke and on the tissue calcium content in SHRSP without the accelerating effect of a dietary salt load.

Seven-month-old male SHRSP were divided into three groups: in one group bilateral parathyroidectomy was performed, the other groups were sham-operated. In one of these groups continuous nimodipine treatment was started at that age; the third group served as a control. All animals, fed standard rat chow containing less than 1% NaCl, were observed for 14 weeks.

Nearly half of the parathyroidectomized rats developed fatal tetany in the first weeks after surgery; the surviving rats developed well until the end of observation. During the last days of observation (10.5 months of age) all the control rats showed neurological symptoms typical for stroke (high irritability, transitory convulsions, hemiplegia or paraplegia). None of the nimodipine-treated or the parathyroidectomized rats showed any neurological symptoms until the end of observation. The final blood pressure of the nimodipine-treated rats was identical with that of the control rats; a small but insignificant decrease was measured at the end of observation in parathyroidectomized rats [11,15].

Since all the controls showed symptoms of stroke in week 14 of observation, the experiment was terminated. Brains and kidneys were removed and homogenized, and the calcium content was measured by atomic absorption. In addition, the calcium content of organs from adult normotensive Wistar-Kyoto rats as well as from young SHRSP was investigated. The brain calcium content of the old control SHRSP was more than 50% higher than that of young SHRSP or adult Wistar-Kyoto animals. The calcium content of brains from both nimodipine-treated and parathyroidectomized rats was only slightly higher than that of young SHRSP at a prehypertensive age, but much lower than that of diseased old SHRSP. In the kidneys the calcium content was considerably elevated in the old SHRSP controls compared with the young SHRSP or Wistar-Kyoto rats. In the old parathyroidectomized rats the renal calcium content remained normal. In the nimodipine-treated rats the increase in renal calcium was partially prevented, but the content was still significantly higher than that of old parathyroidectomized or young control rats [11,15].

The results of this experiment confirm our previous observation that tissue damage in severe hypertensive disease is not solely caused by high blood pressure. Some factors in addition to the high blood pressure are involved which, by increasing the tissue concentration of calcium above a critical threshold, produce tissue damage and death. Nimodipine completely prevented the increase in brain calcium content and the appearance of neurological symptoms of stroke in old SHRSP without affecting their high blood pressure. This protective effect was partially mimicked by surgical parathyroidectomy. In contrast with nimodipine, parathyroidectomy also completely prevented the increase in calcium content in the kidney and the concomitant increase in plasma renin activity.

The protective effect of parathyroidectomy suggests that the activity of the parathyroid gland may be activated, and that an excess of hormone from the parathyroid gland may be at least partly responsible for the calcium overload and tissue damage in advanced hypertension. This hypothesis would explain why nimodipine has a similar effect to removal of the parathyroid gland. It prevents the harmful calcium overload, even in the presence of long-standing, extremely high hypertension.

Conclusions

Our results indicate that the therapeutic effects of dihydropyridine calcium antagonists on hypertensive tissue damage include prevention and reversal of cardiac hypertrophy, vascular lesions, and pathologic neurological symptoms even after doses which have no effect on blood pressure.

In a susceptible rat strain (Dahl S rats) hypertension and tissue damage can be induced by a high-salt diet, both being preventable by calcium-antagonistic dihydropyridine derivatives like nisoldipine.

Therapeutic antihypertensive treatment with nitrendipine in SHR improves renal sodium excretion, lowers plasma renin activity and plasma ANP, and produces a regression in cardiac hypertrophy. By contrast, equally effective blood pressure reduction with the sodium-retaining vasodilator minoxidil results in an aggravation of cardiac hypertrophy and an increase in plasma renin activity and plasma ANP.

The increase in plasma ANP levels in these models of hypertension, and its modulation by antihypertensive treatment with a calcium antagonist or an arteriolar vasodilator, show that the changes in ANP plasma levels are secondary to hypertensive disease and the associated cardiac volume overload.

In SHRSP, a dietary salt load results in brain and renal lesions. Tissue damage is largely prevented and survival is increased by nimodipine without decreasing the high blood pressure. A similar protective effect, independent of blood pressure, is achieved by bilateral parathyroidectomy in salt-loaded SHRSP. The natural manifestation of stroke in SHRSP (even without an additional salt load) correlates with the increase in calcium content in brain and kidney tissue. Nimodipine prevents spontaneous strokes and the increase in brain calcium content without affecting the high blood pressure in SHRSP. Parathyroidectomy also mimics the protective and lifesaving effect of nimodipine. Presumably some calcium-promoting factors are activated in advanced hypertension, resulting in cellular damage. Excess hormones from the parathyroid glands may be one of these factors.

References

1. Burnett JC, Kao PC, Hu DC, Heser DW, Heublein D, Granger JP, Opgenorth TJ, Reeder GS (1986) Atrial natriuretic peptide elevation in congestive heart failure in the human. Science 231:1145–1147
2. Fleckenstein A, Frey M, von Witzleben H (1983) Vascular calcium overload – a pathogenic factor in arteriosclerosis and its neutralization by calcium antagonists. In: Kaltenbach M, Neufeld HN (eds) New therapy of ischemic heart disease and hypertension, 5th international adalat symposium. Excerpta Medica, Amsterdam, pp 36–52

3. Fleckenstein A, Frey M, Zorn S, Fleckenstein-Grün G (1985) Experimental basis of the long-term therapy of arterial hypertension with calcium antagonists. Am J Cardiol 56:3H–14H
4. Garthoff B, Kazda S (1981) Calcium antagonist nifedipine normalizes high blood pressure and prevents mortality in salt loaded DS substrain of Dahl rats. Eur J Pharmacol 74:111–113
5. Garthoff B, Kazda S (1982) Prevention and reversal of salt-induced hypertension in DS-Dahl rats by nisoldipine. Fed Proc 41:1664
6. Garthoff B, Kazda S, Knorr A, Thomas G (1983) Factors involved in the antihypertensive action of calcium antagonists. Hypertension 5 [Suppl II]:II34–II38
7. Jaffé S, Sutherland LE, Barker DV, Dahl LK (1970) Effects of chronic salt ingestion: morphologic findings in kidneys of rats with differing genetic susceptibilities to hypertension. Arch Pathol 90:1–16
8. Johns EJ, Manitius J (1987) The renal actions of nitrendipine and its influence on the neural regulation of calcium and sodium reabsorption in the rat. J Cardiovasc Pharmacol 9 [Suppl 1]:S49–S56
9. Katsube N, Schwartz D, Needleman P (1986) Release of atriopeptin in the rat by vasoconstrictors and water immersion correlates with changes in right atrial pressure. Biochem Biophys Res Commun 133:937–942
10. Kazda S (1986) Effects of nitrendipine on vascular integrity. Am J Cardiol 58:31D–34D
11. Kazda S (1988) The calcium channel and vascular injury. In: Morad M, Nayler W, Kazda S, Schramm M (eds) The calcium channel: structure, function and implications. Springer, Berlin Heidelberg New York, pp 326–334
12. Kazda S, Garthoff B, Dycka J, Iwai I (1982) Prevention of malignant hypertension in salt loaded "S" Dahl rats with the calcium antagonist nifedipine. Clin Exp Hypertens [A] 4:1231–1242
13. Kazda S, Garthoff B, Luckhaus G (1983) Calcium antagonists in hypertensive disease: experimental evidence for a new therapeutic concept. Postgrad Med J 59 [Suppl 2]:78–83
14. Kazda S, Garthoff B, Hirth C, Preis W, Stasch JP (1986) Parathyroidectomy mimics the protective effect of the calcium antagonist nimodipine in salt-loaded stroke-prone spontaneously hypertensive rats. J Hypertens 4 [Suppl 3]:S482–S485
15. Kazda S, Grunt M, Hirth C, Preis W, Stasch JP (1987) Calcium antagonism and protection of tissue from calcium damage. J Hypertens 5 [Suppl 4]:S37–S42
16. Kazda S, Hirth C, Stasch JP (1988) Diuretic effect of nitrendipine contributes to its antihypertensive efficacy: a review. J Cardiovasc Pharmacol 12 [Suppl 4]:S1–S5
17. Klütsch K, Schmidt P, Grosswendt J (1972) Der Einfluss von BAY A 1040 auf die Nierenfunktion des Hypertonikers. Arzneimittelforschung 22:377–380
18. Luckhaus G, Garthoff B, Kazda S (1982) Prevention of hypertensive vasculopathy by nifedipine in salt-loaded Dahl rats. Arzneimittelforschung 32:1421–1425
19. Luckhaus G, Nash G, Garthoff B, Kazda S, Feller W (1985) Prevention of hypertensive vasculopathy by nifedipine in salt-loaded Dahl rats. Arzneimittelforschung 35:115–121
20. Nagaoka A, Shino A, Inotsuka H (1978) Relationship between hypertension and cerebrovascular lesions in stroke-prone spontaneously hypertensive rats. Jpn Heart J 19:604–605
21. Nickerson PA, Conran RM (1981) Parathyroidectomy ameliorates vascular lesions induced by deoxycorticosterone in rat. Am J Pathol 105:185–190
22. Okamoto K, Yamori X, Nagaoka A (1974) Establishment of the stroke prone spontaneously hypertensive rat (SHR). Circ Res 34,35 [Suppl 1]:I143–I153
23. Stasch JP, Kazda S, Hirth C (1986) Effect on hypertension, cardiac hypertrophy and atrial natriuretic peptides of treatment with nitrendipine in SHR. J Hypertens 4 [Suppl 6]:S160–S162

24. Stasch JP, Kazda S, Hirth C (1987) The different effects of a calcium antagonist and a sodium retaining vasodilator on blood pressure, cardiac hypertrophy, and atrial natriuretic peptides in adult spontaneously hypertensive rats. J Hypertens 5 [Suppl 5]:S211–S213
25. Stasch JP, Kazda S, Hirth C, Morich F (1987) Role of nisoldipine on blood pressure, cardiac hypertrophy, and atrial natriuretic peptides in spontaneously hypertensive rats. Hypertension 10:303–307
26. Turek Z, Kubat K, Kazda S, Hoofd L, Rakusan K (1987) Improved myocardial capillarisation in spontaneously hypertensive rats treated with nifedipine. Cardiovasc Res 21:725–729

Nifedipine and Vascular Smooth Muscle Cells in Atherogenesis: In Vivo and In Vitro Studies

P. Pauletto[1], S. Sartore[2], G. Scannapieco[1], A.C. Borrione[2], A.M.C. Zanellato[2], M. Tonello[1], and A.C. Pessina

[1] Istituto di Clinica Medica I, Università di Padova, Via N. Giustiniani, 2, I-35128 Padova, Italy
[2] Istituto di Patologia Generale, Università di Padova, Via N.Giustiniani, 2, I-35128 Padova, Italy

Introduction

Several calcium antagonists have been reported to be effective antiatherogenic agents [1]. Among the calcium antagonists, nifedipine is the foremost and perhaps the most widely studied drug. In experimental animals [2] as well as in humans [3], nifedipine has been found to prevent the development of newly formed lesions and/or the progression of preexisting atheromas. Some conflicting findings regarding the efficacy of nifedipine and other calcium antagonists [1,4] in preventing experimental atherosclerosis are likely to be due to differences in the experimental conditions, the species or breed of animals used, or the method employed for assessing atherosclerosis. On the whole, the results of the majority of studies are consistent with the assumption of an appreciable antiatherogenic effect of at least three calcium antagonists, namely nifedipine, verapamil, and isradipine [1], and indicate that calcium antagonists promise well for clinical use in this field also. Nevertheless, the mechanism or mechanisms by which calcium antagonists may exert their antiatherogenic effect remains poorly understood. Since it is presently well accepted that vascular smooth muscle cells (SMC) are the major contributor to the formation of atherosclerotic plaque [5], we have studied SMC composition in the intimal thickening of hypercholesterolemic rabbits treated with nifedipine. A panel of monoclonal antimyosin antibodies was used to study SMC differentiation during atherogenesis and nifedipine treatment. These antibodies were directed against smooth-muscle or non-muscle myosin isoforms [6]. This approach allowed the identification of different SMC populations in the aortic wall and enabled us to define their modulation during atherogenesis. We also studied the influence of nifedipine on growth rate and incorporation of [3H]thymidine in SMC cultured from the aortic media.

G. Bruschi A. Borghetti (Eds.)
Cellular Aspects of Hypertension
© Springer-Verlag Berlin · Heidelberg 1991

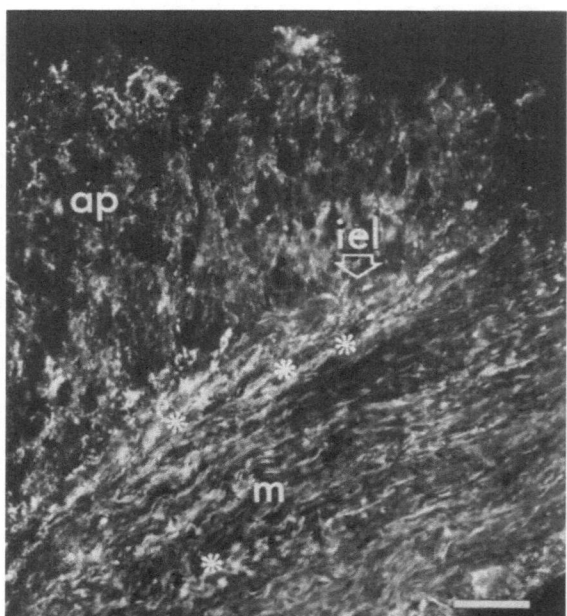

Fig. 1. Indirect immunofluorescence staining with NM-G2 anti-nonmuscle-myosin anti-body on a cryosection from aorta of a rabbit fed a cholesterol-enriched diet for 2 months. Note that the atherosclerotic plaque (*ap*) and a large medial SMC population (*asterisks*) are labeled by this antibody. *m*, medial layer of rabbit aorta; *iel*, internal elastic lamina. Bar: 70 μm

Materials and Methods

Preparation of animals: Sixteen New Zealand White rabbits (mean body weight about 3 kg) were fed a 1% cholesterol-enriched diet for 2 months. At the same time, 8 out of 16 animals were started on nifedipine slow-release treatment (20 mg b.i.d.). Eight other age-matched animals were kept on normal diet and served as controls. Body weight and blood pressure were measured weekly; plasma cholesterol levels were measured in each animal before and after the study period.

Assessment of atherosclerosis: After killing of each animal, the entire aorta was removed up to the bifurcation and fixed under a constant pressure of 90 mmHg in 10% buffered formalin for 15 min. The aortas were then embedded in paraffin and seriate histological sections taken at 5 mm intervals. Sections were then stained with Weigert–Van Gieson stain and analyzed by computerized planimetry, as previously described [7].

Immunocytochemical analysis: After death, the aortic specimens for immunocytochemical analysis were immediately frozen under liquid nitrogen and stored at −70°C until use. Cryosections of aortic tissue from each animal were processed for immunofluorescence according to the procedures in use in our laboratory [8,9]. Two monoclonal antimyosin antibodies, SM-E7 and NM-G2, were used in this study. As previously described [6], the SM-E7 antibody is specific for an epitope shared by myosin heavy chains 1 and 2 present in aortic smooth muscle of adult rabbit; the NM-G2 antibody recognizes a myosin heavy chain epitope present in nonmuscle cell types. These antibodies were applied to cryosections of aorta after appropriate dilution and bound antibodies were revealed by rabbit antimouse IgG coupled with fluorescein (FITC) or rhodamine (RITC) isothiocyanate. Sections were examined by fluorescence microscope and micrographs were taken using the appropriate filters.

Cell cultures: SMC were obtained from the aortic media of adult cattle by the explant method as previously described [10]. Subconfluent cultures of SMC at the second passage were used after being growth-arrested. To avoid binding of nifedipine to plasma proteins, a synthetic serum (CPSR$_1$, Sigma Chem, Co., St. Louis, USA) was added to the culture medium. Nifedipine ($10^{-5}M$, $10^{-6}M$, and $10^{-7}M$) was diluted in alcohol 0.1% (final concentration) and added to the culture medium every 24 h up to 72 h. Quadruplicate cultures were used for each experimental point. Cell number was determined by an electronic cell counter. Moreover, [^{3}H]thymidine 2.5 μCi/ml was added to the culture medium of both control and nifedipine-treated cells. Incorporation of [^{3}H]thymidine into the cells was measured by a β-counter after a 24-h period of incubation.

Results

The cholesterol-enriched diet led to a marked increase in plasma cholesterol levels in both nifedipine-treated and untreated animals (range 1800–2000 mg/100 ml). Before the diet, cholesterol levels ranged between 30 and 60 mg/100 ml, and these remained unchanged in the age-matched controls fed a standard diet. Throughout the study period, no appreciable differences were observed in body weight and blood pressure between the groups. Microscopy of the seriate histological sections showed a striking progression of atherosclerotic lesions in the cholesterol-fed, untreated rabbits (mean intimal area of each section: 1.49 ± 0.53 mm^2). By contrast, a negligible amount of intimal thickening was evident in the cholesterol-fed but nifedipine-treated rabbits as well as in the control rabbits fed a standard diet.

In agreement with our previous observations [8], the immunocytochemical analysis of the aortic sections taken from control animals showed that almost all medial SMC were recognized by the SM-E7 antibody, whereas only a few cells also reacted with the NM-G2 antibody (not shown). These double-labeled SMC, however, represented the major cell type in the atherosclerotic plaque (Fig. 1), where a minority of cells were recognized by the SM-E7 antibody alone and very few reacted only with the NM-G2 anti-

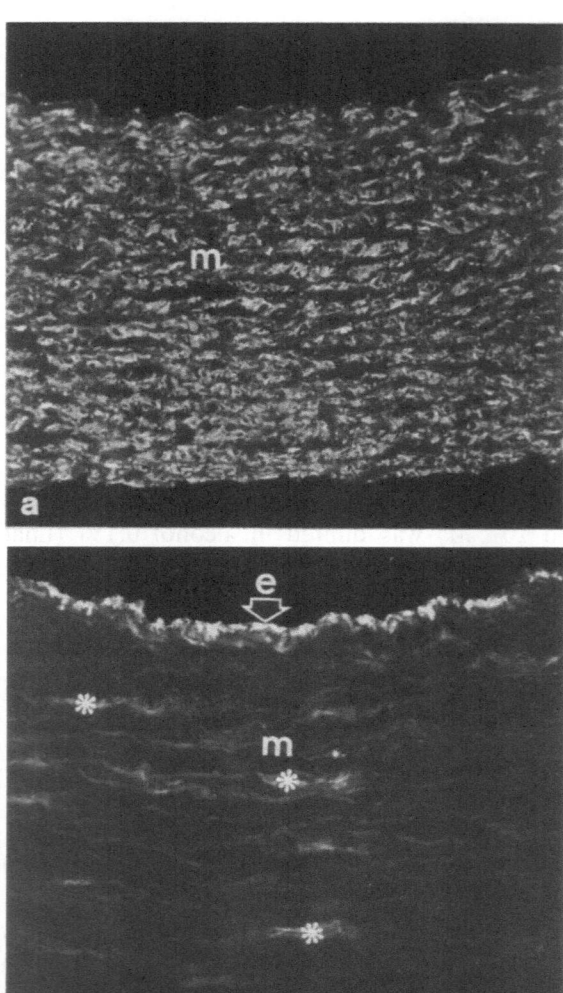

Fig. 2a,b. Indirect immunofluorescence staining with SM-E7 (**a**) and NM-G2 (**b**) anti-myosin antibodies on cryosections from aorta of a rabbit fed a cholesterol-enriched diet for 2 months along with nifedipine treatment. Note the absence of atherosclerotic plaque and the very low number of medial SMC reacting with NM-G2 (*asterisks*). *e*, aortic endothelium; *m*, medial layer of rabbit aorta, Bar: 70 μm

body. Interestingly, in the cholesterol-fed animals the size of the cell population recognized by both antibodies increased in the media underlying the plaque as well, particularly near the internal elastic lamina (Fig. 1). In the cholesterol-fed, nifedipine-treated rabbits in which atherogenesis was suppressed, the pattern of immunoreactivity to the SM-E7 antibody in the media

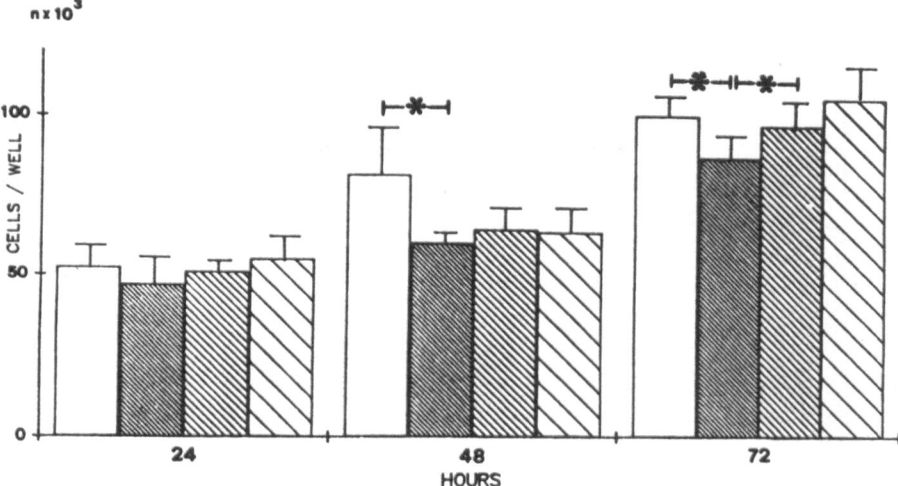

Fig. 3. Effect of nifedipine on cell proliferation in secondary cultures of bovine aortic SMC. Cells were grown in 10% $CPSR_1$ ($\square$) with nifedipine added at a concentration of $10^{-5}M$ (▧), $10^{-6}M$ (▨), and $10^{-7}M$ (▨). At the time points indicated quadruplicate cultures were trypsinized and cell number determined using an electronic cell counter. Mean±SD is given. *,$p < 0.001$ (Student's t test for unpaired data)

was essentially the same as that observed in control animals fed the standard diet (i.e., almost all SMC reacted only with SM-E7 antibody; Fig. 2a). Moreover, in these rabbits the number of medial SMC reacting with both antibodies is much lower than in the cholesterol-fed, untreated rabbits (Fig. 2b) and comparable to that observed in control animals fed the standard diet.

As for the in vitro experiments, nifedipine was able to reduce the growth rate of cultured SMC in a dose-dependent manner (Fig. 3). Incorporation of [3H]thymidine by cultured SMC was also reduced in a dose-dependent manner by the addition of nifedipine to culture medium (Fig. 4).

Discussion

The present study confirms that nifedipine is effective in preventing the development of atherosclerotic lesions in cholesterol-fed rabbits and offers some insights into the cellular mechanisms of atherogenesis and those involved in the prevention of lesion formation. The immunocytochemical approach using antimyosin antibodies is able to define different SMC populations in both the plaque and the arterial media, and to show their relative proportions in different pathophysiological conditions. In previous studies [11], we have shown that the predominant type of SMC present in the arterial media of fetal rabbits is SMC labeled with both the SM-E7 and the

CPM 10 ^{3}/WELL

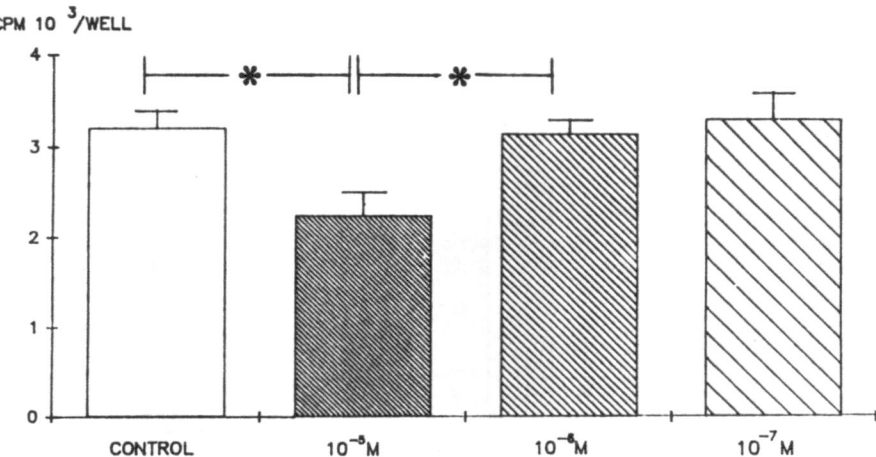

Fig. 4. Effect of different concentrations of nifedipine on [^{3}H]thymidine incorporation in cultured bovine aortic SMC. Mean±SD is given *, $p < 0.001$ (Student's t test for unpaired data)

NM-G2 antibodies; thus we defined as "fetal" myosin pattern the coexpression in SMC of both smooth muscle and nonmuscle myosin. In the aortic media of adult rabbits almost all SMC express only smooth muscle myosin ("adult" myosin pattern); however, a minority of medial SMC still express the "fetal" myosin pattern, and might represent a sort of stem-cell-like compartment. During the development of atherosclerotic lesions induced by cholesterol feeding, we observed a marked increase in medial SMC characterized by the "fetal" myosin pattern. This increase could be due to the activation and subsequent proliferation of the above-described medial SMC population present in the normal media and characterized by the "fetal" myosin pattern. Alternatively, it could be due to phenotypic modulation of some medial SMC shifting from the "adult" to the "fetal" myosin pattern under atherogenic stimuli. Whatever the origin of these "immature" (or "activated") SMC, they migrate and/or proliferate in the plaque, where they represent by far the largest cell population.

Nifedipine treatment resulted in a marked decrease of the double-labeled "immature" SMC present in the arterial media. This decrease could be brought about via two different mechanisms: (a) inhibition of proliferation of the stem-cell-like SMC present in the media; (b) blockade of SMC modulation from a differentiated phenotype to a dedifferentiated one. Either mechanism could explain the suppression of atherosclerotic lesions development observed in nifedipine-treated rabbits through reduction of the "activated" medial SMC available for plaque formation. Indeed, in agreement with previous studies [12], the in vitro experiments showed that nifedipine is able to reduce the growth rate of cultured SMC. It is known that the cellular response to growth factor stimulation, including PDGF [13], is mediated by an elevation of intracellular calcium. It therefore seems reasonable to assume that the inhibitory effects exerted by nifedipine on SMC proliferation and/or phenotypic modulation are mediated by a change in intracellular calcium availability.

References

1. Overturf M, Smith S (1989) Are calcium antagonists effective inhibitors of athero-genesis? In: Dal Palù C, Ross R (eds) Hypertension and atherosclerosis. Excerpta Medica, Amsterdam, pp 98–106
2. Henry PH, Bentley KI (1981) Suppression of atherogenesis in cholesterol-fed rabbits treated with nifedipine. J Clin Invest 68:1366–1369
3. Lichtlen PR, Nellessen U, Rafflenbeul W, Jost S, Hecker H (1987) International nifedipine trial on antiatherosclerotic therapy (INTACT). Cardiovasc Drug Ther 1:71–79
4. Stender S, Stender I, Nordestgaard B, Kjeldsen K (1984) No effect of nifedipine on atherogenesis in cholesterol-fed rabbits. Arteriosclerosis 4:389–394
5. Ross R (1986) The pathogenesis of atherosclerosis – an update. N Engl J Med 314:488
6. Borrione AC, Zanellato AMC, Scannapieco G, Pauletto P, Sartore S (1989) Myosin heavy-chain isoforms in adult and developing rabbit vascular smooth muscle. Eur J Biochem 183:413
7. Pauletto P, Scannapieco G, Vescovo G, Pessina AC, Dal Palù C (1989) Prevention of aortic hypertrophy in spontaneously hypertensive turkeys. J Hypertens 7 [Suppl 6]: S246
8. Pauletto P, Sartore S, Scannapieco G, Borrione AC, Zanellato AMC, Pessina AC, Dal Palù C (1989) Immunocytochemical analysis of myosin isoform distribution in the atherosclerotic lesions of cholesterol-fed rabbits. In: Dal Palù C, Ross R (eds) Hypertension and atherosclerosis. Excerpta Medica, Amsterdam, p 107
9. Zanellato AMC, Borrione AC, Tonello M, Scannapieco G, Pauletto P, Dal Palù C, Sartore S (1988) Myosin heavy chain isoforms in bovine smooth muscle and in smooth muscle cells grown in vitro. A monoclonal antibody study. In: Carraro U (ed) Sarcomeric and non-sarcomeric muscles: basic and applied research prospects for the 90s. Unipress Padova, Padua, p 687
10. Scannapieco G, Pauletto P, Pagnan A, Mattiello A, Biffanti S, Jori G, Dal Palù C (1988) Lipoprotein binding to cultured aortic smooth muscle cells from normo-tensive and hypertensive rats. J Hypertens 6 [Suppl 4]:S269
11. Zanellato AMC (1990) Myosin isoform expression and smooth muscle cell hetero-geneity in normal and atherosclerotic rabbit aorta. Arteriosclerosis 10:996
12. Nilsson J, Sjölund M, Palmberg L, Von Euler AM, Jonzon B, Thyberg J (1985) The calcium antagonist nifedipine inhibits arterial smooth muscle cell proliferation. Atherosclerosis 58:109
13. Moolenaar WH, Tertoolen LGJ, de Laat SW (1989) Growth factors immediately raise cytoplasmic free calcium in human fibroblasts. J Biol Chem 259:8066

Sodium/Calcium Exchange
in Cultured Human Mesangial Cells*

P. Menè, F. Pugliese, and G. A. Cinotti

Cattedra di Nefrologia Medica – 2a Clinica Medica, Università di Roma
"La Sapienza", Policlinico Umberto I, Viale del Policlinico, I-00161 Roma, Italy

Introduction

Mesangial cells are mesenchymal cells of the kidney glomerulus, forming the inner core of the glomerular stalk. They produce an extracellular basement membrane-like matrix, which may be a site for ultrafiltrate formation [1,2,3]. Mesangial contraction may contribute to the regulation of glomerular hemodynamics as well as of surface and/or permeability of the filtration barrier [2,4].

Numerous studies in cultured cells of rat and human origin have established that receptors for hormones, cytokines, and vasoactive agents are coupled to distinct intracellular signal transduction systems regulating mesangial contractility [2]. Recently, an elevation of cytosolic free Ca^{2+} ($[Ca^{2+}]_i$) has been identified as an early signal elicited by several vasoconstrictors [5,6,7,8]. Most agents that reduce the filtration surface area in vivo and contract mesangial cells in culture, stimulate a phosphatidylinositol (4,5)-bisphosphate-specific phospholipase C, with resulting accumulation of inositol phosphates [9,10]. Inositol (1,4,5)-trisphosphate ($InsP_3$) is a Ca^{2+}-gating intracellular messenger in a variety of cell types, releasing internal stores of Ca^{2+}. As a result, $[Ca^{2+}]_i$ is rapidly and transiently elevated [9,10]. On the other hand, little is known about the mechanisms by which mesangial cells achieve fine regulation of their resting $[Ca^{2+}]_i$, and how the intracellular release of Ca^{2+} that follows $InsP_3$ formation is handled. It is of interest that other ion fluxes have been described in activated mesangial cells, including Na^+ influx and H^+ extrusion. Alkalinization induced by Na^+/H^+ exchange activity is balanced by Cl^-/HCO_3^- countertransport, resulting in no net change of cytosolic pH [11–13]. On the other hand, intracellular accumulation of Na^+ may result from its transport across

* These studies were supported by grants from the Ministero della Pubblica Istruzione of Italy (quota 60%), and by a research grant from Janssen Farmaceutici, Rome, Italy (to P.M.).

G. Bruschi A. Borghetti (Eds.)
Cellular Aspects of Hypertension
© Springer-Verlag Berlin · Heidelberg 1991

the plasma membrane, as reported in various serum- or agonist-activated mesenchymal cell types [14]. Na^+/Ca^{2+} countertransport has been identified in certain extrarenal smooth muscle cell types, and may be critical to the regulation of vascular reactivity in various pathophysiologic settings [15–21]. Renal transport epithelia also express Na^+/Ca^{2+} exchange activity [22,23]. To ascertain whether there is any relationship between the intracellular and extracellular concentrations of Na^+ and Ca^{2+} in cultured human mesangial cells, we focused our attention on Na^+/Ca^{2+} exchange as a potential mechanism linking resting and stimulated $[Ca^{2+}]_i$ to the ambient extracellular concentration of Na^+, ($[Na^+]_e$), in cultured mesangial cells.

Methods

Cell Culture

Four human mesangial cell lines were established from glomerular explants as described [7,24] and maintained in culture for up to 15 passages. Glomeruli were obtained from kidneys not suitable for transplantation or histologically normal tissue from nephrectomy specimens. Two human cell lines were kindly provided by Dr. H.E Abboud of the VA Medical Center, Cleveland, Ohio. Characterization of cells was carried out for each line during early passages [7]. Culture medium was RPMI 1640 supplemented with 17% heat-inactivated fetal bovine serum (FBS, Flow Laboratories, Irvine, Scotland), $5\,\mu g/ml$ human recombinant insulin (Novo, Copenhagen, Denmark), $10\,\mu g/ml$ ceftriaxone (Hoffmann-La Roche, Basel, Switzerland). The cultures were maintained at 37°C, in an atmosphere of 95% air and 5% CO_2, and subcultured every 4–7 days.

Measurement of $[Ca^{2+}]_i$

$[Ca^{2+}]_i$ was measured fluorometrically in confluent monolayers of human cells loaded with the Ca^{2+}-sensitive intracellular probe fura-2 [7,25]. Briefly, serum-deprived cells grown on plastic coverslips were loaded with $1\,\mu M$ fura-2 acetoxymethylester (Molecular Probes, Eugene, Oregon) [25] for 40 min in serum-free RPMI 1640, followed by a further 20 min incubation in the same medium without fura-2. Bathing solution was a modified Krebs-Henseleit solution buffered with $20\,mM$ N-2-hydroxyethylpiperazine-N'-2-ethanesulfonic acid (HEPES) and supplemented with 0.2% fatty acid-free bovine serum albumin (BSA). Complete changes of solutions were performed within 3 s by sequential vacuum aspiration and replacement with prewarmed solutions of appropriate $[Na^+]$. Na^+ in replacement solutions was isoosmotically substituted with choline, as the Cl salt. Fluorescence was monitored continuously at 340/500 nm excitation/emission wavelengths

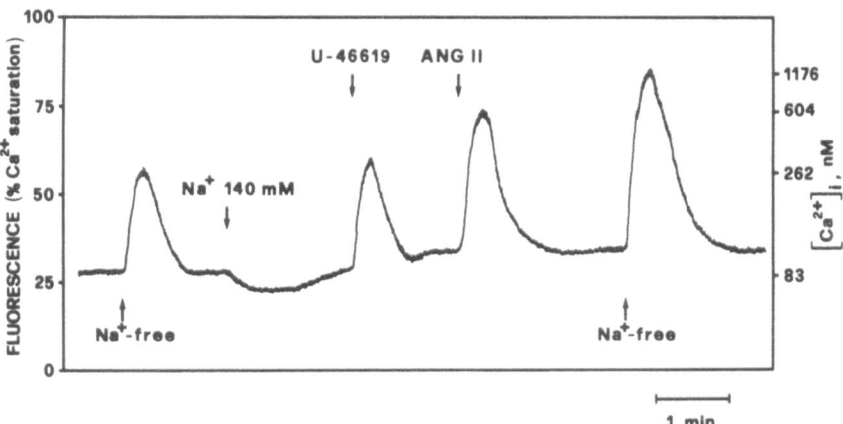

Fig. 1. Effects of removal (*Na$^+$-free*) and readdition (*Na$^+$ 140 mM*) of extracellular Na$^+$ on [Ca^{2+}]$_i$ of fura-2-loaded cultured human mesangial cells. After the first cycle of removal/readdition, the same monolayer is sequentially stimulated with 1 μM U-46619 and angiotensin II (*ANG II*), followed again by Na$^+$ withdrawal. Note enhanced response to Na$^+$ removal after stimulation with vasoconstrictors. Fluorescence tracing representative of *n* = 8 experiments

in a Perkin-Elmer LS5B spectrofluorometer equipped with thermostatically controlled cuvettes and stirring apparatus. Excitation/emission slits were 2.5 and 5 nm, respectively. Calibration of the Ca^{2+}-dependent fluorescence was accomplished by addition of 15 to 40 μM ionomycin, to saturate the dye to maximal fluorescence, followed by 7.5 mM ethyleneglycol-*bis*-(β-aminoethylether)*N, N'*-tetraacetic acid (EGTA) plus 60 mM Tris-HCl, pH 10.5, to release Ca^{2+} from fura-2 and obtain minimal fluorescence. [Ca^{2+}]$_i$ was then calculated according to the formula k_d (F-F_{min}/F_{max}-F) [25], in which F_{min}/F_{max} are the minimal and maximal fluorescence readings during excitation at 340 nm. The k_d of fura-2 for Ca^{2+} is 224 nM [25].

Results

Replacement of the bathing media of quiescent human mesangial cell monolayers with isoosmotic, Na$^+$-free solution resulted in an immediate, transient rise of [Ca^{2+}]$_i$, from resting levels of 68 ± 5 nM (mean ± SE, *n* = 25) to 848 ± 84 nM (*n* = 25). [Ca^{2+}]$_i$ returned to baseline within approximately 45 s (Fig. 1). The cells could then be stimulated with several agonists of phospholipase C, with increased amplitude of responses (see below). When the bathing solution was changed again to 140 mM [Na$^+$]$_e$, [Ca^{2+}]$_i$ promptly decreased by 20 ± 6 nM, followed by slow recovery towards resting levels within 3 min (Fig. 1). Identical responses could be repeatedly elicited in the same monolayer. Peak [Ca^{2+}]$_i$ upon Na$^+$ removal was

Table 1. Effects of graded removal of Na^+ on $[Ca^{2+}]_i$ of cultured human mesangial cells (mean $\pm$ SE from $n = 3-5$ experiments)

$[Na^+]_e$ (mM)	Peak $[Ca^{2+}]_i$ (nM)
140	73 ± 9 (no change)
110	171 ± 36
80	285 ± 48
50	297 ± 26
20	480 ± 100
0	685 ± 74

dependent upon $[Na^+]_e$, as shown in Table 1. Graded reduction of $[Na^+]_e$ from 110 to 0 mM resulted in increasingly larger $[Ca^{2+}]_i$ transients. The $[Ca^{2+}]_i$ response was not an artifact of perturbation of the monolayer by changing the solutions, since no increment of $[Ca^{2+}]_i$ was detected when the same 140 mM $[Na^+]_e$ solution was employed for replacement. The $[Ca^{2+}]_i$ response to Na^+ withdrawal was completely inhibited by pretreatment of the cells with 1 mM amiloride for 5 min prior to Na^+ removal ($n = 5$). Subtraction of amiloride-dependent fluorescence background was required for these experiments.

To identify the source of elevated $[Ca^{2+}]_i$, we employed Na^+-free solutions to which 3 mM EGTA was added to chelate extracellular Ca^{2+}. This maneuver abolished the $[Ca^{2+}]_i$ transient ($n = 6$). Under these conditions, both the thromboxane A_2 mimetic, U-46619, and angiotensin II were still capable of eliciting rapid Ca^{2+} mobilization from intracellular sources [5–8]. Moreover, pretreatment of the cells for 10 min with 10 μM dantrolene, a blocker of the release of Ca^{2+} from intracellular stores, failed to significantly modify the response to Na^+ removal (peak $[Ca^{2+}]_i$ 584 $\pm$ 134 nM, NS vs control Na^+-free solutions, $n = 6$). Taken together, these experiments suggest that extracellular Ca^{2+} entry accounts for the $[Ca^{2+}]_i$ response to extracellular Na^+ removal.

Voltage-gated Ca^{2+} channels activated by membrane depolarization upon abrupt Na^+ withdrawal could theoretically mediate Ca^{2+} influx in mesangial cells. To verify this hypothesis, we preincubated the cells with 10 μM of the dihydropyridine Ca^{2+}-channel blockers verapamil, nifedipine, or nicardipine. No inhibition of the $[Ca^{2+}]_i$ rise upon Na^+ removal was noted under these conditions, implying that voltage-gated Ca^{2+} channels are not involved in this cellular response (Table 2). Moreover, membrane depolarization did not promote substantial Ca^{2+} entry in this preparation, since addition of 75 mM KCl or of the monovalent cation ionophore, gramicidin D (0.2 μM), had no effects on $[Ca^{2+}]_i$ within 3 min of addition (not shown).

The effects of sudden Na^+ removal should be magnified by Na^+ loading of the cells, were they due to facilitated Na^+ efflux in exchange for Ca^{2+}. To

Table 2. Effects of voltage-operated Ca^{2+} channel blockers on the $[Ca^{2+}]_i$ response to Na^+ removal in cultured human mesangial cells (mean $\pm$ SE from 4 experiments)

Condition	Peak $[Ca^{2+}]_i$ (nM)
Control Na^+ removal	440 ± 49
$+10\ \mu M$ verapamil	457 ± 12
$+10\ \mu M$ nifedipine	496 ± 11
$+10\ \mu M$ nicardipine	550 ± 64

All differences NS by unpaired t tests.

verify this point, we employed the Na^+ ionophore monensin, which by itself slowly elevates $[Ca^{2+}]_i$ in these cells, to steady-state levels of 251 59 nM ($n = 6$). This may be taken as evidence of Na^+/Ca^{2+} countertransport in these cells, as internalized Na^+ may be extruded, at least in part, via exchange with Ca^{2+}. To block extrusion of Na^+ by the plasma membrane Na^+, K^+-ATPase, the cells were pretreated for 5 min with 0.1 mM ouabain. Ouabain alone had no effect on $[Ca^{2+}]_i$. The combination of monensin and ouabain markedly increased $[Ca^{2+}]_i$ within 1–2 min. Removal of Na^+ at this time resulted in magnification of amplitude and duration of the $[Ca^{2+}]_i$ transient, from average peak values of 741 ± 77 to 1566 ± 232 nM ($p < 0.05$, unpaired t tests, $n = 6$). This is consistent with enhanced influx of Ca^{2+} in exchange for a larger intracellular Na^+ mass.

Na^+/Ca^{2+} exchange might play a role in modulating the $[Ca^{2+}]_i$ responses to vasoconstrictors. Table 3 shows marked enhancement of the $[Ca^{2+}]_i$ responses to U-46619 or angiotensin II in the absence of extracellular Na^+. This is also graphically illustrated by the sequential studies on the same monolayer of Fig. 1. In these experiments, after initial removal of Na^+ the media was replaced again with a 140 mM $[Na^+]_e$ solution, to which 1 μM U-46619 or angiotensin II was added. A rapid, transient rise of $[Ca^{2+}]_i$ was elicited by these vasoconstrictors. Following return of $[Ca^{2+}]_i$ to baseline, Na^+ was removed again, resulting in markedly larger elevations of $[Ca^{2+}]_i$ than in first part of the experiment.

Table 3. Enhancement of agonist-induced $[Ca^{2+}]_i$ elevation by Na^+ removal in fura-2-loaded confluent monolayers of cultured human mesangial cells (mean $\pm$ SE peak $[Ca^{2+}]_i$, nM)

Basal	U-46619	ANG II	After Na^+ removal	
			U-46619	ANG II
79 ± 9	688 ± 220	986 ± 142	$1152 \pm 123^*$	$1646 \pm 102^*$

$^*p < 0.05$ vs control (1 μM U-46619 or angiotensin II alone, respectively), one-way analysis of variance on $n = 5-10$ experiments. *ANG*, angiotensin.

Discussion

The present investigation demonstrates the presence of a Na^+/Ca^{2+} exchanger in cultured human mesangial cells. This antiporter, similar to other excitable cells or transport epithelia, regulates $[Ca^{2+}]_i$ as a function of Na^+ gradients across the plasma membrane [15–23]. Activity of the antiporter is revealed by rapid withdrawal of $[Na^+]_e$, thus creating an outward gradient for Na^+, whose efflux is balanced by Ca^{2+} influx. The transient nature of the $[Ca^{2+}]_i$ response most likely results from activation of other Ca^{2+} homeostatic mechanisms by the massive elevation of $[Ca^{2+}]_i$ [21]. Additionally, depletion of intracellular Na^+, whose concentration is already kept low by the Na^+, K^+-ATPase, rapidly terminates the Na^+ gradient driving Na^+/Ca^{2+} exchange. On the other hand, increasing $[Na^+]_i$ with ionophores plus ouabain augments the driving force for Na^+ efflux, resulting in internalization of larger amounts of Ca^{2+}. The exchanger is capable of operating in reverse mode, that is, Ca^{2+} out – Na^+ in; this is indicated by the brief reduction of $[Ca^{2+}]_i$ below baseline upon rapid restoring of normal $[Na^+]_e$ from Na^+-free solutions.

Influx of extracellular Na^+ does not occur via voltage-operated divalent cation channels (VOC), since (a) human mesangial cells do not respond to depolarization with detectable increments of $[Ca^{2+}]_i$, (b) VOC blockers do not modify the $[Ca^{2+}]_i$ responses to Na^+ withdrawal, and (c) no depolarization of human mesangial cells occurs in Na^+-free solutions, as fluorometrically assessed in bisoxonol-loaded cells (P. Menè, F. Pugliese, G.A. Cinotti, unpublished observations). On the other hand, it remains to be clarified whether Na^+/Ca^{2+} countertransport contributes to agonist-induced depolarization in these cells. This would be expected if the exchange were electrogenic, as in the case of most excitable cells, in which the apparent stoichiometry is 3 Na^+ for 1 Ca^{2+} [26].

Recently, removal of extracellular Na^+ has been shown to discharge intracellular Ca^{2+} stores via an InsP-dependent process in human skin fibroblasts [27]. This mechanism does not seem to apply to human mesangial cells, in which the release of sequestered Ca^{2+} is ruled out by abrogation of the $[Ca^{2+}]_i$ responses by EGTA, but not by dantrolene. On the contrary, several agonists of phospholipase C produce a typical $[Ca^{2+}]_i$ "spike" even when extracellular Ca^{2+} is chelated by EGTA. Moreover, in the study by Smith et al. [27], depletion of intracellular Ca^{2+} stores with bradykinin blunted the $[Ca^{2+}]_i$ response to Na^+ withdrawal. Opposite behavior was exhibited by our cells, with several stimuli of $[Ca^{2+}]_i$ potentiating the effects of Na^+-free media.

The Na^+/Ca^{2+} exchanger does not seem to play a major role in the basal regulation of $[Ca^{2+}]_i$, since, following the $[Ca^{2+}]_i$ transient, pre-stimulation levels are sharply restored even in Na^+-free solutions by other Ca^{2+} transport mechanisms. On the other hand, the Na^+/Ca^{2+} exchanger is constitutive, since no induction is required for the cells to express countertransport activity, contrary to the findings of previous studies in certain rat smooth muscle preparations [15,16]. The fact that human mesangial cells express such transport activity under resting conditions suggests a possible relevant role of Na^+/Ca^{2+} exchange in the human glomerulus. One possible such function may be the modulation of $[Ca^{2+}]_i$ fluctuations in response to vasoconstrictors. Agonist-induced $[Ca^{2+}]_i$ transients are magnified in the absence of extracellular Na^+. This implies that Na^+ influx in exchange for Ca^{2+} significantly contributes to restoring prestimulation $[Ca^{2+}]_i$, following discharge of internal Ca^{2+} stores by $InsP_3$. However, this is not the sole mechanism counteracting the release of Ca^{2+}, as shown by the lack of prolongation of the $[Ca^{2+}]_i$ transient under Na^+-free conditions. Therefore, reuptake of Ca^{2+} into storage organelles, active extrusion of Ca^{2+}, or other ion exchange mechanisms overcome the lack of external Na^+ in disposing of the immediate cytoplasmic Ca^{2+} excess. On the other hand, the activity of the exchanger is potentiated by phospholipase C agonists, as demonstrated by sequential experiments. The issue of whether a prior rise of $[Ca^{2+}]_i$ or the direct actions of vasoconstrictors through receptor-operated mechanisms are responsible for activation of the exchanger, is still open. It is also possible that vasoconstrictor-stimulated influx of Na^+ allows more efficient exchange with Ca^{2+} upon withdrawal of extracellular Na^+. This situation would be mimicked by the studies with monensin and ouabain. A rise of $[Na^+]_i$ is a known consequence of stimulation of the cells by vasoconstrictors [14]. Potential pathways include activation of Na^+/H^+ exchange, previously documented by us and others in this cell type [13,28,29], Na^+ channel opening [30], and, possibly, the same Na^+/Ca^{2+} exchanger [15–23,26,27].

Other Ca^{2+}-dependent functions, such as secretion or cell proliferation, may be affected by Na^+/Ca^{2+} exchange mechanisms. A putative role of vascular smooth muscle Na^+/Ca^{2+} exchange in the pathogenesis of hypertension is presently being investigated [31,32].

The hypothesis that $[Ca^{2+}]_i$ of resistance vessels is regulated by transmembrane Na^+ fluxes requires evaluation by in vivo and in vitro approaches. The contractile behavior and functional similarities between mesangial cells and smooth muscle suggest a possible extrapolation of our findings to the systemic microcirculation. Our observations indicate that human mesangial cells may represent a useful experimental model for the study of cellular ion transport mechanisms and their regulation.

Acknowledgements. We are indebted to Dr. H.E. Abboud (Veterans Administration Medical Center, Cleveland, Ohio) for providing us with two lines of cultured human mesangial cells. We are also grateful to Drs. F. Cozzi (Institute of Pediatric Medicine, University of Rome, Italy) and A. Ceccamea (Institute of Pathology, University of Rome, Italy) for providing human kidney specimens for cell culture.

References

1. Latta H (1973) Ultrastructure of the glomerulus and juxtaglomerular apparatus, In: Orloff J, Berliner RW, Geiger SR (eds) Handbook of physiology. Renal physiology, sect 8, chap 1. Americal Physiological Society, Washington pp 1–29
2. Menè P, Simonson MS, Dunn MJ (1989) Physiology of the mesangial cell. Physiol Rev 69:1347–1424
3. Latta H, Fliegel S (1985) Mesangial fenestrations, sieving, filtration, and flow. Lab Invest 52:591–598
4. Brenner BM, Dworkin LD, Ichikawa I (1986) Glomerular ultrafiltration. In: Brenner BM, Rector FC (eds) The kidney, vol 1. Saunders, Philadelphia, pp 124–144
5. Bonventre JV, Skorecki K, Kreisberg JI, Cheung JY (1986) Vasopressin increases cytosolic free calcium concentration in glomerular mesangial cells. Am J Physiol 251 (Renal Fluid Electrolyte Physiol 20):F94–F102
6. Hassid A, Pidikiti N, Gamero D (1986) Effects of vasoactive peptides on cytosolic calcium in cultured mesangial cells. Am J Physiol 251 (Renal Fluid Electrolyte Physiol 20):F1018–F1028
7. Menè P, Dubyak GR, Abboud HE, Scarpa A, Dunn MJ (1988) Phospholipase C activation by prostaglandins and thromboxane A_2 in cultured mesangial cells. Am J Physiol 255 (Renal Fluid Electrolyte Physiol 24):F1059–F1069
8. Takeda KH, Meyer-Lehnert H, Kim JK, Schrier RW (1988) Effect of angiotensin II on Ca^{2+} kinetics and contraction in cultured rat glomerular mesangial cells. Am J Physiol 254 (Renal Fluid Electrolyte Physiol 23):F254–F266, 1988
9. Menè P, Simonson MS, Dunn MJ (1989) Phospholipids in signal transduction of mesangial cells. Am J Physiol 256 (Renal Fluid Electrolyte Physiol 25):F375–F386
10. Berridge MJ, Irvine RF (1989) Inositol phosphates and cell signalling. Nature 341:197–205
11. Boyarsky G, Ganz M, Sterzel RB, Boron WF: pH regulation in single glomerular mesangial cells. I. Acid extrusion in absence and presence of HCO_3^- Am J Physiol 255 (Cell Physiol 24):C844–C856
12. Boyarsky G, Ganz M, Sterzel RB, Boron WF (1988) pH regulation in single glomerular mesangial cells. II. Na^+-dependent and -independent Cl^--HCO_3^- exchangers. Am J Physiol 255 (Cell Physiol 24):C857–C869
13. Ganz MB, Boyarsky G, Sterzel RB, Boron WF (1989) Arginine vasopressin enhances pH_i regulation in the presence of HCO_3^- by stimulating three acid-base transport systems. Nature 337:648–651
14. Owen NE (1984) Platelet-derived growth factor stimulates Na^+ influx in vascular smooth muscle cells. Am J Physiol 247 (Cell Physiol 16):C501–C505
15. Vigne P, Breittmayer J-P, Duval D, Frelin C, Ladzunski M (1988) The Na^+/Ca^{2+} antiporter in aortic smooth muscle cells. Characterization and demonstration of an activation by phorbol esters. J Biol Chem 263:8078–8083
16. Smith JB, Zheng T, Smith L (1989) Relationship between cytosolic free Ca^{2+} and Na^+/Ca^{2+} exchange in aortic muscle cells. Am J Physiol 256 (Cell Physiol 25):C147–C154
17. Aaronson PI, Benham CD (1989) Alterations in $[Ca^{2+}]_i$ mediated by sodium-calcium exchange in smooth muscle cells isolated from the guineapig ureter. J Physiol (Lond) 416:1–18
18. van Breemen C, Aaronson P, Loutzenhiser R: Sodium-calcium interactions in mammalian smooth muscle. Pharmacol Rev 30:167–208, 1979
19. Ashida T, Blaustein MP (1987) Regulation of cell calcium and contractility in mammalian arterial smooth muscle: the role of sodium-calcium exchange. J Physiol (Lond) 392:617–635

20. Nabel EG, Berk BC, Brock TA, Smith TW (1988) Na^+/Ca^{2+} exchange in cultured vascular smooth muscle. Circ Res 62:486–493

21. Rasmussen H, Barrett PQ (1984) Calcium messenger system: an integrated view. Physiol Rev 64:938–984

22. Dominguez JH, Macias WL, Rothrock JK, Kwon S (1990) External potassium regulates cell calcium in rat proximal tubules. Kidney Int 37:207 (abstract)

23. Kashgarian M, Ardito T, Giebisch G, Geibel J (1990) Evidence for a Na^+/Ca^{2+} exchanger and a Ca^{2+} channel in the basolateral membrane of the rabbit proximal tubule using confocal microscopy. Kidney Int 37:584 (abstract)

24. Striker GE, Killen PD, Farin FM (1980) Human glomerular cells in vitro: isolation and characterization. Transplant Proc 12:88–99

25. Grynkiewicz G, Poenie M, Tsien RY (1985) A new generation of Ca^{2+} indicators with greatly improved fluorescence properties. J Biol Chem 260:3440–3450

26. Eisner DA, Lederer WJ (1985) Na-Ca exchange: stoichiometry and electrogenicity. Am J Physiol 248 (17):C189–C202

27. Smith JB, Dwyer SD, Smith L (1989) Decreasing extracellular Na^+ concentration triggers inositol polyphosphate production and Ca^{2+} mobilization. J Biol Chem 264:831–837

28. Simonson MS, Menè P, Dubyak GR, Dunn MJ (1988) Identification and transmembrane signaling of leukotriene D_4 receptors in human mesangial cells. Am J Physiol 255 (Cell Physiol 24):C771–C780

29. Ganz MB, Boyarski G, Boron WF, Sterzel RB (1988) Effects of angiotensin II and vasopressin on intracellular pH of glomerular mesangial cells. Am J Physiol 254 (Renal Fluid Electrolyte Physiol 23):F787–F794

30. Hille B (1984) Na^+ and K^+ channels of axons, in ionic channels of excitable membranes. Sinauer, Sunderland, pp 58–75

31. Blaustein MP (1977) Sodium ions, calcium ions, blood pressure regulation, and hypertension: a reassessment and a hypothesis. Am J Physiol 232 (Cell Physiology): C165–C173

32. Abbott A (1988) Interrelationship between Na^+ and Ca^{2+} metabolism in hypertension. Trends Pharmacol Sci 9:111–113

Part III
Membrane Abnormalities: Blood Cell Models

Relation of Cell Permeability to Salt Sensitivity in Hypertension*

S. M. Friedman

Department of Anatomy, Faculty of Medicine University of British Columbia,
2177 Wesbrook Mall, Vancouver, BC, Canada, V6T 1W5

More than 30 years ago we proposed that mineralocorticoid hypertension begins with increased permeability of the vascular smooth muscle cell, followed by the progressive accumulation of sodium, and the restructuring of the cell to cope with the load [1]. Although supportive evidence from many laboratories continued to accumulate over the years, direct evidence for increased sodium permeability and transport as primary events in the hypertensive response to deoxycorticosterone acetate (DOCA) was obtained only 3 years ago [2]. This was followed in the next year by the demonstration that over the first 16 days of treatment with this mineralocorticoid, blood pressure was directly related to cell [Na] and inversely related to the transmembrane Na gradient, as had earlier been predicted [3].

Thus far, however, only an association between blood pressure and sodium distribution had been established. Although there was considerable ancillary information in the literature to suggest that the transmembrane distribution of Na is probably a regulator of blood pressure in vivo, there was, as yet, no evidence for such a causal connection. To prove such a connection required the induction of regular changes in the transmembrane partition of Na in vivo over a range compatible with life, and the demonstration that such changes were precisely and predictably related to blood pressure.

In general, two basic methods were used to explore this problem in the rat [4,5]. First, blood [Na] was raised or lowered over the limited range of ± 15 mM by means of dialysis against an intraperitoneal load of 15% of body weight of modified physiological salt solution (PSS). This range was fully compatible with complete recovery by the following day. Blood and dialysate were monitored over 5 h to cover the initial period of adjustment as well as the sustained period of relative equilibrium. The second procedure involved the same dialysis, but after 30 or 120 min blood pressure was measured, a blood sample taken for [Na], [K] and haematocrit determination, and a few

* This work was carried out with the aid of a grant from the Medical Research Council of Canada.

G. Bruschi A. Borghetti (Eds.)
Cellular Aspects of Hypertension
© Springer-Verlag Berlin · Heidelberg 1991

Table 1. Effect of an intraperitoneal load of high-Na (300 mM), PSS compared with an equivalent osmotic load of sucrose, on blood [Na] and on blood pressure. Values expressed as difference (Δ) from intraperitoneal dialysis with normal PSS

	Δ Blood [Na] (mM)		Δ Systolic BP (mmHg)		Δ Diastolic BP (mmHg)	
	Sucrose	High Na	Sucrose	High Na	Sucrose	High Na
30 min	3.8 ± 0.5	14.7 ± 0.8	11 ± 7	21 ± 5	10 ± 6	22 ± 3
1 h	1.9 ± 0.5	14.9 ± 0.8	-3 ± 5	22 ± 4	3 ± 6	22 ± 3
2 h	-0.2 ± 0.8	12.8 ± 0.6	-8 ± 5	20 ± 6	4 ± 5	23 ± 5
3 h	0.6 ± 0.7	14.2 ± 0.7	-23 ± 7	14 ± 3	-10 ± 8	19 ± 4
4 h	0.3 ± 0.9	13.3 ± 0.9	-22 ± 6	12 ± 5	-12 ± 5	12 ± 5
5 h	-2.7 ± 0.6	13.6 ± 0.9	-15 ± 5	11 ± 3	-7 ± 6	10 ± 2
No. of rats	8	8	8	8	8	8

minutes later the tail artery was rapidly excised and plunged into ice-cold ($<3°C$) lithium-substituted PSS (Li-PSS) for 45 min to exchange extracellular Na for Li. The artery was then taken for the detailed analysis of cell Na, K and water (V_i).

With high-Na dialysates blood [Na] rose rapidly to reach a maximum at about 30 min and remained at about the same level for the 5 h of the study (Table 1). With low-Na dialysates equilibrium values were attained more slowly. Parallel changes in the osmotic load (sucrose) alone, either up or down, produced little effect on $[Na]_o$.

Elevations of $[Na]_o$ were matched by rises in both systolic and diastolic blood pressure. This was evidently a specific effect of Na, since the sustained level of blood pressure was significantly reduced by dialysates of matched hyperosmolarity. This contrary effect of osmolarity was particularly important with subnormal Na loads, which lowered blood pressure when the

Table 2. Transmembrane partition of Na in fresh tail arteries rapidly excised after 2 h intraperitoneal dialysis with high-Na PSS. Arteries were analyzed in groups of 10 after 45 min immersion in Li-PSS ($<3°C$)

Groups by [Na] dialysate (mM)	$[Na]_o$ (mM)	Na_i (mmol/kg d.w.)	$[Na]_i$ (mM)	Groups by Diastolic BP (mmHg)	$[Na]_i$ (mM)
141	141.8 ± 0.4	33.2 ± 1.2	29.2 ± 1.9	63.1 ± 1.3	32.8 ± 2.0
200	147.8 ± 0.5	34.9 ± 1.6	29.9 ± 2.6	71.6 ± 0.6	35.6 ± 2.2
225	150.9 ± 0.8	35.8 ± 1.4	36.1 ± 2.7	76.4 ± 0.4	38.1 ± 2.9
250	154.0 ± 0.9	35.6 ± 0.9	34.3 ± 3.0	80.4 ± 0.5	34.2 ± 4.2
300	160.0 ± 0.7	38.8 ± 0.9	34.5 ± 1.9	87.7 ± 0.9	36.5 ± 2.4
325	163.9 ± 0.6	40.3 ± 1.3	43.7 ± 5.3	97.0 ± 1.1	41.8 ± 3.2
No. of rats	60	60	60	60	60

Table 3. Effect of 30 min intraperitoneal dialysis with high Na PSS (141–325 mM as in Table 2) on transmembrane partition of Na and water in fresh, rapidly excised tail arteries and on BP in rats treated with DOCA-saline for 3 days with or without prior uninephrectomy

Dependent variable	Regression coefficient/mM $[Na]_o$		
	Control	DOCA-saline	DOCA-saline, uninephrect
Δ Diastolic BP (mmHg)	0.81 ± 0.23	$1.43 \pm 0.28^*$	$1.75 \pm 0.31^*$
Δ Na_i (mmol/kg d.w.)	0.14 ± 0.08	$0.21 \pm 0.07^*$	$0.19 \pm 0.06^*$
Δ V_i (ml/kg d.w.)	3 ± 4	$5 \pm 3^*$	$6 \pm 3^*$
Δ $[Na]_i$ $(mM)^\dagger$	0.18	0.27	0.26
Δ E_{Na} $(mV)^\dagger$	-0.004	-0.070	-0.119

Linear regression coefficients calculated using as independent variable the unselected data ($n = 60$) for $[Na]_o$, measured as blood [Na], and the corresponding values for blood pressure, Na_i, and V_i as dependent variables.
† Regression coefficients derived from regressions of Na_1 and V_1.
* $p < 0.02$.

normal osmolarity of dialysate was retained by sucrose replacement, but raised blood pressure when the dialysate was hyposmotic. Thus, in general, the effects of changes in $[Na]_o$ are countered by changes in osmolarity acting in the opposite sense. These effects on blood pressure could not be explained by changes in either blood or extracellular fluid volumes.

Blood [Na], defining the extracellular concentration of sodium, $[Na]_o$, varied in direct proportion to the [Na] in the intraperitoneal dialysate (Table 2). Analysis of tissues excised at 2 h showed that intracellular Na, $[Na]_i$, was directly proportional to $[Na]_o$ at about 0.3 mmol/kg dry weight per millimolar. The entry of Na into cells was incompletely matched by water, so that in terms of concentration, $[Na]_i$, the regression coefficient was close to 0.4 mM/mM. When these data ($n = 60$) were sorted for blood pressure, a direct relation between $[Na]_i$ and blood pressure was observed. The transmembrane Na gradient expressed in terms of the Na equilibrium potential, E_{Na}, declined at about 0.06 mV/mmHg over the same range. The same relation of blood pressure to $[Na]_i$ and to E_{Na} was observed at 30 min.

These data established that a causal relation exists between the transmembrane partition of Na and water on the one hand and blood pressure in vivo on the other. They did not, of course, define the mechanisms underlying this relation. It remained to demonstrate that the increased permeability of the vascular smooth muscle cell induced by a mineralocorticoid such as DOCA was sufficient explanation for salt sensitivity and the progressive rise in blood pressure. The observations of three experiments provide this demonstration. The findings are summarised here and will be reported in full elsewhere.

For the first experiment, six groups of ten normal rats were loaded as before with PSS varying in [Na] from normal to 300 mM. Tissues were excised 30 min later according to the standard procedure described above. The second experiment was identical except that the rats were in the 3rd day of DOCA-saline treatment. The third experiment was the same as the second, but the rats had been uninephrectomized 1 week before the start of treatment. The data obtained in each experiment were sorted according to the values for [Na]$_o$ measured a few minutes before artery excision. The raw data for each variable were plotted separately for each experiment as ordinate values against [Na]$_o$ as abscissa, and intercept and coefficient for the linear regressions determined. The significance of observed differences between treatments was then calculated for each variable by comparing these parameters [6]. The essential data are presented in Table 3.

The rise in blood pressure, both systolic and diastolic, induced by a rise in [Na]$_o$ was significantly greater in both groups exposed to DOCA-saline than in untreated controls. This increased sensitivity to Na tended to be accentuated by prior uninephrectomy, but the difference was marginal. The coefficient for the regression of cell Na was similarly significantly increased in the treated animals. This too was unaffected by prior uninephrectomy. The increased entry of Na in these animals was associated with a small increase in water (V$_i$).

The changes in [Na]$_i$ associated with the rise in [Na]$_o$ over the range of 140–160 mM were calculated using the intercepts and coefficients obtained from the raw data. At 30 min [Na]$_i$ increased at the rate of 0.18 mM/mM [Na]$_o$ in untreated rats, compared with 0.26 in treated animals. This translates into a steep fall in the Na gradient expressed as E_{Na} in the intact DOCA-saline treated rats, and a still steeper fall in those uninephrectomized before treatment.

As described above, a rise or fall in [Na]$_o$ is matched by parallel shifts of Na into or out of cells. These are not fully matched by corresponding movements of water, so that the change in cell Na concentration is always proportionately greater than that of the extracellular environment. Thus, as [Na]$_i$ rises the transmembrane gradient, or equilibrium potential for Na, E_{Na}, falls. Blood pressure is an in vivo expression of this relationship. That is to say, blood pressure is a direct function of [Na]$_i$ and, by the same token, an inverse function of E_{Na}. When the permeability of the cell is increased these changes are exaggerated, and [Na]$_i$ and blood pressure rise more steeply than normal in relation to a step increase in [Na]$_o$. Sensitivity to a salt load is thus a direct expression of increased permeability of the vascular smooth muscle cell to Na.

As pointed out above, a Na load is necessarily hyperosmotic. The effective osmolarity must, however, decline as cells become more permeable to Na. This was readily assessed by recovering the load at the end of the 5-h experiment. The results are shown in Fig. 1. The hyperosmotic effect of a sucrose load equivalent to 300 mM Na was manifest in the withdrawal

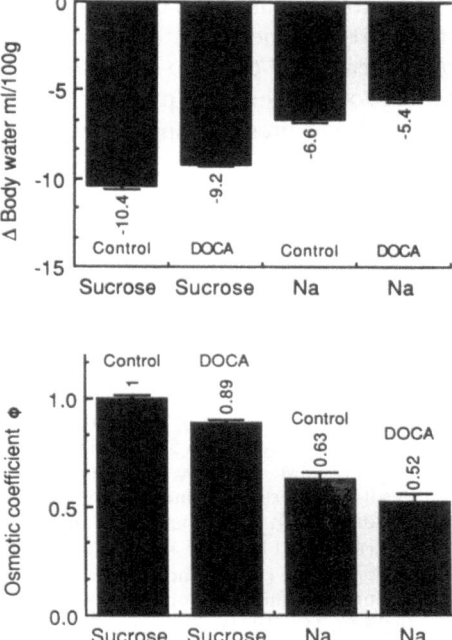

Fig. 1. *Top*: Change in net body weight/100 g original weight, after 5 h intraperitoneal dialysis with hyperosmotic PSS containing either additional Na (300 mM), or osmotically equivalent sucrose. *Bottom*: Osmotic coefficient, ϕ, calculated with sucrose taken as 1. $n =$ 40 for each column

of about 10 ml water/100 g body weight, as previously reported. The corresponding Na load was significantly less hyperosmotic, as expected, since a sizeable proportion of the load enters cells. The osmotic efficiency of both sucrose and Na was diminished significantly, both to about the same degree, in DOCA-treated animals. The conclusion that the increase in permeability induced by mineralocorticoids affects the matrix of the membrane, and is not limited to the specific Na channels, is inescapable. This suggests to us that the defect involves the phospholipids rather than proteins.

An insufficient renal function marginally aggravated the pressor response to [Na]$_o$. The most probable explanation lies in the retardation in the excretion of the Na load. Thus, the rise in blood [Na] in response to an intraperitoneal load was greater, and sustained at a higher level over the 5-h period of observation, in uninephrectomized than in intact rats.

As we have often emphasized, an increase in permeability need not in itself raise blood pressure, so long as homeostatic mechanisms suffice to counter the tendency for [Na] to increase. Thus, an increase of [Na]$_i$ stimulates the Na, K-ATPase to transport extra Na out of the cell. An increase in [Na]$_i$ also stimulates protein synthesis, protein-polysaccharides for export to the paracellular matrix, and proteins for contraction and for the transport enzyme itself. Enhanced renal excretion to reduce the Na load presented to the cell also comes into play. $[Na]_i$, however, rises inexorably as these processes become progressively inadequate to the task, inducing both a rise in blood pressure and restructuring of the cell to cope with this added burden.

References

1. Friedman SM, Friedman CL (1963) Effects of ions on vascular smooth muscle. In: Hamilton WF, Dow P (eds) Circulation. Am Physiol Soc, Washington, pp 1135–1166 (Handbook of physiology, vol II, sect 2)
2. Friedman SM, Tanaka M (1987) Increased sodium permeability and transport as primary events in the hypertensive response to deoxycorticosterone acetate (DOCA) in the rat. J Hypertens 5:341–345
3. Friedman SM, McIndoe RA, Tanaka M (1988) The relation of cellular sodium to the onset of hypertension induced by DOCA-saline in the rat. J Hypertens 6:63–69
4. Friedman SM, McIndoe RA, Tanaka M (1990) The relation of blood sodium concentration to blood pressure in the rat. J Hypertens 8:61–66
5. Friedman SM (1990) The relation of cell volume, cell sodium, and the transmembrane sodium gradient to blood pressure. J Hypertens 8:67–73
6. Jackson TE (1984) Comparison of a class of regression equations. Am J Physiol 246:R271–R276

RFLP Study of the SHR Genome in Relation to Cell Membrane Abnormalities

Y.V. Kotelevtsev and Y.V. Postnov

Central Research Laboratory of the Ministry of Public Health of the USSR, Timoschenko Ave. 23, Moscow 121359, USSR

Investigation of the mode of inheritance of systolic blood pressure as a quantitative trait in families and its distribution in the population has led to two main conclusions: (1) blood pressure is under genetic control and 30%–40% of its variability is determined by the variance in the qenome; (2) there is no single major gene responsible for the variation of blood pressure in a population [1,2]. The variety of clinical forms of primary hypertension suggests a heterogeneity of the disease [3]. This heterogeneity complicates the identification of hereditary factors determining the variability of blood pressure in humans.

Experimental animal strains with the genetic forms of hypertension could be considered as a models of the definite types of human hypertension [4,5], offering a way of investigating homogeneous patterns by complete genetic analysis.

The traditional approach to assessing of the cause of spontaneous hypertension consists of evaluation of physiological, cellular, and biochemical differences between hypertensive and normotensive strains. Most of these differences are quantitative traits themselves, and a simple correlation between them and blood pressure is insufficient to prove causal relations according to paradigm of Rapp [6]. Nevertheless, traits correlating with blood pressure and logically linked with the disease are considered as "intermediate phenotypes" of hypertension [7]. Most investigations try to reveal a chain of intermediate phenotypes from the upper physiological level to protein polymorphism and underlying genome polymorphism [1]. In fact, this is an almost impossible task, due to the great number of variables which influence these phenotypes at the physiological, cellular, and biochemical levels.

In studying monogenic diseases, the reverse approach, based on primary identification of polymorphic genome loci and definition of those cosegregating with the disease phenotype, has proved to be more successful [8]. In the case of spontaneous hypertension the problem should be formulated as follows: polymorphic loci determining elevated blood pressure must be

G. Bruschi A. Borghetti (Eds.)
Cellular Aspects of Hypertension
© Springer-Verlag Berlin · Heidelberg 1991

identified. Afterwards the individual genes responsible for the development of hypertension should be distinguished.

The genome of the rat is approximately equal to the human genome and consists of 3×10^9 nucleotide base pairs or 3000 centimorgan (cM). Genes 20 cM apart have an 80% probability of being inherited together. Botstein proposed dividing the genome into 150 approximately equal linkage groups of 20 cM by means of restriction fragment length polymorphism (RFLP) markers and use them as a genomic markers in linkage studies [8].

Evaluation of blood pressure variability in the progeny strains, F_1 and F_2 generations, permits us to calculate the number of genes determining variability of blood pressure in the progeny strain. For spontaneously hypertensive rats (SHR) this number is about 4 or 5 [2]. More sophisticated analysis carried out on hybrids of SHR with different normotensive strains has indicated epistatic interactions of genes at individual loci [9]. This means that simultaneous inheritance of several genes, each of which has no significant influence on blood pressure, is necessary for complete expression of the trait.

The polygenic nature of the trait and the possibility of epistasis complicates the use of blood pressure itself as a variable in the linkage studies. Genetic analysis of intermediate phenotypes seems to be preferable, at least in the first stage of the study.

SHR are characterized by a number of disturbances in the properties of the plasma membrane relating to cellular ion homeostasis. These plasma membrane abnormalities could probably be linked with hypertension, mainly through an increase in cytoplasmic free Ca^{2+} and the subsequent changes in vascular contractility or enhanced neurotransmitter release [10]. Increased Na^+, K^+ cotransport, Ca^{2+} accumulation, Na^+/H^+ exchange in the erythrocytes and enhanced protein kinase C activity in the brain tissue have been chosen as intermediate phenotypes.

Na^+, K^+ cotransport was measured as ouabain-independent, furosemide-sensitive influx of ^{86}Rb in erythrocytes, which was 1.5-fold higher in SHR than in Wistan-Kyoto rats (WKY). Analysis of 42 (SHR × WKY)F_2 hybrids revealed a positive correlation of the trait with blood pressure. The orrelation coefficient ($r = 0.5$, $p = 0.01$) was estimated between cotransport value and mean blood pressure measured directly under ether anesthesia [11]. It was also shown that the difference in Na^+, K^+ cotransport in the prehypertensive stage excludes augmentation of this parameter as a result of a rise in blood pressure.

The second trait examined was accumulation of Ca^{2+} in erythrocytes in the presence of 5 mM orthovanadate which is twice as high in SHR as in WKY [12]. The positive correlation of this parameter was established with the difference between BP of conscious and ether anesthetised animals.

To examine the involvement of cellular oncogenes in the pathogenetic mechanism of plasma membrane abnormalities in primary hypertension [13], we have studies the polymorphism of cellular oncogenes. Analysis

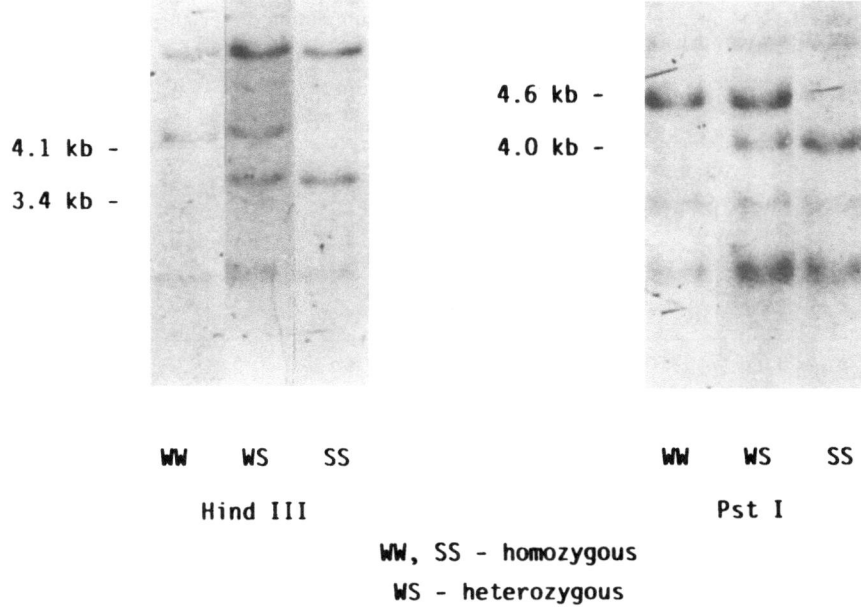

WW   WS   SS   WW   WS   SS

Hind III   Pst I

WW, SS - homozygous

WS - heterozygous

Fig. 1. Southern blot hybridization of DNA of (SHR × WKY)F$_2$ hybrids with v-*src*

of SHR and WKY genomic DNA digested by *Eco*RI, *Sal*I, *Bam*HI, *Pst*I, *Hind*III and *Pvu*II, and using plasmids or incisions containing virus oncogenes including v-*ras*, v-*myc*, v-*sis*, v-*fos*, v-*src*, v-*abl* as probes, revealed polymorphism in the c-*src* locus. Alleles were determined by the position of hybridization bands of Southern blotting after *Hind*III restriction: 3.4 kb for SHR (allele S^{src}) and 4.1 kb for WKY (allele W^{src}). RFLP also was shown with *Pst*I restrictase with the bands 4 kb–S^{src} and 4 kb–W^{src} [14]. Polymorphism of the minor band (4.0 kb) in the c-*fos* locus after *Bam*HI restriction was present only in some of the SHR rats, so it was excluded from further consideration.

In the group of 26 male (SHR × WKY)F$_2$ hybrids above mentioned intermediate phenotypes were measured. Animals were divided into three groups according to their c-*src* genotypes as determined by Southern blotting: homozygous (SS; SHR genotype) and WW (WKY genotype), and heterozygous (WS). The hybridization patterns are shown in Fig. 1. Genotypes determined in the *Hind*III and *Pst*I restrictions coincided in all 26 cases. It appears that mean values for Na$^+$, K$^+$ cotransport in the SS group were equal to those in SHR, those in the WW group to those in WKY, and the WS group showed an intermediate value significantly different from both SS and WW groups (Table 1). This is characteristic for codominant inheritance of the trait and indicates that c-*src* locus cosegregates with the genome locus determining Na$^+$, K$^+$ cotransport. There were

Table 1. Na$^+$, K$^+$ cotransport and calcium accumulation in erythrocytes of (SHR × WKY)F$_2$ hybrids with different *c-src* genotypes

Group	Genotype	Number of animals	Na, K-cotransport (µmol/l cells per h)	Ca accumulation (µmol/l cells)
1	WKY	5	298 ± 36	21.08 ± 1.93
(SHR × WKY)F$_2$ hybrids				
2	WW	8	251 ± 24*	19.08 ± 0.76$^+$
3	WS	14	379 ± 22*	22.22 ± 1.56$^+$
4	SS	4	448 ± 96*	24.14 ± 1.80$^+$
5	SHR	5	431 ± 50	40.71 ± 3.14

*$p_{2,3} < 0.03$; *$p_{2,4} < 0.002$; $^+ p_{2,3} < 0.02$; $^+ p_{2,4} < 0.03$.

no animals having the SS genotype and cotransport values of the WKY range or WW genotypes and cotransport values of the SHR range. Thus, tight linkage between RFLP in *c-src* and cotransport determining locus was demonstrated.

Ca^{2+} accumulation was inherited in a different manner. In the erythrocytes of WW hybrids it was equal to that of WKY, lower than in the WS and SS groups, which in turn had significantly lower levels than SHR (Table 1). This trait is probably under polygenic control with epistatic interaction, but the *c–src* locus has a definite influence upon it.

No correlations between Na$^+$/H$^+$ exchange or protein kinase C activity and the *c-src* genotype were found.

From these data, we conclude that the locus marked by the RFLP in *c-src* contains gene(s) which determine augmented Na$^+$, K$^+$ cotransport and Ca^{2+} accumulation in the erythrocytes of SHR.

This study demonstrates the validity of the RFLP approach in the analysis of quantitative traits forming intermediate hypertensive phenotypes.

Recently Paterson et al. have introduced a very promising approach called quantitative trait loci (QTL) mapping [15]. This approach makes it possible to analyze the probability that the definite loci characterized by flanking RFLP do or do not contribute to the value of a quantitative trait in the back cross of F$_2$ generation. It was successfully applied to the analysis of quantitative traits in tomatoes with a genome of about 1500 cM (12 chromosomes). Using 70 RFLP markers and 237 hybrids, several loci controlling such polygenic quantitative traits as fruit mass (6 QTL), concentration of soluble solids (4 QTL), and fruit pH (5 QTL) were localized on the RFLP map.

Further theoretical analysis of this approach carried out by Lander and Botstein allows evaluation of the number of hybrids and RFLP markers needed and offers optimal cross schemes and methods of selective genome determinations [16]. It seems that an RFLP map of the rat containing 150 evenly distributed markers and analysis of about 500 hybrids would be sufficient for identification of loci determining elevated blood pressure in SHR. The task could be simplified by the use of recombinant inbred strains introduced by Pravenec et al. [17].

The main problem to be solved now is the construction of the RFLP map of the SHR. Methods developed during construction of the RFLP map of humans [18] could be used to facilitate completion of the rat map.

References

1. Camussi A, Bianchi G (1988) Genetics of essential hypertension from the unimodal-bimodal controversy to molecular tochnology. Hypertension 12:620–628
2. Rapp JP (1983) Genetics of experimental and human hypertension. In: Genest J, Kuchel O, Hamet P, Cantin M (eds) Hypertension: physiology and treatment. McGraw-Hill, New York, pp 582–598
3. Laragh JH (1983) Personal views on the mechanisms of hypertension. In: Genest J, Kuchel O, Hamet P, Cantin M (eds) Hypertension: physiology and treatment. McGraw-Hill, New York, pp 615–631
4. Yamori Y (1983) Physiology of the various strains of spontaneously hypertensive rats. In: Genest J, Kuchel O, Hamet P, Cantin M (eds) Hypertension: physiology and treatment. McGraw-Hill, New York, pp 556–581
5. Bianchi G, Ferrari P (1983) Animal models of arterial hypertension. In: Genest J, Kuchel O, Hamet P, Cantin M (eds) Hypertension: physiology and treatment. McGraw-Hill, New York, pp 534–555
6. Rapp JP (1983) A paradigm for identification of primary genetic causes of hypertension in rats. Hypertension 5 [Suppl I]:I198–I203
7. Sing CF, Boerwinkle, E, Turner ST (1988) Genetics of primary hypertension. Clin Exp Hypertens [A] 8:623–651
8. Botstein D, White R, Skolnick M, Dawis RW (1980) Construction of genetic linkage map in man using restriction fragment length polymorphism. Am J Hum Genet 32:314–331
9. Schlager G, Chang-Shin C (1989) The role of dominance and epistasis in the genetic control of blood pressure in rodent models of hypertension. 6th international symposium on SHR and related studies. Iowa City, p 32
10. Postnov YV, Orlov SN (1984) Cell membrane alterations as a source of primary hypertension. J Hypertens 2:1–6
11. Kotelevtsev YV, Orlov SN, Pokudin NI, Agnayev VM, Postnov YV (1987) Genetic analysis of Na^+, K^+-cotransport. Ca^{2+} content in erythrocytes and blood pressure in SHR × WKY F_2 hybrids (in Russian). Biull Eksp Biol Med 103:456–458
12. Orlov SN, Pokudin NI, Postnov YV (1988) Calcium transport in erythrocytes of rats with spontaneous hypertension. J Hypertens 6:829–837
13. Postnov YV (1988) Cell membrane alteration in primary hypertension an approach to its explanation. Acta Physiol Scand 133 [Suppl 571]:175–180
14. Kotelevtsev YV, Brashishkite DA, Spitkovsky DD, Kiseljov FL, Postnov YV (1988) Interstrain restriction fragment length polymorphism of c-fos and c-src oncogene loci in spontaneously hypertensive and normotensive rats. J Hypertens 6:779–781
15. Paterson AH, Lander ES, Hewitt JD, Peterson S, Linkoln S, Tanksley SD (1988) Resolution of quantitative traits into Mendelian factors by using a complete RFLP linkage map. Nature 335:721–726
16. Lander ES, Botstein D (1989) Mapping mendelian factors underlying quantitative traits using RFLP linkage maps. Genetics 121(1):185–199

17. Pravenec M, Klir P, Kren V, Zicha J, Kunes J (1989) An analysis of spontaneous hypertension in SHR by means of new recombinant inbred strains. J Hypertens 7(3):217–221
18. Donis-Keller H, Lander ES et al. (1987) A genetic linkage map of the human genome. Cell 51(2):313–337

Regulation of the Na^+/H^+ Exchanger in Essential Hypertension: Functional and Genetic Abnormalities*

M. Canessa

Endocrinology-Hypertension Division, Laboratory of Cellular Transport,
Brigham and Women's Hospital, Harvard Medical School, 221 Longwood Avenue,
Boston, MA 02115, USA

Introduction

Ten years ago, we initiated studies of the genetic variations of Na^+ transport systems as an approach to the genetics of hypertension. One of these biochemical phenotypes, Na^+/Li^+ exchange (Na/Li EXC), is elevated in a sizable proportion of the essentially hypertensive population. Furthermore, the abnormalities observed have been shown by family studies to be largely genetically determined [1,2,3]. In pursuing the question of what significance these abnormalities might have in the pathophysiology of hypertension, we found that Na^+/Li^+ countertransport has many features in common with the Na^+/H^+ exchanges (Na/H EXC). The Na^+ exchanges system is expressed not only in red blood cells (rbc), but in three target tissues of the hypertensive process: vascular smooth muscle (VSM) [4,5], kidney epithelial [6] and adrenal glomerulosa cells [7]. In these cells Na/H EXC regulates cell pH and volume, and it is modulated by vasocontrictor and growth factor agonist which mobilize Ca^{2+} [4,8,10].

The present paper summarizes our recent findings on the study of rbc Na/H EXC in hypertensive and diabetic patients.

Ion Transport and the Polygenic Model of Hypertension

Even though there is general agreement that heredity plays an important role in blood pressure variability, few biochemical phenotypes that could correlate genetic factors and hypertension are available. During the past decade, we have examined whether or not genetic factors contribute to the abnormalities in Na^+ transport systems in essential hypertension and

* Supported by grants from the National Institute of Health, NHLBI 35664 and by a specialized Center of Research Award in Hypertension (1P50HP-36568).

G. Bruschi A. Borghetti (Eds.)
Cellular Aspects of Hypertension
© Springer-Verlag Berlin · Heidelberg 1991

have documented that: (a) the Na^+ exchange and Na^+-K^+-Cl^- cotransport systems display genetic differences in their expression in humans and in particular in hypertensive patients [1,2,3]; (b) these Na^+ transport systems are expressed not only in rbc, but more importantly, in target tissues of the hypertensive process such as VSM [4,5], kidney epithelial [6] and adrenal glomerulosa cells [7]; and (c) both ion transporters are modulated with tissue specificity by vasoconstrictor agonists and growth factors which mobilize cellular Ca^{2+} [4,8,10].

It was also foreseen when initiating these studies that at least some of these transport phenotypes might prove to be genetically transmitted traits, their expression predating the onset of clinical hypertension. Such traits would be useful markers for determining the primary genetic abnormalities which result in the observed phenotype and, ultimately, hypertension. The goals of these studies have been largely met. The observation that hypertensive subjects and their first-degree relatives have elevated Na/Li EXC suggested that genetic factors are involved [1,2,3,11,12]. Furthermore, it was shown that rbc Na/Li EXC is family aggregated, possesses high heritability, exhibits racial differences [13–15], and is elevated in "nonmodulating hypertensives" [20]. These findings indicate that shared genes, rather than shared environment, are principal determinants of interindividual differences. A segregation analysis of Mormon families in Utah concluded that a model of polygenic inheritance and/or a recessive major gene can account for 80% of the variability of Na/Li EXC in the human population [16,17]. Similar conclusions were reached in studies performed in Rochester and Seattle [18,19].

Relationship of the Na/Li Exchange Abnormalities to Hypertension

The Na^+/Li^+ countertransport heritability provided a tool to define the role of one genetic locus for Na^+ transport in hypertension. However, it does not provide insight into how this transport abnormality could be involved in the pathophysiology of hypertension. To address this question, our strategy was to determine the cellular function of the Na/Li EXC, investigating whether it is a mode of operation of the Na/H EXC. We and others [2,6] have noted the striking similarities of the Na/Li EXC and the ubiquitous Na/H EXC system. To this end, we documented the presence of Na/H EXC in human rbc and studied its kinetic properties [21–23], providing an analytical tool to examine its functional properties in hypertensive patients [24]. As expected, we found that Na/H EXC can transport Li^+ with higher affinity than Na^+ and both can operate in different hetero/homo exchange modes. The interaction of Li^+ with the Na/H EXC and H^+ with Na/Li EXC, [25–27] indicates that both transport processes can be driven by the same system with some differences (Fig. 1). Rbc Na/H EXC is partially ATP-dependent, amiloride-sensitive, and operates at $pH_i < 7.0$, while Na/Li and Na/Na EXC

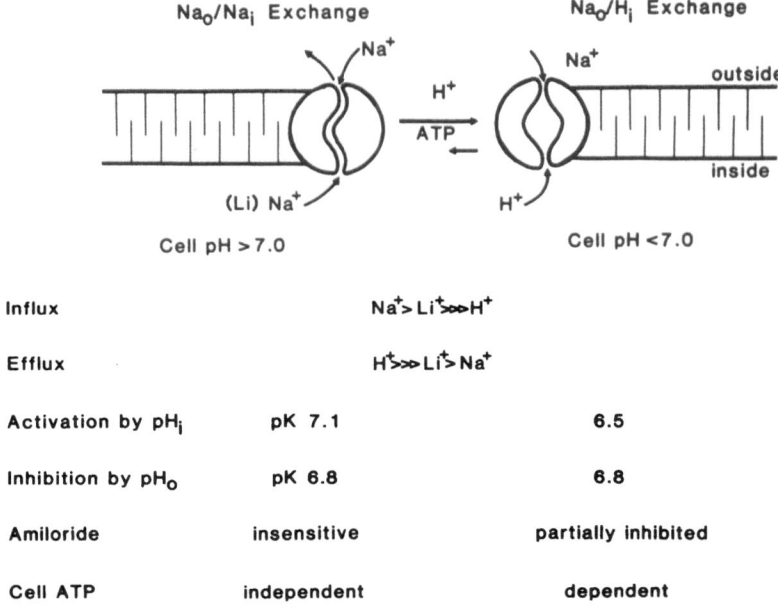

Fig. 1. Conformational model to account for the properties of Na/Li and Na/H exchange

are ATP-independent, amiloride-insensitive and can operate at pH$_i$ < 7.0 [21–23]. These findings led us to propose a two-conformation model for the Na/H EXC with the differences in amiloride sensitivity and ATP dependence reflecting distinctively phosphorylated forms [22–27] (Fig. 1).

To clarify the functional role of NaH EXC in hypertension, we have studied the activity of this system in culture vascular smooth muscle cells of the spontaneously hypertensive (SHR) and Milan hypertensive (MHS) rat strains [28,29] and found it altered in the SHR [28]. We have also established that the kinetic mechanism of Na/H EXC activation by angiotensin II (Ang II) is modulated at the Na$^+$ site [5]. The role of Na/H EXC in modulating Ang II-stimulated aldosterone secretion is also being examined [7]. Considering the presence of this transport system in the many tissues and the relationship of Na/H EXC to Na/Li EXC in hypertensives, we have focused our investigations on the biochemical and kinetic properties of Na/H EXC in blood cells of normal and hypertensive subjects.

Kinetic Abnormalities of the Red Cell Na/H EXC in Hypertensive Patients

In recent years, we have investigated the kinetic properties and of the Na/H EXC phenotype in blood cells of hypertensive patients. The central hypothesis being tested states that an important pathophysiological event in

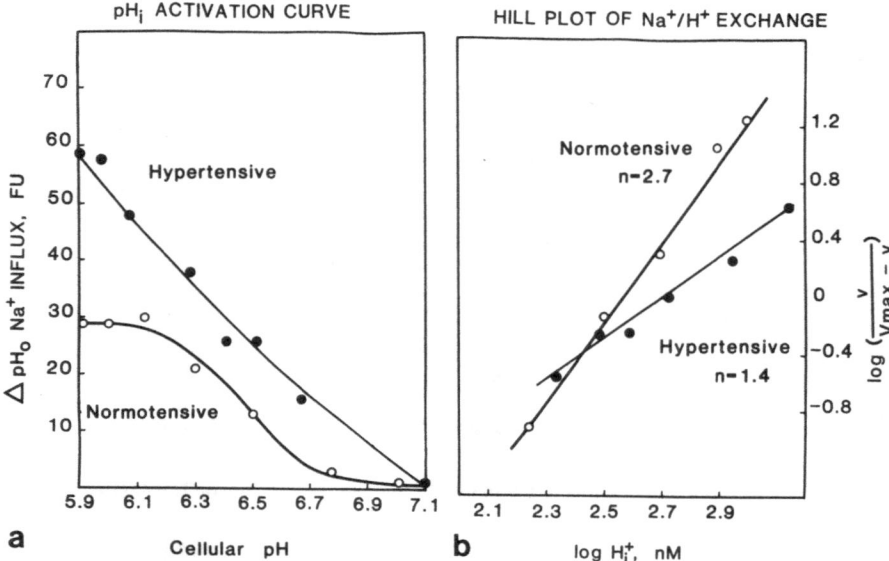

Fig. 2. **a** Activation by cell pH of net Na^+ influx into rbc of a normotensive and a hypertensive subject. ΔpH_o determines Na^+ influx driven by a outward H^+ gradient and was calculated as the difference between Na^+ influx at pH_o 8.0 and pH_o 6.0 [23]. **b** Hill plot of the rate of Na/H EXC versus cellular H^+ concentration; notice that the different slopes determine a Hill coefficient (n_{app}) lower for the hypertensive patient than for the normotensive subject

hypertension is an abnormal regulation of Na/H EXC in several cell types. This abnormal regulation may lead to: (a) increased vascular reactivity, because it potentiates vasoconstrictor effects and favors vascular hyperplasia; (b) abnormal Na^+ reabsorption in the early proximal tubule modulated by and (c) abnormal aldosterone secretion because it modulates the Ang II-stimulated pathway.

Na/H EXC was measured in Na^+-depleted, acid-loaded, DIDS-treated cells by determining Na^+ influx driven by an outward H^+ gradient [23]. In a normotensive subject, net Na^+ influx at pH_o 6.0 increases linearly with cell acidification and is significantly lower than at pH_o 8.0. When an outward H^+ gradient is imposed (pH_o 8.0), net Na^+ influx increases sigmoidally with cell acidification. Subtracting the influx at pH_o 6.0, this component (ΔpH_o) represents the Na^+ driven by Na/H EXC, with Na^+ transport reaching saturation at low pH_i (Fig. 2a). The dependence of Na/H EXC (ΔpH_o) on internal H^+ was analyzed by a Hill plot (Fig. 2b) which gave a Hill coefficient (n_{app}) of 2.7 and pK 6.4. In 46 normotensive subjects, the V_{max} was 47 ± 4 (SE) mmol/l cell h; pK 6.4 ± 0.03; and n_{app}, 2.56 ± 0.14. In a hypertensive patient (Fig. 2a), the activation of Na/H EXC by cell acidification was not sigmoidal, but linear. Notice that n_{app} is lower and V_{max} is higher than in the normotensive patients (Fig. 2b).

In 30 untreated hypertensives, the V_{max} of Na$^+$/H$^+$ was higher (62 ± 6), the n_{app} lower (1.61 ± 0.12) and the pK similar (6.24 ± 0.05) than in normotensive subjects [30]. The decreased n_{app} indicated that cooperative interactions of the H$_i^+$ ligand for Na/H EXC are defective in these patients. These properties could reflect defective control by H$_i^+$, which may be partially compensated by an increased number of transport units. The frequency distribution of the V_{max} and n_{app} for pH$_i$ activation of Na/H EXC in red cells of normotensive and hypertensive patients showed marked skewness to the right, as previously reported for Na/Li EXC.

What are the Mechanisms Involved in the Abnormalities of Na/H EXC

Li$^+$, as well as Na$^+$, ions are asymmetrically transported by Na/H EXC [22–27], the rate of Na$^+$ influx being 100 times higher than that of Na$^+$ or Li$^+$ efflux. The functional asymmetry of the antiporter is decreased in hypertensive patients [24]: At physiological cell pH, more Li$^+$ (or Na$^+$) than H$^+$ is outwardly transported in exchange for external Na$^+$. Furthermore, the degree of activation by the H$_i^+$ regulatory site of Na/H EXC is decreased ($n_{app} < 2.0$) in most hypertensive patients (80%) [24–30]. In many hypertensive subjects, an elevated V_{max} of Na/H EXC is observed in rbc and, even more markedly, lymphocyte Na/H EXC (unpublished observations).

However, our data indicate that there is no a tight linear relationship between elevated Na/Li EXC and Na/H EXC. This apparent discrepancy can be explained if one considers that Na/Li EXC measured at pHi 7.4 provides an estimation of the number of antiporter sites, but it does not determine the activity of Na/H EXC which is determined by Ca$^+$-dependent and independent phosphorylation or another regulatory process. Given the relationship between these two systems and the ubiquity of the Na/H EXC, the alterations in Na/Li and Na/H EXC can be of pathophysiological significance for hypertension.

Aronson has proposed the presence of an internal H$^+$ regulatory site in the Na/H EXC to explain the kinetic behavior of the Na$^+$/H$^+$ antiporter when activated by cell H$^+$ [31]. The H$^+$-regulatory site is likely to be an allosteric site, not a transport site. The decreased n_{app} for H$_i^+$ activation of Na/H EXC in rbc of hypertensive patients might be caused by differences in the transporter molecular structure, or by post-translational modification such as phosphorylation. On the other hand, the concomitant elevation of Na/Li EXC may reflect abnormalities in the regulation of its expression. We are examining several hypotheses to account for the kinetic and functional abnormalities of this antiporter in hypertensive individuals: (a) differences in the antiporter gene encoding the protein or the regulation of its expression; (b) differences in the translation of the antiporter gene; (c) post-translational differences, i.e., glycosylation or phosphorylation controlled by post-receptor messages such as cytosolic Ca^{2+}, protein kinase C, calmodulin. Several experimental approaches are being used to assess the contribution of these possibilities.

Is Na/H EXC the Major Gene Responsible for Variations in Na/Li Exchange?

As described above, there is strong evidence that the Na/Li EXC system is mediated by the Na/H antiporter, and several studies have concluded that the observed variation in countertransport is genetically determined, with a substantial part of this variation determined by a major gene [16–19]. One obvious candidate gene for the abnormal activity of the Na/Li EXC is the Na/H EXC gene. This gene has recently been cloned from human DNA by complementation of a mouse fibroblast cell line deficient in antiporter activity [32]. We requested the collaboration of Drs Sardet and Pouyssegur to investigate the role of Na/H gene in hypertension. I encouraged Dr. Richard Lifton from our Division to pursue studies on the Na/H EXC gene. Lifton et al. [33] have cloned and mapped a 90-kb region spanning the human amiloride-sensitive Na/H antiporter on overlapping cosmid clones. This cloned region extends 27 kb upstream of the 5' end of the longest antiporter cDNA and reveals a primary transcription unit of at least 60 kb. Using these genomic clones, Lifton et al. identified two polymorphisms within an intron of the Na/H EXC transcription unit. A linkage analysis using other markers on human chromosome 1 precisely located the antiporter gene 3 cM proximal to the rhesus blood group and 4 cM distal to the anonymous DNA marker CMM8 [33].

Further on, Lifton et al. [34] have used linkage analysis to test the hypothesis that the Na/H EXC is the major gene responsible for variation in Na/Li EXC levels. The study was performed in the Utah families previously studied by Williams et al. [16–17]. Lifton et al. [34] determined genotypes at the Na/H EXC locus and flanking loci in pedigrees supporting major gene segregation of elevated Na/Li EXC and excluded linkage between them with an odds ratio of almost 1 000 000:1 against linkage. By analysis of 93 hypertensive sibling pairs, Lifton et al. further demonstrated that the Na/H EXC gene locus explains none of the variance in Na/Li EXC in hypertensives.

However, the translation of the antiporter gene using Northern blots has not yet been examined. We can interpret the variations in the V_{max} of Li/Na EXC as variations in the number of Na/H antiporter sites, but not its modulated activity (Fig. 1). The Li/Na EXC can operate at pHi 7.4, at which the H_i^+ regulatory site is not occupied, and it is ATP-independent and amiloride-insensitive. Furthermore, the findings of Lifton et al. [34] do not exclude the role of other genes coding for isoforms of the antiporter not yet identified. From the physiological point view, antiporter isoforms may perform a variety of functions in different cells or in the same cell. It is

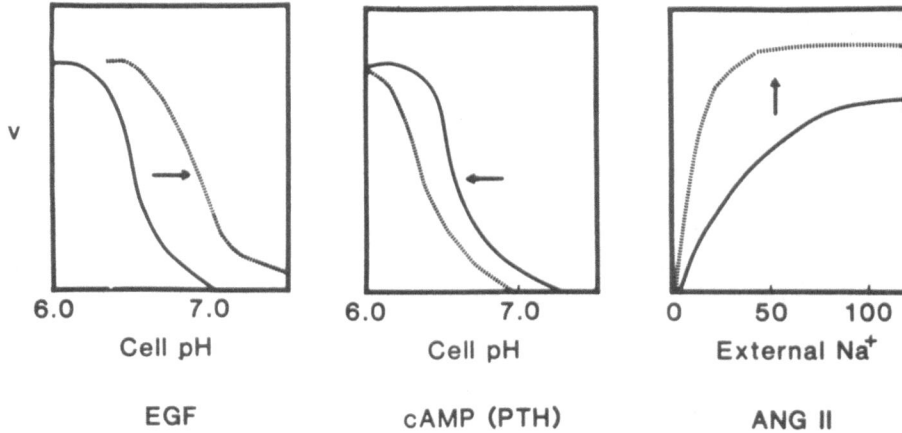

EGF   cAMP (PTH)   ANG II

Fig. 3. Mechanisms of regulation of Na/H EXC by several agonists. Growth factors such as epidermal growth factor reduce the K_m for cellular H$^+$ to activate Na/H EXC [41]. Several studies have shown that parathyroid hormone (PTH) and its second messenger cAMP increases the K_m for cellular H$^+$ [43,44]. Angiotensin activates the antiporter, decreasing the K_m for external Na$^+$ ions and increasing its turnover rate (V_{max}/K_m) [5]. Activation by ang II can occur through PKC-dependent and -independent pathways [51]

possible that Na$^+$/H$^+$ antiporter isoforms could be differentially regulated by post-translational modifications such as phosphorylation. Sardet et al. have recently reported phosphorylation of the Na$^+$/H$^+$ antiporter encoded by chromosome 1P by epidermal growth factor, thrombin, and phorbol esters [35].

Are Na/H Exchange Abnormalities Caused by Post-translational Modifications?

We reported several years ago that rbc Na/H EXC is modulated by an increase in cytosolic Ca^{2+} elicited by Ca^{2+} ionophores [21]. We have previously discussed how the relationship between Na$^+$ and Ca^{2+} ions does not take place only through a Na$^+$/Ca^{2+} exchanger, but through the Ca^{2+} modulation of Na$^+$ transport [2]. Cytosolic Ca^{2+} has been consistently reported to be elevated in platelets of hypertensive patients [36–39]. It has been shown that Ca^{2+}-mobilizing, vasoconstrictor, and growth factor agonists (Ang II, PDGF, thrombin) in VSM cells [9,10] modulate the antiporter by some post-receptor event, i.e., phospholipase C (PLC) activation, release of inositol triphosphate, increased cytosolic Ca^{2+} and diacylglycerol, protein kinase C (PKC) activation. Several kinetic mechanisms of Na/H EXC modulation have been established (Fig. 3) [40] Growth factors decreases the K_m for H$_i^+$ [41] cAMP increases the K_m for H$_i$ [42,43] and Ang II decreases the K_m for Na$_o^+$ [5]. However, the mechanism of Na/H

EXC activation by increased cytosolic Ca^{2+} has not yet been defined. The study of the kinetics of Ca^{2+} activation of Na/H EXC may shed light into the nature of the abnormalities observed in hypertensive patients. Increased Ca^{2+} may increase V_{max}, increase n_{app}, or decrease the K_m for H_i^+. The reduced n_{app} observed in hypertensives may imply that the modulation by cytosolic Ca^{2+} is impaired even though the number of sites is elevated (increased Na/Li EXC). Numerous studies have determined that PKC is a modulator of Na/H EXC [40] and enhanced PKC levels have been reported in rbc of a small number of hypertensive patients [44]. It is important to clarify which PKC isoform is present in human rbc and in other cell types before its activity can be correlated with increased antiporter activity [45].

Alicia Rivera is studying in my laboratory the role of cytosolic Ca^{2+} on Na/H EXC activity in human rbc and lymphocytes. Measurements of cytosolic Ca^{2+} in lymphocytes of 15 normal subjects have given mean values of $70 \pm 4 nM$. In 15 hypertensive patients there was a trend to higher values ($102 \pm 9 nM$). A reduction of cytosolic Ca^{2+} (using BAPTA or external Ca = 0) from $54 nM$ to $10 nM$ produces a 80% reduction of Na^+-dependent H^+ efflux. Preliminary results indicate that in human rbc, Na^+/H^+ activity falls 50% after reduction of cytosolic Ca^{2+}. These studies indicate that a large fraction of the Na/H EXC activity is determined by normal cytosolic Ca^{2+} levels. Thus, elevation of cytosolic Ca^{2+} is not required to account for increased activity of the antiporter, as proposed by Aviv [46]. We have found that in VSM of the Milan hypertensive strain, elevation of cytosolic Ca^{2+} is not accompanied by elevation of Na/H EXC, as previously observed in VSM of the SHR strain [9,47].

What are the Abnormalities of Na/H EXC in Diabetes?

An important clue in the clarification of the mechanism involved in the abnormal regulatory properties of Na/H EXC in human rbc seems to evolve from studies performed by my laboratory on patients with insulin-dependent diabetes, (IDDM), IDDM with nephropathy (IDDM + NP), and in hypertensive blacks with insulin-resistant diabetes (IDDM-II).

Studies in collaboration with Drs Marianna Bak and Andrew Krowleski from the Joslin Clinic, Boston have indicated that in IDDM patients, the V_{max} of Na/Li EXC is normal but the V_{max} of Na/H EXC is elevated and the kinetics of cell pH activation have a normal Hill coefficient [48]. As previously found in essentially hypertensive patients, in IDDM + NP patients the V_{max} of Na/Li and Na/H EXC are elevated but the Hill coefficient is lower than 2.0. Both groups of patients therefore exhibit a widespread abnormality in the pH_i regulation of Na/H EXC accompanied by an increased number of sites or normal number of sites. The significant elevation of the V_{max} of Na/H in IDDM patients in comparison to normotensive whites suggests that insulin receptors might be involved in the regulation of

Na/H EXC. Because these patients are treated with exogenous insulin, the hormone might have a long-term effect on the expression or a short-term modulatory action on the Na/H antiporter.

Studies in collaboration with Dr. Bonita Falkner from the Hahneman School of Medicine have indicated that the V_{max} of Na/H EXC in rbc of normotensive and hypertensive nonobese young blacks is significantly lower than normotensive whites [23,24]. This finding (even though preliminary) is in agreement with previously reported racial differences in the Na/Li EXC. Moreover, the young black hypertensives determined to be insulin-resistant as determined by the insulin clamp technique have elevated Na/H EXC.

These findings indicate that post-insulin-receptor events appear to be an important determinant of the V_{max} of Na/H EXC in human rbc in diabetes and hypertension [49,50]. The mechanism involved in the regulation of Na/H exchange by the insulin receptor in human rbc remains to be determined. Because rbc do not have a insulin-related glucose transporter, nor glycogen or lipid metabolism modulated by insulin, other mechanisms should be investigated. Does the insulin receptor directly phosphorylate the antiporter through its tyrosine-kinase activity? Or, alternatively, does the insulin receptor regulate the antiporter indirectly through phosphorylation of the calmodulin-activated Ca^{2+} pump which sets the cytosolic Ca^{2+} levels?

References

1. Canessa M, Adragna N, Solomon HS, Connolly TM, Tosteson DC (1980) Increased sodium-lithium countertransport in red cells of patients with essential hypertension. N Engl J Med 302:772–776
2. Canessa M, Brugnara C, Escobales NE (1987) The Li-Na exchange and Na-K-Cl cotransport in essential hypertension. Hypertension 10:4–10
3. Canessa M (1988) Genetic variants of Na transport systems in human red cells. In: Nagel R (ed) Genetically abnormal red cells Vol 2. CRC Press, Boca Raton. pp 131–148
4. Berk BC, Brock TA, Gimbrone MA, Alexander RW (1987) Early agonist-mediated ionic events in cultured vascular smooth muscle cells: calcium mobilization is associated with intracellular acidification. J Biol Chem 262:5065–5072
5. Vallega G, Canessa ML, Berk CB, Brock TA, Alexander RW (1988) Vascular smooth muscle Na$^+$/H$^+$ exchange kinetics and its activation by angiotensin II. Am J Physiol 254:C751–758
6. Mahnesmith RL. Aronson PS (1985) The plasma membrane sodium/hydrogen exchange and its role in physiological and pathophysiological processes. Circ Res 56:773–788
7. Conlin P, Vallega G, Canessa M, Williams RW (1990) Na$^+$/H$^+$ exchange in adrenal glomerulosa cells and its activation by angiotensin II. Endocrinology 127:236–244
8. Mitsuhashi T, Ives HE (1988) Intracellular Ca^{2+} requirement for activation of the Na/H exchanger in vascular smooth muscle cells. J Biol Chem 263:8790–8795
9. Berk B, Canessa M, Vallega G, Griesling C, Alexander RW (1988) Agonist-mediated changes in intracellular pH: role in vascular smooth muscle cell function. J Cardiovasc Pharmacol 12:S104–114

10. Hatori N, Fine BP, Cragoe E, Aviv A (1987) Angiotensin II effect on cytosolic pH in cultured rat vascular smooth muscle cells. J Biol Chem 262:5073–5078
11. Hilton PJ (1986) Cellular sodium transport in essential hypertension. N Engl J Med 314:222–229
12. Blaustein MP (1984) Sodium transport in hypertension: where are we going? Hypertension 6:445–453
13. Canessa M, Solomon H, Falkner B, Adragna N, Tosteson DC. Familial aggregation of sodium countertransport and essential hypertension. In: Villarreal H, Sambhi MP (eds) Topics of pathophysiology of hypertension. Nijhoff, Boston, pp 78–87
14. Lewitter FI, Canessa M (1985) Red cell transport studies in adult twins. Am J Hum Genet 36:172
15. Canessa M, Spalvins A, Adragna M, Falkner B (1984) Red cell sodium countertransport and cotransport in normotensive and hypertensive blacks. Hypertension 6:344–351
16. Dadone MM, Hasstedt SJ, Hunt SC, Smith JB, Ash O, Williams RR (1984) Genetic analysis of sodium-lithium countertransport in 10 hypertension-prone kindreds. Am J Med Genet 17:565–577
17. Hasstedt SJ, Wu LL, Owen-Ash K, Kuida H, Williams RR (1988) Hypertension and sodium-lithium countertransport in Utah pedigrees: evidence for major-locus inheritance. Am J Hum Genet 43:14–22
18. Boerwinkle E, Turner ST, Weinshilboum R, Johnson M, Richelson E, Sing CF (1986) Analysis of the distribution of erythrocyte sodium lithium cuntertransport in a sample representation of the general population. Genet Epidemiol 3:365–378
19. Motulsky AG, Burke W, Billings PR, Ward RH (1987) Hypertension and the genetics of red cell membrane abnormalities. Ciba Found Symp 130:150–166
20. Redgrave J, Canessa M, Gleason R, Hollenberg NK, Williams GH (1989) Erythrocyte countertransport in non-modulating essential hypertension. Hypertension 13:721–726
21. Escobales NE, Canessa M (1985) Ca-activated Na^+ fluxes in human red cells: amiloride-sensitivity. J Biol Chem 260:11914–11923
22. Escobales N, Canessa M (1986) Amiloride-sensitive Na transport in human red cells: evidence for a Na/H exchange system. J Membr Biol 90:21–28
23. Semplicini A, Canessa M (1989) Kinetic and stoichiometry of the red cell Na+/H+ exchange. J Membr Biol 107:219–228
24. Morgan K, Canessa M, Goldzer R, Moore T, Williams GH (1988) Red cell Na^+/H^+ exchange has a defective H^+ regulatory site in hypertensive patients with elevated Na^+/Li^+ exchange. Clin Res 36:430a
25. Canessa M, Spalvins A, Escobales N (1986) Li/H and Li/Na exchange in human red cells: effect of proton gradients. Biophys J 49:141a
26. Canessa M, Spalvins A (1987) Kinetic effect of internal and external H^+ on Li/H and Li/Na exchange in human red cells. Biophys J 51:567a
27. Canessa M, Morgan K, Semplicini A (1988) Genetic differences in lithium-sodium exchange and regulation of the sodium-hydrogen exchanger in essential hypertension. J Cardiovasc Pharmacol 12:S92–98
28. Berk BC, Vallega G, Muslim AJ, Gordon HM, Canessa M, Alexander RW (1989) Spontaneously hypertensive rat vascular smooth muscle cells in culture exhibit increased growth and Na/H exchange. J Clin Invest 83:822–829
29. Vallega G, Atkinson WW, Tsai E, Torielli L, Canessa M (1989) Milan hypertensive (MHS) rat vascular smooth muscle cells exhibit increased growth and Na-K-Cl cotransport. FASEB J 3:A1187
30. Canessa M, Morgan K, Goldzer R, Moore TJ, Spalvins (1991) Kinetic abnormalities of the red cell sodium-proton exchange in hypertensive patients. Hypertension 17:340–348

31. Aronson PS (1985) Kinetic properties of the plasma membrane Na+/H+ exchanger. Annu Rev Physiol 47:545–560

32. Sardet C, Franchi A, Pouyssegur J (1989) Molecular cloning, primary structure, and expression of the human growth factor-activatable Na$^+$/H$^+$ antiporter. Cell 56:271–280

33. Lifton RP, Sardet C, Pouyssegur J, Lalouel JM (1990) Cloning of the human genomic amiloride-sensitive NA/H antiporter gene; identification of genetic polymorphism and localization on the genetic map of chromosome 1P. Genomics 7: 131–135

34. Lifton RP, Hunt SC, Williams RR, Poyssegur J, Lalouel JM (1991) Exclusion of the Na$^+$/H$^+$ antiporter as a candidate gene in essential hypertension. Hypertension 17: 8–14

35. Sardet C, Counillon L, Franchi A, Pouyssegur J (1990) Growth factors induce phosphorylation of the Na$^+$/H$^+$ antiporter, a glycoprotein of 110D. Science 247: 723–726

36. Resink TJ, Dimitrov D, Zschauer A, Erne P, Tkachuk VA, Buhler FR (1986) Platelet calcium-linked abnormalities in essential hypertension. Ann NY Acad Sci 488:252–263

37. Erne P, Bolli P, Burgisser E, Buhler FR (1984) Correlation of platelet calcium with blood pressure. N Engl J Med 310:1084–1088

38. Oshima T, Matsuura H, Kido K, Matsumoto K, Fujii H et al. (1988) Intra-lumphocytic sodium and free calcium and plasma renin in essential hypertension. Hypertension 12:26–31

39. Quan Sang KHL, Benlian P, Kanawati C, Montenay-Garestier, Meyer P, Devyinck MA (1985) Platelet cytosolic free calcium concentration in primary hypertension. J Hypertens 3:S33–36

40. Grinstein S, Rothstein A (1986) Mechanisms of regulation of the Na/H exchanger. J Membr Biol 90:1–12

41. Moolenar WH, Tsien RY, Van der Saag PT, de Laat SW (1983) Na/H exchange and cytoplasmic pH in the action of growth factors in human fibroblasts. Nature 304:645–648

42. Miller RT, Pollock AS (1987) Modification of the internal pH sensitivity of the Na$^+$/H$^+$ antiporter by parathyroid hormone in a cultured renal cell line. J Biol Chem 262:9115–9120

43. Kahn AM (1985) Parathyroid hormone and dibutyril cAMP inhibit Na$^+$/H$^+$ exchange in renal brush border vesicles. Am J Physiol 248:F212–218

44. Kravtsov GM, Dulin NO, Postnov YV (1988) Activity of protein kinase C in erythrocytes in primary hypertension. J Hypertens 6:853–857

45. Huang Kuo-Pink (1989) The mechanism of protein-kinase C activation. Trends Neurosci 12:425–450

46. Aviv A (1988) The link between cytosolic Ca^{2+} and Na$^+$/H$^+$ antiporter; a verifying factor for essential hypertension. J Hypertens 6:685–691

47. Socorro I, Vallega G, Moore T. Canessa M (1990) Vascular smooth muscle cells from the Milan hypertensive rat exhibit decreased functional angiotensin II receptors. Hypertension 15:591–599

48. Bak MI, Canessa ML, Warramm JH and Krowleski AS (1989) Differences in the activities of Na/H transport system in red blood cells of individuals with and without diabetic nephropathy. Clin Res 37:551a

49. Semplicini A, Canessa M, Mozzato MG, Ceolotto G et al. (1989) Red blood cell Na/H exchange in subjects with essential hypertension. Am J Hypertens 2:903–908

50. Livne A, Veitch R, Grinstein S, Balfe JW, Marquez-Julio A, Rothstein A (1987) Increased platelet Na$^+$/H$^+$ exchange rates in essential hypertension: application of a novel test. Lancet I: 533–536

Regulation of the Na^+/H^+ Exchanger by Protein Kinase C and Its Implications in Hypertension: The Platelet Model*

A. A. Livne, S. Markus, M. Afargan, and O. Aharonovitz

Department of Biology, Ben-Gurion University of the Negev, P.o.b. 653, 84105 Beer-Sheva, Israel

The Na^+/H^+ Exchange System

The Na^+/H^+ exchange system is ubiquitous, reported to be present in plasma membranes of essentially all mammalian cell types. The exchanger catalyzes an amiloride-sensitive, electroneutral countertransport of Na^+ for H^+. Entry of Na^+ in exchange for intracellular H^+ is the predominant transport mode under physiological conditions, although the mechanism is reversible [1]. The Na^+/H^+ exchange has been implicated in a variety of functions, including transepithelial movement of salt and water, the regulation of the cytoplasmic pH and Na^+ concentration, the control of cell volume, and the initiation of growth and proliferation [2].

Activity of Na^+/H^+ exchange is regulated by several different pathways. Mechanisms reported are primarily allosteric modulation by cytoplasmic pH (pH_i), implicating a modifier site [3–7], and covalent protein modification through kinases (for review see [8]). The operation of the modifier is expressed as the exchange activity is stimulated by lowering pH_i below a certain threshold; furthermore, the exchange activity becomes practically quiescent above this threshold or "set point." This set point is close to the normal physiological pH_i of the cell, consistent with a central role of the Na^+/H^+ exchanger in pH_i homeostasis [8]. A variety of stimuli can modulate the exchange activity by shifting the allosteric control of transport to a different setting. For a series of growth factors, hormones, phorbol esters, and diacylglycerol, the shift of the set point is to a more alkaline setting, while for some modulators (parathyroid hormone, cAMP) a shift to a more acidic pH has been observed. The primary structure of the Na^+/H^+ antiporter molecule was recently elucidated by cloning its gene [9], and the

* A grant from the Chief Scientist, Ministry of Health, Isreal, is acknowledged. This research was supported by the Fund for Basic Research administered by The Israel Academy of Sciences and Humanities.

mechanism for the control of the transporter by covalent modification can now be approached on a molecular level.

The Na$^+$/H$^+$ Exchange in Essential Hypertension

After activation by a receptor-mediated process, the phosphoinositide signalling system leads to contraction in vascular smooth muscle cells, with the participation of protein kinase C (PKC) and Na$^+$/H$^+$ exchange (for review see [10]). PKC and phosphoproteins affected by PKC apparently play the major role in regulating smooth muscle tone during the sustained phase of smooth muscle contraction [11]. PKC is characterized by molecular heterogeneity with distinct tissue expression and intracellular localization; different PKC subspecies probably have distinct functions in the processing and modulation of a variety of physiological and pathological responses to external signals [12].

It has been proposed that elevated rates of Na$^+$/H$^+$ exchange in membranes of blood vessels and of kidney tubules play a role in the pathophysiology of essential hypertension [2,13]. In the vascular smooth muscle, increased intracellular Na$^+$ may lead to increased Ca^{2+} concentration [14] and vascular tone. Higher tone may also be affected by intracellular pH changes imposed by the Na$^+$/H$^+$ exchange [2]. Furthermore, the enhanced exchange may lead to the development of vessel hypertrophy and increased vessel resistance associated with hypertension. Overactivity of the Na$^+$/H$^+$ exchange in the luminal membrane of the proximal renal tubular cell would enhance Na$^+$ reabsorption and lead to a defect in renal salt excretion. Hypertension might then be expressed via two consequences: (a) the renal defect would require that the kidney be perfused at an arterial pressure greater than normal to maintain salt balance [15], and (b) the secondary release of natriuretic hormonal factors could lead to an increase in peripheral vascular resistance [16,17].

The proposed involvement of elevated Na$^+$/H$^+$ exchange in the pathophysiology of essential hypertension is supported by a series of studies.

1. Several cell types of essential hypertensives and of hypertensive rats, including vascular smooth muscle cells and renal brush border membranes, exhibit increased Na$^+$/H$^+$ exchange activity. Studies are summarized in Table 1. In addition, it has been reported that Li$^+$ clearance, a measure of proximal tubular reabsorption of Na$^+$ is decreased in patients with essential hypertension and in normotensive subjects with a family history of hypertension [18]. This study, however, was not corroborated by another investigation [19]. Of particular interest is a recent report that endothelin-induced vasoconstriction of human peripheral resistance vessels is mediated in part by stimulation of Na$^+$/H$^+$ exchange [20].

Table 1. Increased Na$^+$/H$^+$ exchange activity in hypertension

	Cells	Ref.
Essential hypertension	Platelets	Livne et al. [35]
	Platelets	Schmouder & Weder [36]
	Leukocytes	Ng et al. [25, 37]
	Erythrocytes	Semplicini et al. [19]
	Erythrocytes	Orlov et al. [38]
Hypertensive rats	Lymphocytes	Feig et al. [39]
	Vascular smooth muscle cells	Berk et al. [40]
	Renal brush border (membranes)	Morduchowicz et al. [41]
	Erythrocytes	Orlov et al. [38]

2. PKC activity is elevated in platelets of hypertensive rats [21] and erythrocytes of essential hypertensives [22].
3. The cytosolic pH is more alkaline in resistance vessels [23] and platelets of hypertensive rats [24], as well as in leukocytes [25] of essential hypertensives. Furthermore, inhibition of PKC results in greater reduction of pH$_i$ in platelets of hypertensive rats than controls. In contrast, reduced pH$_i$ in lymphocytes from the spontaneously hypertensive rat has been reported [26]. pH$_i$ is a steady-state value, representing the outcome of several factors including metabolic generation of protons, Na$^+$/H$^+$ exchange, buffer capacity, and Cl$^-$/HCO$_3^-$ exchange. Thus, even if Na$^+$/H$^+$ exchange is elevated, alkalinization must not necessarily be coupled to it.

Regulation of Na$^+$/H$^+$ Exchange by PKC

Hormones and other agents that stimulate Na$^+$/H$^+$ exchange are known to activate protein kinases, primarily tyrosine kinase and protein kinase C [8]. The present study aims to clarify the following questions: (a) Which of these kinases, tyrosine kinase or PKC, predominantly regulates Na$^+$/H$^+$ exchange in platelets? (b) How tight is the regulation? (c) Are the rate and set point of the exchange closely linked or differentially affected by factors affecting the kinase activity? Platelets were chosen for the study in view of their similarity to the smooth muscle cell, the advantages they offer for hypertension research and the studies already performed [18,27].

The experimental procedure for the fluorimetric measurements of Na$^+$/H$^+$ exchange was as described elsewhere [28–30], with the following modifications. Platelets were loaded with bis(carboxyethyl)carboxy fluorescein (BCECF) by incubation of platelet-rich plasma with the parent acetoxymethyl ester for 30 min at 22°C and then gel-filtered as described, with the Sepharose column equilibrated and eluted with the NaCl medium, but at pH 6.8. The eluted platelets were supplemented with 1 mM CaCl$_2$, 1 mM MgCl$_2$, and

Table 2. Na^+/H^+ exchange rate and pHi set point of human platelets as affected by PMA and staurosporine

	(n)	Na/H exchange[a]		Correlation	
		Rate[b]	pHi set point	Rate vs set point	Confidence
Control	(6)	0.031 ± 0.008	7.27 ± 0.03	0.96	>99%
PMA, $0.1 \mu M$	(5)	0.053 ± 0.008	7.64 ± 0.03	0.48	<80%
Staurosporine, $0.5 \mu M$	(6)	0.006 ± 0.002	7.04 ± 0.02	0.85	>95%

[a] Mean ± SE.
[b] ΔpH/9 s at pH 7.0.

$0.2 \, mM$ probenecid. An aliquot ($10-20 \mu l$) of the gel-filtered platelets was rapidly mixed with $1.8 \, ml$ of the assay medium composed of $60 \, mM$ Na propionate, $80 \, mM$ NaCl, $10 \, mM$ glucose, $5 \, mM$ KCl, $1 \, mM$ $MgCl_2$, $1 \, mM$ $CaCl_2$, $20 \, mM$ HEPES, pH 7.3, adjusted to 285 mosmol. All tested agents [phorbol myristate acetate (PMA), staurosporine, or dihexanoyl glycerol] were added to the assay medium dissolved in $1.8 \mu l$ dimethyl formamide, prior to the platelets. The solvent was added equally to the control. Fluorimetric tracings were recorded for 90 s, starting within 3 s after the addition of the platelets. The Na^+/H^+ exchange rate was determined by the change in pH_i, which is sensitive to 5-(N-methyl-N-isobutyl) amiloride and dependent on external Na^+. The set point was calculated by determining the rates of the Na^+/H^+ exchange at several pH_i values (at least seven measurements) and extrapolating the curve (log rate vs pH_i) to pH_i, for which the rate is ≤ 0.001 pH/9 s.

Figure 1 presents measurements relating to Na^+/H^+ exchange in platelets from six normotensive individuals (four female, two male; average age: 40 ± 3 years; average weight: 68 ± 3 kg), showing the exchange rate and pH_i set point, the data are summarized in Table 2.

The pH_i set point of the control is close to the pH_i values recorded earlier [28–30] in platelets under the same conditions, indicating the major role played by the Na^+/H^+ exchange in the control of intracellular pH. In all subjects PMA caused an increase in the exchange rate coupled with an alkaline shift of the pH_i set point. Conversely, staurosporine reduced the rate with a coinciding acidic shift of the pH_i set point.

Overall, the range of the pH shifts is 0.6 pH units. The exchange rate of staurosporine-treated platelets is approximately 10% of the maximum rate, obtained by PMA treatments. The rate and pH_i set point values are closely correlated, particularly for the control and staurosporine treatment. Taken together, these results point to a tight control of the Na^+/H^+ exchange in platelets by PKC, in regard to both the rate and set point.

The observed shifts in the pH_i set point coupled with changes in the exchange rate and the apparent involvement of protein kinase can be con-

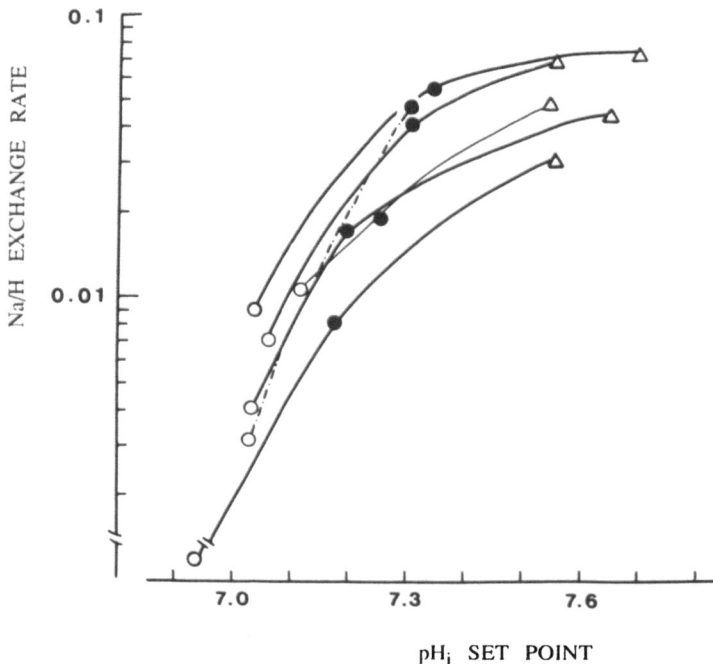

Fig. 1. Na$^+$/H$^+$ exchange rate ΔpH/9s at pH 7.0 and pHi set point in human platelets as affected by PMA and staurosporine. *Open circles*, staurosporine 0.5 μM; *closed circles*, control; *triangles*, PMA 0.1 μM. The *lines* link the experimental points for each individual. In one individual (-.-.) no PMA assay point was performed

solidated into a unifying hypothesis as already presented elsewhere [8]: it is assumed that phosphorylation at or near the modifier site is responsible for the shifts in the set point as affected by PMA and staurosporine. Furthermore, the extent of phosphorylation is an explession of the balance between a kinase and a phosphatase. The phosphorylation could lead to a conformational change of the protein or to an increased local proton concentration created by the negative charges of the phosphate group. In support of this hypothesis, the role of phosphorylation in the activation of the Na$^+$/H$^+$ exchanger has been recently demonstrated: mitogenic activation of resting cells stimulated phosphorylation of the Na$^+$/H$^+$ antiporter, with a concomitant rise in pH$_i$ [31].

Several lines of evidence indicate that PKC is the kinase that predominantly regulates Na$^+$/H$^+$ exchange in platelets:

1. Stimulation of PKC by PMA would be expected to elevate the phosphorylation of the Na$^+$/H$^+$ exchanger and thus lead both to an increased exchange rate and an alkaline shift of the pH$_i$ set point. Both of these are indeed apparent (Fig. 1, Table 2).

2. Like PMA, dihexanoyl glycerol (DHG), known to activate PKC in intact platelets [32], affects both the rate and set point of the exchange. $K_{0.5}$ for stimulation of the rate by DHG is approx. $4 \mu M$, with rate increases of over 220% relative to control apparent when $40 \mu M$ DHG is added.
3. Staurosporine, a potent kinase inhibitor that more potently inhibits PKC than other kinases, markedly affects the Na^+/H^+ exchange parameters. Ki for staurosporine in inhibition of the Na^+/H^+ exchange rate is $16 \, nM$. This is even lower than the Ki of $100 \, nM$ observed for the inhibition of PKC in intact platelets [33], when analyzed on the basis of the phosphorylation of 47-kDa protein (pleckstrin), a specific substrate for PKC.
4. The effects of DHG (2 above) are abolished by staurosporine.
5. Using other agents as well to affect Na^+/H^+ exchange, it was concluded in earlier studies that PKC activates Na^+/H^+ exchange in platelets [34].

Conclusion

1. Na^+/H^+ exchange rate and pH_i set point are highly correlated and apparently closely linked.
2. Na^+/H^+ exchange is tightly regulated by PKC.
3. The apparent mechanism for the regulation is phosphorylation at or near the transporter modifier site that causes a conformation change or increased local $[H^+]$.
4. The tight control of Na^+/H^+ exchange by PKC and the heterogeneity of PKC already demonstrated apparently lead to polymorphic expression of Na^+/H^+ exchange. Such polymorphism may play an important role in the manifestation of hypertension.

Acknowledgements. We wish to thank the Oelbaum family, Barrie Rose, and Toby and Joey Tanenbaum, all from Toronto, Canada, for their generous aid.

References

1. Aronson PS (1985) Kinetic properties of the plasma membrane Na-H exchanger. Annu Rev Physiol 47:545–460
2. Mahhensmith RT, Aronson PS (1985) The plasma membrane sodium-hydrogen exchanger and its role in physiological and pathophysiological processes. Circ Res 57:573–588
3. Aronson PS, Nee J, Suhn MA (1982). Modifier role of internal H^+ in activiting the Na^+-H^+ exchanger in renal microvillus membrane vesicles. Nature 299:161–163
4. Paris S, Pouyssegur J (1984) Growth factors activate the Na^+-H^+ antiporter in quiescent fibroblasts by increasing its affinity for intracellular H^+. J Biol Chem 259:10989–10994
5. Grinstein S, Cohen S, Goetz JD, Rothstein A, Gelfand EW (1985) Characterization of the activation of the Na^+-H^+ exchange by phorbol esters: changes in cytoplasmic pH dependence of the antiport. Proc Natl Acad Sci USA 82:1429–1433

6. Grinstein S, Rothstein A, Cohen S (1985) Mechanism of osmotic activation of Na$^+$/H$^+$ exchange in rat thymic lymphocytes. J Gen Physiol 85:765–785
7. Montose MH, Murer H (1986) Regulation of intracellular pH in LLC-PK, cells by Na$^+$/H$^+$ exchange. J Membr Biol 93:33–42
8. Grinstein S, Rothstein A (1986) Mechanism of regulation of the Na$^+$/H$^+$ exchanger. J Membr Biol 90:1–12
9. Sardet C, Franchi A, Pouyssegur J (1989) Molecular cloning, primary structure and expression of the human growth factor-activatable Na$^+$/H$^+$ antiporter. Cell 56: 271–280
10. Heagerty AM, Ollenshaw JD (1987) The phosphoinositide signalling system and hypertension. J Hypertens 5:515–524
11. Rasmussen H, Takuwa Y, Park S (1987) Protein kinase in the regulation of smooth muscle contraction. FASEB J 1:177–185
12. Nishizuka A (1988) The molecular heterogeneity of protein kinase C and its implication for cellular regulation. Nature 334:661–665
13. Siefter JL, Aronson PS (1986) Properties and physiological role of the plasma membrane sodium-hydrogen exchanger. J Clin Invest 78:859–864
14. Blaustein MP (1984) Sodium transport and hypertension. Hypertension 6:445–453
15. Guyton AC, Coleman TG, Cowley AN Jr, Manning RD Jr, Norman RA Jr, Ferguson JD (1974) A systemic analysis approach to understanding long-range arterial blood pressure control and hypertension. Circ Res 35:159–176
16. Haddy FJ (1980) Mechanism, prevention and therapy of sodium-dependent hypertension. Am J Med 69:746–758
17. de Wardener HE, MacGregor GA (1970) Dahl's hypothesis that a saluretic substance may be responsible for a sustained rise in arterial pressure: its possible role in essential hypertension. Kidney Int 18:1–9
18. Weder AD (1986) Red-cell lithium-sodium countertransport and renal lithium clearance in hypertension. N Engl J Med 314:198–201
19. Semplicini A, Canessa M, Mozzato MG, Ceolotto G, Mazola M, Buzzaccarin F, Casolino P, Pessina AC (1989) Red blood cell Na$^+$/H$^+$ and Li$^+$/Na$^+$ exchange in patients with essential hypertension. Am J Hypertens 2:903–908
20. Richards NT, Poston L, Goldsmith DJA, Cragoe EJ, Hilton PJ (1989) Endothelin-induced contraction of human peripheral resistance vessels is partly dependent on stimulation of sodium-hydrogen exchange. J Hypertens 7:777–780
21. Takaori K, Ito S, Kanayama Y, Takeda T (1986) Protein kinase C activity in platelets from spontaneously hypertensive rats (SHR) and normotensive Wistar Kyoto rats (WKY). Biochem Biophys Res Commun 141:768–773
22. Kravtsov GM, Dulin NO, Postnov YV (1988) Activity of protein kinase C in erythrocytes in primary hypertension. J Hypertens 6:853–857
23. Izzard SA, Heagety AM (1989) The measurement of internal pH in resistance arterioles: evidence that intracellular pH is more alkaline in SHR than WKY animals. J Hypertens 7:173–180
24. Inariba et al. (1988) 12th international congress of hypertension (abstract)
25. Ng LL, Dudley C, Bomford J, Hawley D (1989) Leucocyte intracellular pH and Na$^+$/H$^+$ antiport activity in human hypertension. J Hypertens 7:471–475
26. Battle DC, Saleh A, Rombola G (1990) Reduced intracellular pH in lymphocytes from spontaneously hypertensive rat. Hypertension 15:97–103
27. Buhler FR, Resink TJ (1988) Platelet abnormalities and the pathophysiology of essential hypertension. Experientia 44:94–97
28. Livne A, Grinstein S, Rothstein A (1987) Characterization of Na$^+$/H$^+$ exchange in platelets. Thromb Haemost 58:971–977
29. Funder J, Hershco L, Rothstein A, Livne A (1988) Na$^+$/H$^+$ exchange and aggregation of human platelets activated by ADP: the exchange is not required for aggregation. Biochim Biophys Acta 938:425–433

30. Argaman A, Livne A (1988) Angiotensin II and Na^+/H^+ exchange in human blood platelets. J Hum Hypertens 2:161–166
31. Sardet C, Counillon L, Franchi A, Pouyssegur J (1990) Growth factors induce phosphorylation of the Na^+/H^+ antiporter, a glycoprotein of 110-KD. Science 247:723–726
32. Lapetina EG, Reep B, Ganong BR, Bell RM (1985) Exogenous sn-1,2 diacylglycerols containing saturated fatty acids function as bioregulators of protein kinase C in human platelets. J Biol Chem 260:1353–1361
33. Watson SP, McNally J, Shipman LJ, Godfrey PP (1988) The action of the protein kinase C inhibitor, staurosporine, on human platelets. Biochem J 249:345–350
34. Siffert W, Scheid P (1986) A phorbol ester and 1-oleoyl-2-acetylglycerol induce Na^+-H^+ exchange in human platelets. Biochem Biophys Res Commun 141:13–19
35. Livne A, Balfe JW, Veich R, Marquez-Julio A, Grinstein S, Rothstein A (1987) Increased platelet Na^+-H^+ exchange rates in essential hypertension: application of a novel test. Lancet i:533–536
36. Schmouder RL, Weder AB (1989) Platelet sodium-proton exchange is increased in essential hypertension. J Hypertens 7:325–330
37. Ng LL, Harker M (1988). Sodium influx in human leucocytes: a novel technique for assessment of sodium-proton antiport activity. Clin Sci 75:179–184
38. Orlov SN, Postnow IY, Pokudin NI, Kukharenko VY, Postnov YV (1989) Na^+-H^+ exchange and other ion-transport systems in erythrocytes of essential hypertensives and spontaneously hypertensive rats: a comparative analysis. J Hypertens 7:781–788
39. Feig PU, D'Occio MA, Boylan JW (1987) Lymphocyte membrane sodium-proton exchange in spontaneously hypertensive rats. Hypertension 9:282–288
40. Berk BC, Vallega G, Muslin AJ, Gordon HM, Canessa M, Alexander RW (1989) Spontaneously hypertensive rat vascular smooth muscle cells in culture exhibit increased growth and Na^+/H^+ exchange. J Clin Invest 83:822–829
41. Morduchowicz GA, Sheikh-Hammad D, Jo OD, Nord EP, Lee DBN, Yanagawa N (1989) Increased Na^+/H^+ antiport activity in the renal brush border membrane of SHR. Kidney Int 36:576–581

Phosphoinositide Metabolism in Hypertension

P. Marche and D. L. Zhu

Department de Pharmacologie, Inserum U7, Hôpital Necker, 161, rue de Sèvres,
F-75730 Paris Cedex, France

In essential hypertension, an increase in peripheral resistance is associated
with hypertrophy/hyperplasia of smooth muscle cells of the media as well
with cell hypercontractility. As a result, the wall-to-lumen ratio decreases
and both the basal tone and the response to vasoconstrictor agents increase.
To gain insight into the pathogenesis of hypertension, the mechanisms
underlying both the smooth muscle hypertrophy and the enhanced reactivity
of these cells to vasoactive agents have been investigated. There is ample
evidence that free cytosolic calcium plays an important role in the development of tension in vascular smooth muscle. An altered cell calcium
homeostasis, which can be involved in the cellular hyperresponsiveness and
hyperproliferation, has been reported in various cell types of patients with
essential hypertension and of rats with genetic or experimental hypertension
(reviewed in [1]). Recently it has been demonstrated that agonists or hormones that elicit physiological responses within cells by modifying the internal
free calcium concentration exert their action by activating, through specific
enzymes, the metabolism of inositol-containing phospholipds [2]. These
phospholipids, also called phosphoinositides, include phosphatidylinositol,
phosphatidylinositol 4-phosphate, and phosphatidylinositol 4,5-bisphosphate
(PI-P$_2$). Thus, the metabolism of phosphoinositides, acting as a signaling
system, has been investigated in various cell types and in various models
of experimental hypertension; reviews on this topic have been already
published [3,4]. The scope of this paper is therefore to summarize and to
update the results so far obtained and to examine how such a lipid metabolism
can be involved in the development of hypertension.

The Phosphoinositide Signaling System

Occupation of the receptors that transmit the signal by increasing free Ca^{2+}
content in the cytoplasm activates a phosphoninositide-specific phospholipase
C (PLase C). Although much progress has been made recently in the

G. Bruschi A. Borghetti (Eds.)
Cellular Aspects of Hypertension
© Springer-Verlag Berlin · Heidelberg 1991

purification, characterization, and cDNA cloning of multiple isoforms of the enzyme, the specific functions/activities that can be ascribed to each of the isoenzymes are still unclear [5]. Nevertheless, a body of evidence indicates that following activation of PLase C, the first lipid to be hydrolyzed is PI-P_2. This results in the release of two compounds, namely inositol 1,4,5-trisphosphate (IP_3) and diacylglycerol (DG), with different second messenger functions [2].

IP_3 migrates in the cytoplasm and is responsible for the rapid mobilization of intracellular Ca^{2+} by activating Ca^{2+} efflux from a subpopulation of the endoplasmic reticulum. The cytoplasmic rise in Ca^{2+} further induces cellular responses through various reactions such as (a) the formation of a complex with calmodulin which activates calcium/calmodulin-dependent enzymes (such as the myosin light chain kinase), (b) the direct activation of phospholipase A_2, which results in the production of eicosanoids (some of them having been reported as endowed with second messenger functions), and (c) the activation of protein kinase C (see below).

DG is retained within the membrane, where its primary signaling role is the activation of a calcium- and phospholipid-dependent protein kinase called protein kinase C (PKC). PKC, which is the cellular receptor for tumor promoters such as phorbol esters, can provide the cell with both positive and negative feedback control mechanisms [6]. Thus the enzyme has been demonstrated to possess a large variety of substrates that include receptors, membrane proteins, contractile and cytoskeletal proteins, enzymes, etc. and to be involved in multiple functions such as hormone and neurotransmitter release, secretion, muscle contraction and relaxation, glucose transport, glycogenolysis; its activity appears also to be linked to that of Na^+/H^+ exchanger and to the modulation of ion channel activity [6,7]. PKC therefore participates in the modulation of membrane functions and in the regulation of cell growth and proliferation.

This rapid survey of IP_3 and DG functions demonstrates that the two second messenger pathways described above are interconnected. As with cell-surface receptors that utilize adenylate cyclase, there is some evidence that PLase C activity is triggered through guanine nucleotide binding proteins regarded as receptor-effector couplers.

Role of Phosphoinositide Metabolism

Cell Reactivity and Muscle Contraction

In various cell types, agonist-induced secretion involves PLase C hydrolysis of PI-P_2 as its first step and then proceeds through the biochemical events triggered by IP_3 and DG. Likewise, vascular smooth muscle contraction stimulated by angiotensin II, vasopressin, platelet derived growth factor (PDGF), α_1-adrenergic drugs, and, more recently, endothelin proceeds through a common mechanism which involves IP_3 and DG as second messengers. In this muscle, the degree of phosphorylation of the myosin light chain (due mainly to myosin light chain kinase but also, although to a lesser extent, to PKC) correlates with the degree of tension development. Phorbol esters, acting through PKC activation, induce sustained contraction of arterial strips. The fact that such contractions can be blocked by verapamil (a Ca^{2+} channel antagonist) points out the role of Ca^{2+} influx. The entry of Ca^{2+} from the extracellular space into the cytoplasm is mediated by partly voltage-dependent Ca^{2+} channels and is related to transmembrane Na^+ flux which is promoted through the activation of an amiloride-sensitive Na^+/H^+ antiport. This exchanger is partly under the control of PKC. On its own, PKC can also modulate Ca^{2+} channels. In the vascular smooth muscle, IP_3-induced Ca^{2+} mobilization occurs from the sarcoplasmic reticulum and the resulting Ca^{2+} rise is of sufficient magnitude to initiate tension development via calcium/calmodulin-mediated activation of myosin light chain kinase. The importance of phosphoinositide metabolism in vasoconstriction has also been confirmed by the observation that the elevation in cellular cyclic guanosine monophosphate – which is known to inhibit hydrolysis of phosphoinositides – appeared to be the mechanism whereby many vasodilator drugs exert their effects.

Cell Growth and Proliferation

As with various cell types, vascular smooth muscle cell proliferation results from a variety of intracellular changes that occur following receptor activation [8]. The phosphoinositide metabolism is undoubtedly involved in the control of cell proliferation [9]. One of the earliest events to occur after the binding of various mitogens to their receptors is an increase in the fluxes of Na^+, K^+, and H^+ across the plasma membrane. Thus the occupation of receptors by a mitogen, such as PDGF, stimulates a rapid influx of Na^+ into cells via an amiloride-sensitive Na^+/H^+ antiport. This increases the cell content of Na^+ and causes cytoplasmic alkalinization. Since the activity of the Na^+/K^+ pump is regulated by intracellular Na^+, there is secondary stimulation of the Na^+/K^+ pump activity, which increases K^+ and restores the electrochemical gradient for Na^+. Changes in monovalent ion fluxes are accompanied by a rapid mobilization of Ca^{2+} from intracellular stores, which leads to a transient increase in cytosolic Ca^{2+}. In addition, PDGF stimulates DG and IP_3 formation as well as phosphorylation of an endogenous 76 kDa protein, a specific substrate of PKC [9]. Moreover, the PDGF stimulation of DNA synthesis can be abolished by long-term exposure to phorbol esters, a treatment known to desensitize PKC. Thus, activation of PKC represents one of the pathways through which some growth factors can initiate cell proliferation [6,9]. In addition to its role in stimulating cell division, PKC activation can also inhibit PLase C and the consequent Ca^{2+} mobilization by negative feedback control [6]. PKC appears therefore to be important in coordinating the network of early events triggered by the receptor activation. On the other hand, growth factors such as PDGF and epidermal growth factor stimulate tyrosine phosphorylation of PLase C and appear therefore capable of modulating the inositol phospholipid signaling system. In the aforementioned example, the PDGF-induced activation of PLase C, and hence the hydrolysis of PI-P$_2$, can therefore be considered as one of the earliest biochemical components initiating the series of events which finally result in the cell physiological responses.

Phosphoinositide Metabolism in Hypertension

Impaired Ca^{2+} transport and handling, associated with enhanced free Ca^{2+} content, has been reported in various cell types (not only those of the vascular bed) of patients with essential hypertension and of hypertensive animals [1,10]. This led us to consider whether, in primary hypertension, abnormal cell membrane structure and function of genetic origin may participate in the pathogenesis of hypertension. The discovery and elucidation of roles played by the phosphoinositide metabolism-related and PKC-related biochemical events in the cell responsiveness and proliferation (see above) prompted us to investigate the possible involvement of inositol lipid metabolism in hypertension.

The first studies, which concerned the shuttle responsible for the interconversion between phosphatidylinositol and its phosphorylated derivatives, have indicated that the phosphoinositide system operated differently in hypertensives than in normotensives [11]. This prompted investigators to study phosphoinositide metabolism in activable cells, in other words to investigate in hypertensive subjects the reactivity of cells toward agonists that activate PLase C. Blood platelets have been extensively studied in this respect because (a) they are readily accessible in patients, in whom they have been demonstrated to exhibit hyperactivability (see below) and (b) they display many analogies with vascular smooth muscle cells, in that in both cell types contraction operates through similar Ca^{2+}-dependent protein machinery [12]. Both cell types have in common many receptor types, such as those for serotonin, epinephrine, prostaglandin, vasopressin, angiotensin II. Catecholamines acting through β-receptors induce a rise in cyclic AMP and relax smooth muscle; likewise, an increase in cyclic AMP diminishes further platelet activation. Both smooth muscle cells and platelets appear to behave as bidirectional control systems with regard to receptor activation processes: the signal that induces the turnover of phosphoinositides promotes the activation of cellular functions, whereas the signal that produces cyclic AMP usually antagonizes such activation. As a consequence, it is the balance between the cellular levels of cyclic AMP and Ca^{2+} which determines the state of myosin light chain phosphorylation.

In hypertensive patients and in spontaneously hypertensive rats (SHR), platelets have been shown to display an increased sensitivity towards various agonists as measured by agonist-triggered physiological responses such as shape change, adhesion/aggregation, thromboxane formation, and release reaction ([13,14] and references therein). Since platelet functions are under the control of variations in cellular Ca^{2+}, which, as detailed above, is largely dependent on phosphoinositide turnover, platelets appear to be an attractive model for studying the impaired calcium metabolism that occurs in cells from hypertensives. For example, the hyperresponsiveness of platelets to thrombin is probably the result of enhanced thrombin-induced PLase C-mediated cytosolic Ca^{2+} elevation, phosphoinositide turnover, and protein phosphorylation [14,15]. On the other hand, the intrinsic activity of PKC was found to be similar in SHR and in normotensive controls [15]. Recently, it has also been shown [16] that (a) in deoxycorticosterone acetate (DOCA)/ Na hypertensive rats, platelet physiological responses and PLase C activity were similar to those in control animals, and (b) in Dahl salt-sensitive rats fed a NaCl-rich diet, these variables, were enhanced compared to those in the animals fed a NaCl-poor diet. A similar influence of NaCl could not be observed in Dahl salt-resistant rats [16]. The platelet functional hyper-responsiveness and the PLase C hyperactivity are therefore not secondary to either the high blood pressure or the high NaCl intake. The hyperactivity of platelet PLase C is likely to be of genetic origin and can, in Dahl rats, be regulated by factors that influence the blood pressure [16]. Furthermore, since in platelets the Na^+/H^+ exchanger is dependent on PKC, it is likely that enhanced PLase C activity accounts – through an increase in PKC activity – for the enhanced Na^+/H^+ exchange activity observed in platelets of SHR and of patients with essential hypertension [17,18].

In a number of blood vessels, the enhancement of vascular muscle reactivity and of calcium sensitivity in response to hormones or neuro-transmitters has been described in established hypertension [10]. Increased activities of Plase C and/or of the Na^+/H^+ exchanger have also been reported in the vascular wall and in cultured smooth muscle cells and fibroblasts isolated from the aorta of SHR, compared to controls [19–22]. In addition to their increased responsiveness, aortic myocytes from SHR grow faster than those from Wistar-Kyoto (WKY) [21,23–25], and this cannot be ascribed to a difference in PKC activity [20]. Since PLase C is involved in the cell proliferation process [9], an impairment of phosphoinositide metabolism appears likely to participate in the abnormal growth of SHR-derived cells. However, the different responses to various stimuli (either growth factors or contractile agonists) may occur at different sites in the biochemical pathways leading to DNA synthesis and cell division, as listed in [24]. In this respect, investigations of the receptor density have shown that differences could exist between SHR and WKY, but result were not always consistent ([20,21,23,25] and references therein).

As might be expected with increased PLase C activity, the resulting increase in DG generation results in greater activation of PKC; this is in fact what can be observed in tissues of hypertensive animals (see [10] for review). However, as discussed above, the role of Ca^{2+} influx through potential-operated channels in PKC-mediated contraction in hypertension, could be accounted for by direct phosphorylation of contractile proteins. In addition, since both PLase C and the Na^+/H^+ exchanger participate in the proliferation of smooth muscle cells, one may postulate that increased activities of these biochemical systems are involved in the vascular hypertrophy/hyperplasia associated with hypertension [8]. In agreement with the assumption of a genetically determined overactivity of PLase C in SHR, the abnormal proliferating ability of vascular smooth muscle cells from SHR can also be observed in cells from young, still normotensive animals.

The cellular mechanisms underlying the altered activity of PLase C in cells from hypertensives remain to be established. An impaired PLase C activity provides a molecular basis for further studies of the mechanisms responsible of the high blood pressure. However, still at the membrane level, additional signal-transduction processes may also be involved in the cell hypersensitivity. These include the adenylate cyclase, the tyrosine kinase and other, still unidentified, signaling pathways, some of which have already been shown to be modified in primary hypertension [26–28].

Along with a decrease in the plasma membrane-associated calmodulin-dependent Ca^{2+} ATPase activity and in the sarcoplasmic reticulum-located cAMP-stimulated Ca^{2+} ATPase activity [3,10], the changes in phosphoinositide metabolism and related events obviously participate in an increased intracellular Ca^{2+} content. These impairments, which eventually lead to smooth muscle hypertrophy/hyperplasia and contractility, are likely to be responsible for the increase in the peripheral resistance observed in hypertension.

References

1. Postnov Y, Orlov SN (1985) Ion transport across plasma membrane in primary hypertension. Physiol Rev 65:904–945
2. Berridge MJ, Irvine RF (1984) Inositol trisphosphate, a novel second messenger in cellular signal transduction. Nature 312:315–321
3. Heagerty AM, Ollerenshaw JD (1987) The phosphoinositide signalling system and hypertension. J Hypertens 5:515–524
4. Marche P (1989) Membrane phosphoinositide metabolism in hypertension. News Physiol Sci 4:230–233
5. Crooke ST, Bennett CF (1989) Mammalian phosphoinositide-specific phospholipase C isoenzymes. Cell Calcium 10:309–323
6. Nishizuka Y (1986) Studies and perspectives of protein kinase C. Science 233:305–312
7. Shearman MS, Sekuguchi K, Nishizuka Y. Modulation of ion channel activity: a key function of the protein kinase C enzyme family. Pharmacol Rev 41:212–235

8. Owens GK (1989) Control of hypertrophic versus hyperplastic growth of vascular smooth muscle cells. Am J Physiol 257:H1755–H1765
9. Whitman M, Cantley L (1988) Phosphoinositide metabolism and the control of cell proliferation. Biochim Biophys Acta 948:327–344
10. Sharma RV and Bhalla RC (1988) Calcium and abnormal reactivity of vascular smooth muscle in hypertension. Cell Calcium 9:267–274
11. Remmal A, Koutouzov S, Girard A, Meyer P, Marche P (1988) Defective phosphoinositide metabolism in primary hypertension. Experientia 44:133–137
12. Erne P, Resink TJ, Bürgin M, Bürgisser E, Bühler FR (1985) Platelets and hypertension. J Cardiovasc Pharmacol 7 [Suppl 6]:S103–S108
13. De Clerck F (1988) Review: blood platelets in human essential hypertension. Agents Actions 18:563–580
14. Marche P, Koutouzov S, Girard A, Barbier P, Meyer P (1989) Hyperresponsiveness of platelet phospholipase C in essential hypertension. J Vasc Med Biol 1:137–141
15. Koutouzov S, Limon I, Marche P (1990) Receptor-dependent and independent protein phosphorylation in platelets- of spontaneously hypertensive rats. Thromb Res 59:475–487
16. Limon I, Blanc J, Koutouzov S, Knorr A, Meyer P, Marche P (1990) Platelet phospholipase C activity in salt-dependent hypertension. Hypertension 15:381–387
17. Ek TP, Deth RC (1988) Elevated phospholipase C and Na^+/H^+ exchange activity in spontaneously hypertensive rats. Hypertension 12:331–332
18. Livne A, Balfe JW, Veitch R, Marquez-Julio A, Grinstein S, Rothstein A (1987) Increased platelet Na^+/H^+ exchange rates in essential hypertension: application of a novel test. Lancet i:533–536
19. Ueahra Y, Ishii M, Ishimitsu T, Sugimoto T (1988) Enhanced phospholipase C activity in the vascular wall of spontaneously hypertensive rats. Hypertension 11:515–524
20. Resink TJ, Scott-Burden T, Baur U, Bürgin M, Bühler FR (1989) Enhanced responsiveness to angiotensin II in vascular smooth muscle cells from spontaneously hypertensive rats is not associated with alterations in protein kinase C. Hypertension 14:283–303
21. Paquet JL, Baudouin-Legros M, Brunelle G, Meyer P (1990) Angiotensin II-induced proliferation of aortic myocytes in SHR. J Hypertens 8:565–572
22. Zhu DL, Durant S, Marche P (1990) Phospholipase C activity in cultured aortic fibrobalsts of hypertensive and normotensive rats. J Vasc Med Biol 2:26–31
23. Berk BC, Muslin AJ, Gordon HM, Canessa M, Alexander RW (1989) Spontaneously hypertensive rat vascular smooth muscle cells in culture exhibit increased growth and Na^+/H^+ exchange. J Clin Invest 83:822–829
24. Hadrava V, Tremblay J, Hamet P (1989) Abnormalities in growth characteristics of aortic smooth muscle cells in spontaneously hypertensive rats. Hypertension 13:589–597
25. Scott-Burden T, Resink TJ, Baur U, Bürgin M, Bühler FR (1989) Epidermal growth factor responsiveness in smooth muscle cells from hypertensive and normotensive rats. Hypertension 13:295–304
26. Remmal A, Koutouzov S, Marche P (1988) Enhanced turnover of phosphatidylcholine in platelets of hypertensive rats. Possible involvement of a phosphatidylcholine-specific phospholipase C. Biochim Biophys Acta 960:236–244
27. Hamet P, Tremblay J (1989) Abnormalities of second messenger systems in hypertension. In: Meyer P, Marche P (eds) Blood cells and arteries in hypertension and atherosclerosis. Raven, New York, pp 171–187
28. Marche P, Limon I, Blanc J, Girard A (1990) Platelet phosphatidylcholine turnover in experimental hypertension. Hypertension 16:190–193

Lymphocyte Membrane Abnormalities in Hypertension

P. B. Furspan and D. F. Bohr

7706 Medical Sciences II, Department of Physiology, University of Michigan, Ann Arbor,

During the past 15 years extensive research in experimental and clinical hypertension has indicated that there are cell membrane abnormalities associated with chronically elevated arterial pressure. Many aspects appear to be altered: cation concentration, cation permeability, active, co-, and countertransport, and calcium binding are some processes that have been examined. These abnormalities have been reported in a number of cell types, e.g., erythrocytes, leukocytes, lymphocytes, platelets, cardiocytes, nerve cells, and hepatocytes, leading to the conclusion that this is a generalized phenomenon in hypertension. Of course, it is recognized that an abnormality in the membrane of these cells may not play a role in the pathogenesis of an elevated arterial pressure. The membranes of these cells are studied because of their possible use as markers for the hypertensive process and because they may reflect similar abnormalities in vascular smooth muscle [1] and regulatory centers in the central nervous system [2] that do influence blood pressure.

Research utilizing the lymphocyte has contributed numerous and important insights into the relationship between a generalized cell membrane abnormality and hypertension. Our own research is supportive of the theory that the primary defect responsible for the membrane abnormalities associated with hypertension is a lack of membrane stability that results from altered calcium binding to the cell membrane surface. At the level of the vascular smooth muscle cell, the generalized increase in membrane permeability caused by decreased binding of calcium may well be responsible for an increase in vascular smooth muscle sensitivity that causes the increase in vascular resistance of hypertension. In this review we will summarize our work and that of others using lymphocytes and other cell types in relation to this hypothesis.

G. Bruschi A. Borghetti (Eds.)
Cellular Aspects of Hypertension
© Springer-Verlag Berlin · Heidelberg 1991

Cation Contents and Fluxes

One of the most consistent findings in relation to lymphocyte membrane abnormalities associated with essential hypertension has been an increased intracellular sodium concentration (for review, see [3]). In 1985, we reported that increased intralymphocytic sodium concentration is also a characteristic of several forms of experimental hypertension: genetic, +15% (stroke-prone substrain of the spontaneously hypertensive rat, SHRSP); mineralocorticoid, +23% (deoxycorticosterone acetate-treated, DOCA); and renal, +30% (one kidney, one clip, 1K-1C). As was found for essential hypertension [4,5], potassium concentration was not significantly altered in any of these forms of experimental hypertension [6]. In this study, we were also interested in whether cation permeability is altered in these forms of experimental hypertension. We measured net sodium influx and net potassium efflux at 4°C indexed to dry cell weight. Lymphocytes from SHRSP and DOCA-treated rats exhibited similarly elevated net sodium fluxes as compared with those in their controls (43% and 35%, respectively). Net potassium flux in lymphocytes from SHRSP was also elevated compared with that in controls (70%). Sodium and potassium fluxes in lymphocytes from control and 2K-1C rats did not differ significantly. An increased sodium and potassium permeability has also been reported for red blood cells [7,8] and vascular smooth muscle from SHR and DOCA-treated rats [9,10]. Friedman and Friedman [10], however, did not observe an elevated sodium permeability in the tail artery from rats with established 2K-1C renal hypertension.

Establishment of an association between a particular abnormality, membrane or otherwise, is not evidence of a causal relationship. Accordingly, we next conducted a study to determine whether these lymphocyte traits (intracellular sodium and potassium contents and net sodium and potassium fluxes) in SHRSP were under the control of the same genetic mechanism that regulates blood pressure [11]. In order to define the relationship between these abnormal membrane traits and arterial pressure elevation, their distributions were studies in F_1, F_2, and backcross progeny of the cross of SHRSP and Wistar-Kyoto rats (WKY). In the F_1 generation, arterial pressure and potassium efflux values resembled those of the WKY parents, suggesting that these traits are dominant in the normotensive rat; sodium influx values resembled those of the SHRSP parents. There was a moderate, but not statistically significant correlation ($r = 0.45$) between blood pressure and sodium influx in the F_2 generation. There was a high and statistically significant correlation between blood pressure and potassium efflux ($r = 0.86$) in the F_2, suggesting that these two traits may be under the control of a common genetic mechanism. We hypothesized that the altered potassium efflux demonstrated in lymphocytes from SHRSP did not represent a primary defect of the cell membrane but rather was related to the effect of an elevated intracellular concentration of calcium on calcium-activated potassium channels. This hypothesis was based on our observation that the intra-

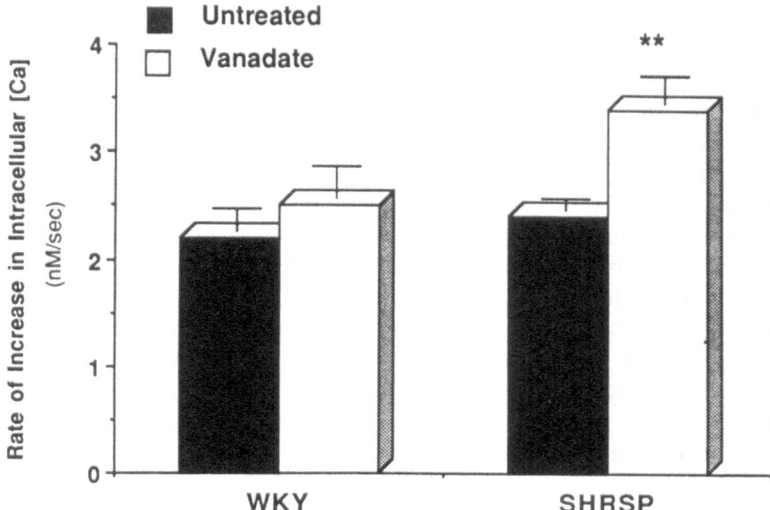

Fig. 1. Rate of increase in intracellular calcium concentration in untreated and vanadate-treated lymphocytes after $1\,mM$ extracellular calcium was added back to calcium-free medium. The rate of increase was significantly greater in the vanadate-treated cells from SHRSP than in those WKY ($p < 0.05$, t test, $n = 3$ for each group)

cellular free calcium concentration in lymphocytes from SHRSP is 50% higher than in those of WKY [12]. The increased intracellular free calcium concentration may be a consequence of an increased membrane permeability to calcium. We used the intracellular fluorescent dye fura 2 to determine whether lymphocyte membranes from SHRSP were leakier to calcium than those from WKY rats [13]. In lymphocytes treated with vanadate (a relatively specific inhibitor of the calcium pump) and incubated in a calcium-free medium (to reduce intracellular calcium concentration), calcium concentration increased at a faster rate in lymphocytes from SHRSP than those from WKY rats when calcium was added back to the medium (see Fig. 1). Bruschi et al. reported similar results using quin 2 and lymphocytes from SHR and WKY rats [14]. Rapp performed a genetic analysis of vascular smooth muscle contraction to cobalt and found that an increased sensitivity to cobalt segregates with an increment in blood pressure [15]. He suggests that increased sensitivity to cobalt may result from the same abnormality that allows calcium to leak into the cell and increase vascular tone.

If increased membrane permeability to calcium was the primary defect associated with hypertension and was responsible for the increased intracellular free calcium concentration and, consequently, the elevated net potassium efflux, it followed that reducing membrane permeability to calcium should have an ameliorating effect on all these factors. When we treated SHRSP and WKY rats with the dihydopyridine calcium channel antagonist felodipine in an attempt to reduce membrane permeability to calcium, we found that blood pressure, intralymphocytic free calcium concentration, and net potassium efflux were reduced to normal or near-normal levels in SHRSP [16].

In a parallel study, Bruner and Webb [17] determined the effect of felodipine on oscillatory contractions in response to norepinephrine in isolated tail arteries from the same animals. Previous studies had indicated that the mechanism for these oscillations involves altered membrane calcium and/or potassium handling and that this vascular change is genetically associated with hypertension in SHRSP [18–20]. Oscillatory activity was markedly reduced in tail arteries from felodipine-treated SHRSP compared with control SHRSP. Felodipine also inhibited oscillatory activity when added directly to bath. The authors postulated that felodipine was acting to lower blood pressure, at least in part, by correcting the genetic defect responsible for oscillatory activity.

Calcium-Activated Potassium Channels

Using the patch clamp technique, we were able to identify and characterize a calcium-activated potassium channel in the plasma membrane of lymphocytes from SHRSP and WKY rats [12]. Potassium currents were recorded that had a slope conductance of 18.1 ± 1.49, and 18.5 ± 1.44, in WKY rat and SHRSP lymphocytes, respectively. Calcium sensitivity of the channels was similar; maximum activation occurred at $700\,nM$ free calcium concentration. Ionomycin, a calcium ionophore, caused a concentration-dependent and proportional increase in net potassium efflux and intracellular free calcium concentration in lymphocytes from both strains of rat. Based on the relationship between net potassium efflux and intracellular free calcium concentration established with ionomycin, and the relationship between potassium channel activity and free calcium concentration established with the patch clamp, the resting net potassium efflux of lymphocytes from SHRSP is greater than would be predicted on the basis of the resting intracellular free calcium concentration. We speculated that the difference was related to the membrane-stabilizing effect of external calcium, i.e., the increased membrane permeability to potassium is caused by a reduced calcium binding to the plasma membrane. The generalized increase in membrane permeability caused by less calcium being bound to the membrane could not only account for the increased net potassium efflux, but would explain the increased net sodium efflux and calcium influx as well.

Membrane Stabilization and Cation Permeability

Decreased calcium binding has been described for the erythocyte membrane in SHR and/or humans with essential hypertension [21,22]. Devynck et al. [23] have also reported fewer calcium binding sites on plasma membranes of heart, nerve, and liver cells from SHR compared to those from WKY. Kwan

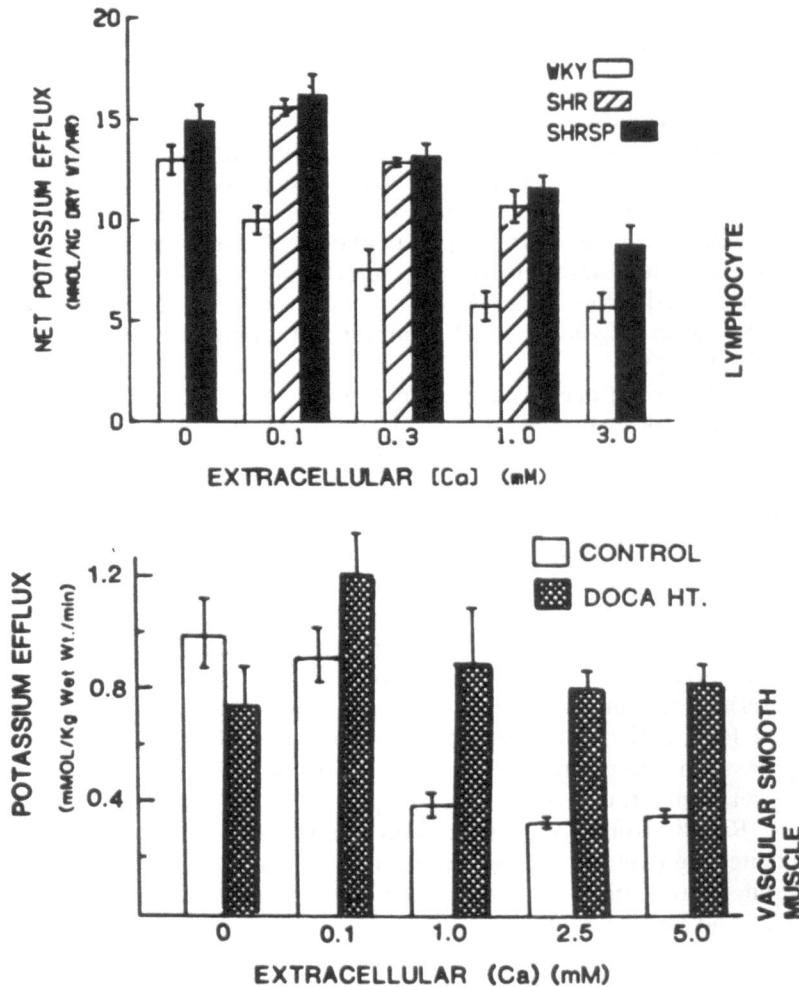

Fig. 2. *Upper graph*: Effects of varying external calcium concentration on net potassium efflux in lymphocytes from WKY, SHR, SHRSP. Number of rats used: WKY, 7; SHR, 4; SHRSP, 6. Bars indicate standard error of the mean. (Reproduced from [31]). *Lower graph*: Similar study of the effect of varying external calcium concentration on potassium efflux in aortic cells from normotensive and DOCA-hypertensive rats. (Plotted from data presented by Jones and Hart [32])

et al. [24] found that the membranes of mesenteric artery cells from SHR bind less calcium than do those from WKY.

At a molecular level, Kowarski et al. [25] found significant reductions of an "integral membrane calcium-binding protein" in various tissues of the SHR as compared to those from WKY. The affinity of this protein for calcium, however, was so high that it would be completely saturated at physiological concentrations of extracellular calcium.

The effect of extracellular calcium on membrane stability may result from the interaction of this cation with the lipid bilayer rather than with a specific membrane protein. Recent reviews have summarized an expanding body of evidence which indicates that the function of membrane proteins can be affected by the lipid composition and state, e.g., fluidity, of the lipid bilayer [26,27]. In a recent study, Bialecki and Tulenko reported an elevated influx of calcium under basal and norepinephrine-stimulated conditions in rabbit carotid arteries whose cell membranes had been enriched with cholesterol in vitro [28]. Tulenko et al. have also observed a similar effect of cholesterol-enrichment on rubidium efflux from rabbit carotid arteries [29]. Wei et al. have reported an alteration in the acyl group composition of synaptosomal membrane phospholipids in SHR compared to WKY rats [30].

Evidence of decreased calcium binding to the membrane of lymphocytes from SHRSP was provided by a study in which we examined the effect of various external calcium concentrations on net potassium efflux and net sodium influx [31]. Net potassium efflux was greater in lymphocytes from SHRSP than in those from WKY at external calcium concentrations of 0.1, 0.3, 1.0, and 3.0 mM, but not at 0 mM (Fig. 2, upper graph). The slightly enhanced potassium efflux from the cells at 0.1 compared to that at 0 mM calcium in the hypertensive tissues may reflect calcium entry into the cell and activation of calcium-activated potassium channels. Net sodium influx in lymphocytes from SHRSP was greater than in those from WKY at all external calcium concentrations tested (0, 0.1, 1.0, and 3.0 mM). Jones and Hart [32] noted a similar effect of external calcium concentration on potassium efflux from vascular smooth muscle from normotensive and hypertensive rats (Fig. 2, lower graph).

Results from a recent study in our laboratory suggest that even small changes in external calcium concentration may have an effect on cell membrane permeability. We examined the effect of dietary calcium (normal, 1%, and high, 2.5%) on blood pressure and lymphocyte membrane characteristics in SHRSP and WKY rats. Increasing dietary intake of calcium from 1% to 2.5% led to a decrease in blood pressure, intralymphocytic free calcium concentration, and net potassium influx in SHRSP but not WKY rats [33]. These changes were associated with a small but significant increase (0.24 mmol/l) in serum ionized calcium. We offer as a speculative hypothesis that this small increase in extracellular calcium concentration over the time course of the experiment (10 weeks) may have an effect on membrane stability equivalent to an acute, but larger increase in extracellular calcium concentration, i.e., increased membrane-bound calcium and reduced membrane cation permeability. In a recent study, Dominiczak et al. [34] have reported that the hypercalcemia (26% higher than control group) of patients with hyperparathyroidism is associated with a significantly lower platelet cytosolic free calcium concentration (93 vs 81.5 nmol/l). After parathyroidectomy and the elimination of hypercalcemia, platelet cytosolic free calcium concentration tended to rise. These results are consistent with our hypothesis that small changes in extracellular calcium over long periods of time may affect membrane stability and, consequently, membrane permeability.

An ability of small changes in serum calcium concentration to affect blood pressure and cell membrane characteristics is consistent with and supportive of the theory that variations in the intake of dietary calcium may play a role in the pathogenesis and maintenance of hypertension [35–38]. It remains to be demonstrated whether small changes in external calcium concentration occurring over prolonged periods of time can affect cell membrane characteristics in an in vitro system, e.g., cell culture.

Although we have primarily discussed changes in passive transport properties as secondary consequences of altered calcium-binding capacity of cell membranes in hypertension, abnormalities of other transport systems may also be secondary to this defect or its consequence, e.g., compensatory adjustments in active transport systems for increased cell membrane permeability. The results of a recent study in our laboratory provide evidence of just such a possibility [13]. When lymphocytes loaded with fura 2 are placed in a calcium-free medium, the intracellular calcium concentration of lymphocytes from SHRSP decreases at a faster rate than that of those from WKY rats. We interpret this result to indicate that the calcium extrusion pump (Ca-ATPase) of lymphocytes from SHRSP exhibits greater activity than that of those from WKY rats. This enhanced activity may be a compensatory development in response to elevated calcium permeability, which, in turn, is the result of decreased membrane binding of calcium.

Conclusion

The lymphocyte has proven to be a useful and reliable cell with which to study altered membrane properties associated with hypertension. The relevance of the data obtained with this cell type to the understanding of pathogenesis of hypertension is born out by parallel findings in other cell types, including vascular smooth muscle. We believe the results of our research can best be explained by and are supportive of the hypothesis that, in hypertension, there is a decrease in the amount of calcium bound to the cell membrane. In the lymphocyte, the resulting less "stabile," i.e., more permeable, cell membrane leads to increased sodium, potassium and calcium fluxes and increased intracellular sodium and calcium concentrations. At the level of the vascular smooth muscle, these alterations of ion permeabilities lead to an increase in vascular smooth muscle sensitivity that causes the increase in vascular resistance and consequent elevation in blood pressure of hypertension.

References

1. Jones AW (1981) Kinetics of active sodium transport in aortas from control and deoxycorticosterone hypertensive rats. Hypertension 3:631–640
2. Bohr DF (1981) What makes the pressure go up? A hypothesis. Hypertension 3 [Suppl II]:II160–II165
3. Ambrosioni E, Costa FV, Borghi C, Baschi S, Mussi A (1986) Cellular and humoral factors in borderline hypertension. J Cardiovasc Pharmacol 8 [Suppl 5]:S15–S22
4. Edmondson RPS, Thomas RD, Hilton PS, Patrick J, Jones NF (1975) Abnormal leucocyte composition and sodium transport in essential hypertension. Lancet 2: 1003–1005
5. Araoye MA, Khatri IM, Yao LL, Freis ED (1978) Leucocyte intracellular cations in hypertension: effect of antihypertensive drugs. Am Heart J 96:731–738
6. Furspan PB, Bohr DF (1985) Lymphocyte abnormalities in three types of hypertension in the rat. Hypertension 7:860–866
7. Harris AL, Guthe CC, van 't Veer F, Bohr DF (1984) Temperature dependence and bidirectional cation fluxes in red blood cells from spontaneously hypertensive rats. Hypertension 6:42–48
8. Friedman SM, Nakashima M, McIndoe RA (1977) Glass electrode measurements of net Na^+ and K^+ fluxes in erythrocytes of the spontaneously hypertensive rat. Can J Physiol Pharmacol 55:1302–1310
9. Jones AW (1981) Kinetics of active sodium transport in aortas from control and deoxycorticosterone hypertensive rats. Hypertension 3:631–640
10. Friedman SM, Friedman CL (1976) Cell permeability, sodium transport, and the hypertensive process in the rat. Circ Res 39:433–441
11. Furspan PB Jokelainen PT, Sing CF, Bohr DF (1987) Genetic relationship between a lymphocyte membrane abnormality and blood pressure in spontaneously hypertensive stroke prone and Wistar-Kyoto rats. J Hypertens 5:293–297
12. Furspan PB, Bohr DF (1990) Calcium sensitivity of a Ca^{2+}-activated K^+ channel in SHRSP. Hypertension 15 [Suppl I]:I97–I101
13. Furspan PB, Bohr DF (1987) Membrane permeability to calcium and calcium pump activity of thymocytes from SHRSP and WKY rats. Fed Proc 46:522A
14. Bruschi G, Bruschi ME, Cavatorta A, Borghetti A (1986) The mechanism of Ca increase in blood cells of spontaneously hypertensive rats. J Cardiovasc Pharmacol 8 [Suppl 8]:S139–S144
15. Rapp JP (1982) A genetic locus (Hyp-2) controlling vascular smooth muscle response in spontaneously hypertensive rats. Hypertension 4:459–467
16. Furspan PB, Bohr DF (1988) Effect of felodipine on blood pressure and lymphocyte membrane characteristics in spontaneously hypertensive stroke-prone rats. J Hypertens 6:S236–S238
17. Bruner CA, Webb RC (1989) Effect of felodipine on blood pressure and vascular reactivity in stroke-prone spontaneously hypertensive rats. J Hypertens 7:31–35
18. Myers JH, Lamb FS, Webb RC (1985) Norepinephrine-induced phasic activity in tail arteries from genetically hypertensive rats. Am J Physiol 248:H419–H423
19. Lamb FS, Myers JH, Hamlin MN, Webb RC (1985) Oscillatory contractions in tail arteries from genetically hypertensive rats. Hypertension 7 [Suppl I]:I25–I30

20. Bruner CA, Myers JH, Sing CF, Jokelainen PT, Webb RC (1986) Genetic association of hypertension and vascular changes in stroke-prone spontaneously hypertensive rats. Hypertension 8:904–910

21. Devynck M-A, Pernollet M-G, Nunez A-M, Meyer P (1981) Analysis of calcium handling in erythrocyte membranes of genetically hypertensive rats. Hypertension 3:397–403

22. Postnov YV, Orlov SN, Pokudin NJ (1980) Decrease of calcium binding by red blood cell membrane in spontaneously hypertensive rats and in essential hypertension. Pflugers Arch 385:191–195

23. Devynck M-A, Pernollet M-G, Nunez A-M, Aragon I, Montenay-Garestier T, Helene C, Meyer P (1982) Diffuse structural alterations in cell membranes of spontaneously hypertensive rats. Proc Natl Acad Sci USA 79:5057–5060

24. Kwan CY, Belbeck L, Daniel EE (1979) Abnormal biochemistry of vascular smooth muscle plasma membrane as an important factor in the initiation and maintenance of hypertension in rats. Blood Vessels 16:259–268

25. Kowarski S, Cowen LA, Schachter D (1986) Decreased content of integral membrane calcium calcium-binding protein (IMCAL) in tissues of the spontaneously hypertensive rat. Proc Natl Acad Sci USA 83:1097–110

26. Carruthers A, Melchior DL (1989) How bilayer lipids affect membrane protein activity. TIBS 11:331–335

27. Yeagle PL (1989) Lipid regulation of cell membrane structure and function. FASEB J 3:1833–1842

28. Bialecki RA, Tulenko TN (1989) Excess membrane cholesterol alters calcium channels in arterial smooth muscle. Am J Physiol 257:C306–C314

29. Tulenko TN, Bialecki R, Gleason M, D'Angelo G (1990) Ion channels, membrane lipids and cholesterol: a role for membrane lipid domains in arterial function. In: TJ, Colatsky (ed) Potassium channels: basic function and therapeutic aspects. Liss, New York, pp 187–203

30. Wei JW, Yang LM, Sun SH, Chiang CL (1987) Phospholipids and fatty acid profile of brain synaptosomal membrane from normotensive and hypertensive rats. Int J Biochem 19:1225–1228

31. Furspan PB, Bohr DF (1986) Calcium related abnormalities in lymphocytes from genetically hypertensive rats. Hypertension 8 [Suppl II]:II123–II126

32. Jones AW, Hart RG (1975) Altered ion transport in aortic smooth muscle during doexycorticosterone acetate hypertension in the rat. Circ Res 37:333–341

33. Furspan PB, Rinaldi G, Hoffman K, Bohr DF (1989) Effect of dietary calcium on a cell membrane abnormality in genetic hypertension. Hypertension 13:727–730

34. Dominiczak AF, Lyall F, Morton JJ, Dargie HJ, Boyle IT, Tune TT, Murray G, Semple PF (1990) Blood pressure, left ventricular mass and intracellular calcium in primary hyperparathyroidism. Clin Sci 78:127–132

35. Bukoski R, Lucas P, Drüeke T, McCarron D (1986) Theoretical mechanisms of dietary calcium's antihypertensive action. Adv Exp Med Biol 208:389–396

36. Walczyk MH, McCarron DA (1987) Calcium and hypertension. Compr Ther 13:10–16

37. McCarron DA, Morris CD (1985) Blood pressure response to oral calcium supplementation in mild to moderate hypertension. Ann Intern Med 103:825–831

38. Ayachi S (1979) Increase dietary calcium lowers blood pressure in the spontaneously hypertensive rat. Metabolism 12:1234–1238

Structural and Functional Alterations of Platelet Membrane in Essential Hypertension

K.-H. Le Quan Sang, C. Astarie, M. Mazeaud, and M.-A. Devynck

Department de Pharmacologie, CNRS I 61670, Faculté de Medicine Necker –
Enfants Malades, 156, rue de Vaugirard, F-75015 Paris, France

Introduction

Several functions of cell membrane have been reported to be altered in
essential and/or experimental hypertension. These include transport systems
for various ions (Na^+, K^+, Ca^{2+}, H^+, Cl^-), enzymatic activities, and
characteristics of receptors for hormones or neurotransmitters (for review,
see [1,2]). These widespread alterations may reflect the adjustment of cell
metabolism to high blood pressure, but they may also result either from the
interdependent regulatory mechanisms adapting the multiple membrane
functions to the cell activity or from an underlying alteration of cell membrane structure.

This last hypothesis has already been assessed in a few studies using
electron spin resonance (ESR) or fluorescence depolarization of probes
inserted into the membrane bilayer, in which the information obtained
depends on the localization of the probe in the membrane. Regarding
essential hypertension, the two studies performed on erythrocyte membrane
reported increases in the rate of pyrene lateral diffusion [3] and in values of
outer hyperfine splitting and order parameter of the ESR spectra for 5-
nitroxystearate [4], both indicating enhanced membrane microviscosity. In a
preliminary study performed on blood platelets from only four hypertensive
patients. Naftilan et al. observed that the fluorescence polarization of
diphenylhexatriene (DPH) was enhanced at temperatures higher than 20°C,
also suggesting an elevated local microviscosity [5]. In contrast, no changes
were observed in patients with secondary hypertension [3,4].

In order to investigate further the membrane structural alterations
possibly associated with essential hypertension, the steady-state fluorescence
depolarization of the DPH trimethyl amino derivative (TMA-DPH) was
studied in blood platelets. In contrast to DPH, which is embedded in the
lipid core of cell membranes and may have access not only to plasma
membrane but to all intracellular membranes, TMA-DPH, in intact cells,
is specifically incorporated into the plasma membrane at the water-lipid

G. Bruschi A. Borghetti (Eds.)
Cellular Aspects of Hypertension
© Springer-Verlag Berlin · Heidelberg 1991

interface, presumably anchored to the phospholipid polar heads by its positively charged group [6]. Since the activities of various membrane proteins, including ion transport systems, have been reported to be modulated by their lipid environment [7], the local mobility of TMA-DPH, estimated by its steady-state fluorescence anisotropy, was measured in parallel with two platelet variables, the cytosolic free Ca^{2+} concentration and the intracellular pH, taken as global indices of membrane ion transport activities. In order to avoid the changes in membrane properties associated with platelet activation or aggregation, these measurements were performed on unstimulated platelets ex vivo.

Materials and Methods

Twenty-four subjects were included in this study: nine healthy normotensive subjects free of any medication and 15 patients with mild to moderate essential hypertension without associated disease. Their characteristics are given in Table 1. Hypertension was diagnosed from supine diastolic blood pressure (Korotkoff phase V). Five hypertensive patients had never received treatment and the others had their antihypertensive treatment interrupted for at least 5 weeks before blood sampling.

Forty milliliters of venous blood were collected in tubes containing 2.73% citric acid, 4.48% trisodic citrate, and 2% glucose as anticoagulant (1/10 vol.). The platelet-rich plasma (PRP) obtained by centrifugation at 530g max for 5 min at 20°C was divided into four parts.

For membrane fluidity studies, platelets were diluted five times with a medium containing (mmol/l) NaCl 145, KCl 5, $Ca(NO_3)_2$, $MgSO_4$ 1,

Table 1. Biological characteristics and platelet variables in normotensive and hypertensive subjects (mean ± SEM)

	Normotensive subjects ($n = 9$)	Hypertensive patients ($n = 15$)	
Male/female	5/4	9/6	
Age (years)	37.3 ± 4.6	45.2 ± 3.2	ns
Systolic blood pressure (mmHg)	124 ± 4	159 ± 4	$p < 0.001$
Diastolic blood pressure (mmHg)	74 ± 3	100 ± 3	$p < 0.001$
Platelet TMA-DPH anisotropy	0.288 ± 0.003	0.275 ± 0.003	$p < 0.01$
Platelet cytosolic Ca^{2+} (nmol/l)	164 ± 11	190 ± 10	ns
Platelet cytosolic pH[a]	7.19 ± 0.03	7.37 ± 0.05	$p < 0.02$

Differences between normotensive subjects and hypertensive patients were calculated with two-tailed Student's t tests.

[a] Platelet cytosolic pH was determined in only 7 normotensive subjects and 12 hypertensive patients.

Na_2HPO_4 0.5, glucose 5, and HEPES 10, pH 7.4 at 37°C, then centrifuged at 270g for 15 min at 20°C and resuspended in the same medium for fluorescence polarization measurements. The platelets were adjusted at a concentration of $5 \pm 1\ 10^7$ cells/ml ($n = 7$) by Rayleigh signal, then labelled with 1 μmol/l TMA-DPH (Molecular Probes) with incubation time of 10 min at 37°C. Under these conditions, aggregation induced by ADP was observed not to be significantly altered by TMA-DPH incorporation.

The fluorescence measurements were performed in quartz suprasil microcuvettes (5 × 5 mm), thermostated at 37°C, with a Perkin-Elmer LS–5B spectrofluorometer. Excitation and emission wavelengths were fixed at 350 and 430 nm respectively, both with 5 nm bandwidth. Fluorescence intensities were corrected by the signals from unlabelled cells. Under vertically polarized excitation light the fluorescence anisotropy is defined as:

$$r = \frac{(1_{//})_v - \theta\,(1_\perp)_v}{(1_{//})_v + 2\theta\,(1_\perp)_v} \text{ with a correcting factor } \theta = \frac{(1_{//})_H}{(1_\perp)_H}$$

where the subscripts V and H indicate vertical or horizontal polarization of the excitation light and the subscripts // and $\perp$ refer to components of the emission light parallel and perpendicular to the direction of polarization of the excitation beam.

Cytosolic free Ca^{2+} concentration was determined with the fluorescent indicator quin 2 and intracellular pH was quantified with the fluorescent indicator 2′,7′-bis(carboxyethyl)-5(6)-carboxyfluorescein (BCECF) as previously reported [8,9].

Results

The interassay and intra-assay variabilities for the fluorescence anisotropies of TMA-DPH embedded in blood platelets were 1% ($n = 4$) and 0.5% $n = 9$), respectively.

The fluorescence anisotropy of TMA-DPH was significantly decreased in hypertensive patients in comparison to normotensive subjects (Table 1 and Fig. 1). This indicates that, in hypertension, plasma membrane has a lowered

Table 2. Correlations between steady-state anisotropy of TMA-DPH embedded in platelet membrane and biological variables

	n	Correlation coefficient	Significance
Sex	24	0.010	ns
Age	24	−0.418	ns
Systolic blood pressure	24	−0.584	$p = 0.003$
Diastolic blood pressure	24	−0.431	$p = 0.033$
Platelet intracellular pH	18	−0.670	$p = 0.003$

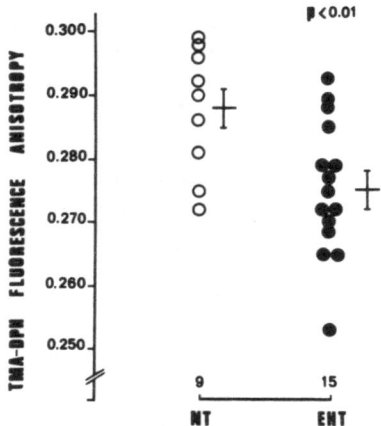

Fig. 1. Individual values of the steady-state fluorescence anisotropy of TMA-DPH embedded in platelet membrane from normotensive subjects (*open circles*) and hypertensive patients (*closed circles*). The significance of the difference was calculated using Student's *t* test

microviscosity at the water-lipid interface. Fluorescence anisotropy of TMA-DPH correlated inversely with systolic and diastolic blood pressures but did not appear to vary with sex or age of the donor (Table 2).

In agreement with previous results [9], cytosolic pH was observed to be significantly higher on average in hypertensive patients than in normotensive subjects (Table 1). The cytosolic Ca^{2+} concentrations also tended to be higher in the hypertensive patients than in the normotensive group (Table 1). The individual values of platelet TMA-DPH fluorescence anisotropy correlated inversely with the corresponding cytosolic pH values (Table 2), indicating that the platelets with decreased TMA-DPH anisotropy were also characterized by raised cytosolic pH. This relationship remained very significant at constant blood pressure or age ($r' = -0.750$, $p < 0.001$ and $r' = 0.772$, $p < 0.001$, respectively).

Discussion

Blood platelets from hypertensive patients are characterized by various functional alterations, including enhanced sensitivity to various aggregating agents such as thrombin, epinephrine, or adenosine diphosphate [2], low serotonin content and uptake [10], increased cytosolic concentration of Ca^{2+} ions [8,11], increased activity of the Na^+/H^+ antiport [12,13], and cytosolic alkalinization [9]. Some of these modifications have been observed to be associated in vivo with plasma cholesterol concentration or can be induced in vitro by changes in membranes lipid composition [14–17]. A pathological role of changes in cell membrane lipid composition, resulting in an altered membrane fluidity, may therefore be suspected [18,19].

In rats with spontaneous hypertension, the microviscosity of plasma membrane has indeed been described as increased in cells from various tissues both in the hypertensive and prehypertensive states, and similar findings were observed in the Sabra hypertension-prone rats compared to the original Sabra strain [18,20]. In addition, salt loading, which is known to further raise blood pressure, also increased the DPH polarization values [21]. However, an increase in the plasma membrane microviscosity cannot be considered as a general feature of high blood pressure. This is indicated by the decreased microviscosity observed in the prehypertensive SHR in intestinal brush border membranes and by the absence of increased DPH fluorescence polarization in erythrocyte and platelet membranes from female rats with genetic hypertension [20,21].

The present observation of decreased TMA-DPH anisotropy in blood platelets from hypertensive patients suggests that structural membrane abnormalities associated with high blood pressure are located mainly in the plasma membrane. As platelet activation and shape change have been reported to induce a rise in TMA-DPH anisotropy [22], the lowered values observed in hypertensive patients are not likely to result from such an in vivo or ex vivo activation.

The precise roles of these structural changes in the TMA-DPH local environment are unknown, but it is interesting to note that the platelets with the highest intracellular pH were those with the lowest TMA-DPH anisotropy, indicating that these structural changes may be physiologically relevant. However, the changes in TMA-DPH anisotropy only averaged 4% and the physiological relevance of such a small change may be questioned. In a study of the lipid fluidity of rat colonic brush border membrane, Brasitus et al. have demonstrated that an increase in fluidity of 5–6% was associated with a 36% increase in the maximal velocity of the amiloride-sensitive sodium-stimulated proton efflux, which is far from negligeable [23]. The rise in the Na^+/H^+ exchange activity observed in platelets from hypertensive patients [12,13] may thus result from an alteration of membrane structure similar to that reported in the present study. A rise in Na^+/H^+ exchange maximal velocity may also participate in the cytosolic alkalinization observed in these patients.

The activity of the Na^+/Li^+ countertransport, suggested to partly reflect that of the Na^+/H^+ exchange, has also been reported to vary with plasma lipids and membrane fatty acids [24]. The altered phase transition temperature of erythrocyte Na^+/Li^+ countertransport in hypertensive patients and their relatives [25] also concords with the assumption of a cell membrane lipid disturbance in essential hypertension.

Another finding of the present study is that the higher the blood pressure, the more evident the decrease of TMA-DPH anisotropy. In order to establish the pathogenetic importance of this membrane alteration, it is crucial to verify whether or not it only results from high blood pressure itself, and if it is also present in endothelial and vascular smooth muscle cells.

References

1. Ives HE (1989) Ion transport defects and hypertension. Where is the link? Hypertension 14:590–597
2. De Clerk F (1986) Blood platelets in human essential hypertension. Agents Actions 18:563–580
3. Orlov SN, Postnov YV (1982) Ca^{2+} binding and membrane fluidity in essential and renal hypertension. Clin Sci 63:281–284
4. Masuyama Y, Tsuda K, Shima H, Ura M, Takeda J, Kimura K, Nishio I (1988) Membrane abnormality of erythrocytes is highly dependent on salt intake and renin profile in essential hypertension: an electron spin resonance study. J Hypertens 6 [Suppl 4]: S267–S268
5. Naftilan AJ, Dzau VJ, Loscalzo J (1986) Preliminary observations on abnormalities of membrane structure and function in essential hypertension. Hypertension 8: 11174–11179
6. Prendergast FG, Haugland RP, Callahan PJ (1981) 1-[4-(Trimethylamino)phenyl]-6-phenylhexa-1,3,5-triene: synthesis, fluorescence properties, and use as a fluorescence probe of lipid bilayers. Biochemistry 20:7333–7338
7. Shinitzky M (1984) Membrane fluidity and cellular functions. In: Shinitzky M (ed) Physiology of membrane fluidity. CRC Press, Boca Raton, pp 1–52
8. Le Quan Sang KH, Devynck MA (1986) Increased platelet cytosolic free calcium concentration in essential hypertension. J Hypertens 4:567–574
9. Astarie C, Levenson J, Simon A, Meyer P, Devynck MA (1989) Platelet cytosolic proton and free calcium concentration in essential hypertension. J Hypertens 7: 485–491
10. Kamal LA, Le Quan Sang KH, Meyer P (1984) Decreased uptake of ^{3}H-serotonin and endogenous content of serotonin in blood platelets in hypertensive patients. Hypertension 6:568–573
11. Erne P, Bolli P, Burgisser E, Buhler FR (1984) Correlation of platelet calcium with blood pressure. Effect of an hypertensive therapy. N Engl J Med 310:1084–1088
12. Livne A, Balfe JW, Veicht R, Marquez-Julio A, Grinstein S, Rothstein A (1987) Increased Na^+-H^+ exchange rates in essential hypertension: application of a novel test. Lancet 8552:533–536
13. Schmouder RL, Weder AB (1989) Platelet sodium-proton exchange is increased in essential hypertension. J Hypertens 7:325–330
14. Shattil SJ, Anaya-Galindo R, Bennett J, Colman RW, Cooper RA (1975) Platelet hypersensitivity induced by cholesterol incorporation. J Clin Invest 55:636–645
15. Guicheney P, Devynck MA, Cloix JF, Pernollet MG, Grichois ML, Meyer P (1988) Platelet 5-HT content and uptake in essential hypertension: role of endogenous digitalis-like factors and plasma cholesterol. J Hypertens 6:873–879
16. Locher R, Knorr M, Edmonds D, Neyses L, Vetter W (1986) Influence of cholesterol on intracellular free calcium concentration in human platelets. J Hypertens 4 [Suppl 6]:S358–S360
17. Le Quan Sang KH, Levenson J, Simon A, Meyer P, Devynck MA (1987) Platelet cytosolic free Ca^{2+} concentration and plasma cholesterol in untreated hypertensives. J Hypertens 6 [Suppl 5]:S251–S254

18. Devynck MA, Pernollet MG, Nunez AM, Aragon I, Montenay-Garestier T, Helene C, Meyer P (1982) Diffuse structural alterations in cell membranes of spontaneously hypertensive rats. Proc Natl Acad Sci USA 79:5057–5060

19. Bing RF, Heagerty AM, Thurston H, Swales JD (1986) Ion transport in hypertension: are changes in the cell membrane responsible? Clin Sci 71:225–230

20. Aragon I, Montenay-Garestier T, Devynck MA, Pernollet MG, Nunez AM, Ben-Ishay D, Meyer P, Helene C (1982) Diffuse structural abnormalities in cell membranes from genetically hypertensive rats: a fluorescence polarization study. Clin Sci 63:49s–52s

21. Lau K, Langman CB, Bafter U, Dudeja PK, Brasitus TA (1986) Increased calcium absorption in prehypertensive spontaneously hypertensive rat. J Clin Invest 78: 1083–1090

22. Kubina M, Lanza F, Cazenave JP, Laustriat G, Kuhry JG (1987) Parallel investigation of exocytosis kinetics and membrane fluidity changes in human platelets with the fluorescent probe trimethylammonium-diphenylhexatriene. Biochim Biophys Acta 901:138–146

23. Brasitus TA, Dudeja PK, Worman HJ, Foster ES (1986) The lipid fluidity of rat colonic brush-border membranes vesicles modulates Na^+-H^+ exchange and osmotic water permeability. Biochim Biophys Acta 855:16–24

24. Duhm J, Behr J (1986) Role of exogenous factors in alterations of red cell Na^+-Li^+ exchange and Na^+-K^+ cotransport in essential hypertension, primary aldosteronism and hypokalemia Scand J Clin Lab Invest [Suppl]180:82–95

25. Levy R, Livne A (1984) The erythrocyte membrane in essential hypertension. Characterization of the temperature dependence of lithium efflux. Biochim Biophys Acta 769:41–48

Platelet Abnormalities in Human Hypertension

A. Lechi[1], C. Lechi[2], C. Lauciello[2], P. Guzzo[2], E. Arosio[1],
P. Minuz[1], and M. Zatti[2]

[1] Istituto di Clinica Medica, Università di Verona, Via Delle Menegone,
  I-37134 Verona, Italy
[2] Istituto di Chimica Clinica, Università di Verona, Via Delle Menegone,
  I-37134 Verona, Italy

Platelet Functional Abnormalities

The importance accorded to platelet function in the pathogenesis of hypertension and its complications rests on many considerations. Platelets may be argued to play a patho-physiological role, since they produce vasoconstrictors (e.g., thromboxane, TXA_2, and serotonin, 5-HT) and growth factors (platelet-derived growth factor, PDGF), as well as participating actively in the development of thromboembolic complications. On the other hand, vasoconstrictor factors, such as epineprine, angiotensin II, and vasopressin, are capable to activate or sensitize the platelets by acting on specific receptors. In hypertension, increased shear stress on the vessel wall and the resulting endothelial damage can also activate the platelets, with important consequences in the onset and continuation of the atherosclerotic process. Finally, if the hypothesis is accepted that essential hypertension is characterized by a widespread genetically determined cellular abnormality, this could affect not only the smooth muscle cells of the blood vessels, but also other cells such as platelets.

The functional similarities between platelets and smooth muscle cells are remarkable: they share (a) an adenylcyclase system activated by α_2-adrenoceptors, (b) receptors for angiotensin II, vasopressin, and serotonin, (c) similar intracellular Ca^{2+} deposits (sarcoplasmic reticulum for muscle cells and dense tubular system for platelets), and (d) similar Ca^{2+}-dependent contractile systems. For these reasons, and because of their availability, platelets have been seen by many authors as a suitable model for in vitro study of cellular anomalies in hypertension.

A considerable number of platelet abnormalities (Table 1) has been described in essentially hypertensive patients [1]. Some studies have demonstrated enhanced platelet aggregation induced by various agonists in hypertensive patients compared with control subjects, but other studies have not confirmed this finding. Such an apparent contradiction could be connected with the differences in the study populations as to age and duration and

G. Bruschi A. Borghetti (Eds.)
Cellular Aspects of Hypertension
© Springer-Verlag Berlin · Heidelberg 1991

Table 1. Platelet abnormalities observed in essential hypertension

A. *Functional abnormalities*
– Increased adhesion
– Increased aggregatory response
– Increased cell size
– Increased release reaction
– Reduced 5-HT content and uptake
– Hypersensitivity to epinephrine
B. *Calcium handling and membrane abnormalities*
– Increased intracellular free Ca^{2+}
– Partial membrane depolarization
– Defective Ca^{2+} binding to the plasma membrane
– Reduced calmodulin stimulation of Ca^{2+}-ATPase and higher capacity for Ca^{2+} transport
– Hypersensitivity of phospholipase C and increased phosphoinositide breakdown
– Increased Na^+/H^+ exchange rate and increased pH_i

severity of disease. In fact, as recently observed by Nyrop and Zweifler [2], hyperaggregability with adenosine diphosphate (ADP) and epinephrine seem, from the date in the literature, to be present only in patients with more severe hypertension (MBP > 120 mmHg), not in those with milder hypertension.

Some authors have found that hypertensive patients show increased platelet size [3,4] and release reaction, determined from plasma β-thromboglobulin or platelet factor 4 levels [3–5]. Reduced platelet 5-HT content and uptake has also been reported in essential hypertension [6,7], together with increased dopamine uptake [8].

A series of abnormalities related to uptake, efflux rate, and functional response of platelets to epinephrine has been described in essential hypertension (for a review, see [9]). On the whole, these studies indicate that hypertensives seem to show hypersensitivity to epinephrine.

Platelet Calcium Abnormalities

Intracellular free Ca^{2+} concentration, $[Ca^{2+}]_i$, has been found to be higher in hypertensive patients [10]. This finding has been subsequently confirmed by most authors [11–16]. Other groups, on the contrary, were not able to find any difference in platelet free Ca^{2+} in experimental [17,18] and human [19] hypertension. A correlation between $[Ca^{2+}]_i$ and blood pressure [10,13,20] or between $[Ca^{2+}]_i$ and intracellular Na^+ [14] has also been demonstrated in some cases (Table 2).

It has been observed that, after stimulation with certain specific agonists, platelet $[Ca^{2+}]_i$ increases more in hypertensives than in normotensives [13]. This can be seen by using a strong agonist like thrombin and also with a

Table 2. Intracellular free calcium in platelets measured with fluorescent indicators in essentially hypertensive patients (EHP) and in rats (SHR, WKY, DOCA-salt)

Erne et al. [10]	EHP	q	increased[a]
Bruschi et al. [11]	EHP, SHR	q	increased
Le Quan Sang et al. [13]	EHP	q	increased[a]
Zimlichman et al. [17]	SHR, WKY	q	no difference
Murakawa et al. [18]	SHR, DOCA-salt	q	no difference
Lechi et al. [12]	EHP	q, f	increased
Cooper et al. [14]	EHP	f	increased[b]
Ueno et al. [15]	EHP	q	increased
Hvarfner et al. [16]	EHP	q	increased
Dominczak et al. [19]	EHP	q	no difference
Pritchard et al. [20]	EHP	q	a

[a] Significant correlation with blood pressure.
[b] Significant correlation with $[Na^+]_i$.
q, Quin 2; f, fura 2.

weak agonist like ADP, both in the presence of extracellular Ca^{2+} or, to a lesser extent, in its absence (Table 3).

The importance of these observations, also confirmed in other types of blood cells [11], lies in the fact that increased free cytosolic Ca^{2+} levels, both at rest and after stimulation, could indicate, although not conclusively, a greater susceptibility to specific stimuli. The functional response to stimuli is characteristic of each type of cell; in the case of smooth muscle cells, an abnormal level of excitability could be closely related to the pathogenesis of hypertension.

Increased $[Ca^{2+}]_i$ could be explained by an intrinsic cellular defect involving one or more regulatory mechanisms (Table 1): defective Ca^{2+} binding to the plasma membrane and increased Ca^{2+} influx, reduced Ca^{2+}-calmodulin stimulation of Ca^{2+}-ATPase and higher capacity for Ca^{2+} transport, hypersensitivity of phospholipase C and increased phosphoinositide breakdown, or altered Ca^{2+} efflux or sequestration.

It is also possible that increased $[Ca^{2+}]_i$ may be related to intracellular Na^+ increase induced by Na^+, K^+-ATPase inhibition caused by a hypothetical plasma ouabain-like factor [21]. Using quin 2 as an intracellular indicator, we were not able to demonstrate a rise in $[Ca^{2+}]_i$ in platelet incubated with ouabain or with plasma extracts containing ouabain-like factor [22]. This may depend on the buffering action of quin 2, since, if fura 2 is used as intracellular dye, a small but significant ouabain-induced rise in platelet $[Ca^{2+}]_i$ can be demonstrated [23].

Reduced Ca^{2+} binding to the plasma membrane has been found in hypertension in various types of cells [24,25] as well in platelets [26], and it has been suggested that this may favor membrane depolarization, with opening of potential-operated Ca^{2+} channels. However, this hypothesis is still to be proved, since partially hyperpolarized platelets respond equally to

Table 3. Platelet cytosolic free Ca^{2+} in patients with essential hypertension (EHP) and in normotensive controls (NTC) in unstimulated platelets, platelets stimulated with thrombin (measured using quin 2), and platelets stimulated with ADP (measured using fura 2). Values are means ± SEM

	NTC	EHP
$[Ca^{2+}]_i$ in unstimulated platelets (nM)	126.6 ± 5.1	144.9 ± 6.4**
	(n = 65)	(n = 64)
$[Ca^{2+}]_i$ after 25 mU/ml thrombin (nM)	218 ± 18	278 ± 15*
	(n = 11)	(n = 11)
$[Ca^{2+}]_i$ after 50 mU/ml thrombin (nM)	373 ± 30	627 ± 135*
	(n = 11)	(n = 11)
$[Ca^{2+}]_i$ after 20 μmol/l ADP with	306 ± 11	406 ± 24**
1 mM extracellular Ca^{2+} (nM)	(n = 10)	(n = 10)
$[Ca^{2+}]_i$ after 20 μmol/l ADP without	250 ± 21	318 ± 16*
extracellular Ca^{2+} (nM)	(n = 10)	(n = 10)

* $p < 0.05$; ** $p < 0.01$.

agonists with increased $[Ca^{2+}]_i$ [27]. Moreover, Ca^{2+} antagonists, at least at therapeutic doses, do not significantly inhibit an agonist-induced rise in $[Ca^{2+}]_i$ [28]. The concentrations of Ca^{2+} antagonist capable of inhibiting a rise in $[Ca^{2+}]_i$ are some orders of magnitude greater than the concentrations required to inhibit voltage-dependent Ca^{2+} channels in vascular smooth muscle. Therefore, in human platelets membrane potential variations responsible for triggering Ca^{2+} influx seem unlikely.

It has been suggested that in some types of cells in essential hypertension the interaction between calmodulin and membrane Ca^{2+}-ATPase [29] is altered. The latter mechanism is responsible for regulating $[Ca^{2+}]_i$ through Ca^{2+} efflux or, above all, its resequestration into deposit pools. In platelets, Resink et al. [29] found increased capacity for Ca^{2+} transport and reduced calmodulin stimulation, interpreting these data as indicative of the inability of the system to effect adequate normalization of $[Ca^{2+}]_i$ increase.

Platelet Activation Systems

Specific platelet agonists like thrombin, TXA_2 and platelet activating factor (PAF) (so-called strong agonists), activate membrane phosphoinositide hydrolysis with formation of inositol triphosphate (IP_3) and diacylglycerol (DG) by a phospholipase C (PLC). IP_3 increases $[Ca^{2+}]_i$ by releasing it from deposits. DG activates protein kinase C (PKC), which leads to phosphorylation of a 47-kDa protein connected with secretion processes and induces expression of fibrinogen receptors. PKC also activates Na^+/H^+ exchange.

Phosphoinositide turnover thus plays a key role in processes of platelet activation as a second messenger of the stimulus-response coupling process.

In human and experimental hypertension phospholipid turnover has been studied in the erythrocytes [30] and in muscular cells [31]. Platelets in essential hypertension have been described as having enhanced phosphoinositide turnover, which could be responsible for increase of $[Ca^{2+}]_i$ [32].

An important role in stimulus-response coupling is also played by phospholipase A_2 (PLA$_2$). Platelet PLA$_2$ is regulated by increase in $[Ca^{2+}]_i$, by Na^+/H^+ exchange, and by the inhibitor lipocortin [33]. PLA$_2$ is responsible for release of arachidonic acid (AA) which is the main source of platelet eicosanoids (PGG$_2$, PGH$_2$, and TXA$_2$) and of the lysophospholipid substrate for platelet-activating factor formation. These are potent agonists which bind to specific receptors and, through a G protein, which acts as a signal transducer, they directly activate PLC and phosphoinositide breakdown.

This system, via PKC and Na^+/H^+ exchange, regulates intracellular pH and is responsible for the cytoplasmic alkalinization which occurs in the platelets after agonist stimulation. A more alkaline cytosolic pH stimulates the activity of various enzymes, including PLA$_2$, and seems to be a prerequisite for AA release [34]. On the other hand, inhibition of Na^+/H^+ exchange reduces PLC activation by ADP and epinephrine, the so-called weak agonists. It has been suggested that activation of Na^+/H^+ exchange, in cooperation with $[Ca^{2+}]_i$, activates in turn PLA$_2$, with liberation of AA and, subsequently, TXA$_2$ [35].

Arachidonic Acid and Thromboxane A$_2$

There is plentiful evidence of altered cell membrane in hypertension [25]. Some authors have reported an abnormal composition of cell membrane fatty acids in human hypertension [36] and some studies seem to attribute a certain importance to the biological effects of polyunsaturated fatty acid supplementation on the vascular system [37] and platelets [38].

Studies of PLA$_2$-eicosanoids in the platelets of hypertensive patients are still few in number. In hypertensive patients, the AA concentration required to induce irreversible platelet aggregation has been found to be lower than in controls [39], but in vitro TXB$_2$ production has not been shown to be significantly different [40].

In a study which compared a group of essentially hypertensive patients without vascular complications to matched control subjects, we found that the in vitro production of TXB$_2$ and urinary excretion of 2,3-dinor-TXB$_2$, which is the most reliable index of in vivo platelet aggregation, was higher in the hypertensive patients, but not to a significant level (Table 4). However, in these patients platelets show significantly greater aggregation and higher $[Ca^{2+}]_i$ increase after stimulation with high doses of ADP capable of inducing platelet TXA$_2$ production and irreversible aggregation. After blockade with

Table 4. Agonist-induced platelet aggregation in patients with essential hypertension (EHP) and in normotensive controls (NTC). Thromboxane A_2 platelet production was evaluated in vitro as TXB_2 production after $10\,\mu M$ ADP and in vivo as 2,3-dinor TXB_2 in 24 h urinary excretion. Means $\pm$ SD

	NTC ($n = 12$)	EHP ($n = 12$)
ADP, $20\,\mu M$	74 ± 12	89 ± 12**
ADP, $20\,\mu M$ + ASA	74 ± 13	80 ± 19
ADP, $10\,\mu M$	67 ± 20	82 ± 16*
ADP, $10\,\mu M$ + ASA	70 ± 13	70 ± 20
Arachidonic acid, $5\,\mu M$	93 ± 7	96 ± 6
Arachidonic acid, $5\,\mu M$ + ASA	1 ± 0.5	2 ± 2
U 46619, $1\,\mu M$	97 ± 11	97 ± 15
U 46619, $1\,\mu M$ + ASA	95 ± 9	98 ± 10
TXB_2 produced in vitro (pg/ml)	573 ± 105	658 ± 101
urinary 2,3-dinor TXB_2 (pg/mg creat.)	92 ± 22	146 ± 27

$*p < 0.05$; $**p < 0.01$.

cyclooxygenase, the difference between hypertensives and normotensives is no longer detectable. By contrast stimulation with AA and the endoperoxide analogue U 46619 produced no significantly different results in the two groups.

We do not know if these data, as well as the in vitro aggregability observed by various authors with ADP and other agonists like epinephrine [1,2], depend on environmental, humoral, and/or mechanical factors, or on some intrinsic platelet abnormality. These date, observed in vitro on washed platelets, direct attention more to intrinsic alteration of the platelets than to environmental factors.

The fact that cyclooxygenase blockade eliminates the difference between normotensives and hypertensive patients does not, in itself, imply that the eicosanoid pathway is altered in the platelets of hypertensive patients. It may well also be explained by more plausible hypotheses. For example, it could indicate that the platelets of hypertensives are characterized by an anomalous increase in Na^+/H^+ exchange and thus of PLA_2 activation. Moreover, we cannot rule out that the inhibition of AA metabolism eliminates an important amplification pathway involving phosphoinositide breakdown.

Na^+/H^+ Exchange

Recently published studies seem to show enhanced activity of Na^+/H^+ exchange in cells of hypertensive patients. Ng et al. [41] demonstrated that the activity of the Na^+/H^+ antiporter is higher in the leukocytes of hyper-

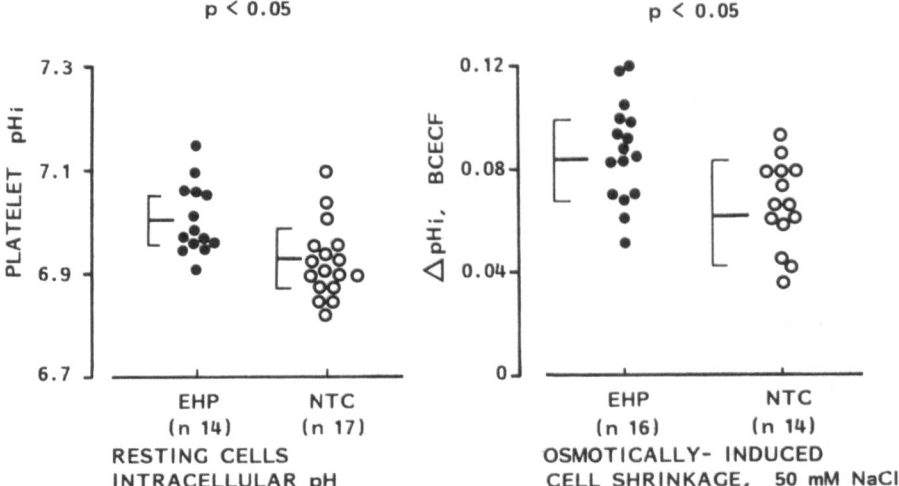

Fig. 1. Intracellular pH (pH_i) of resting platelets and variations in pH_i induced by osmotic cell shrinkage with $50\,mM$ NaCl (means $\pm$ SD). Platelets were incubated with $5\,\mu M$ BCECF-AM for 30 min at 37°C and washed twice before fluorescence measurements (C. Lechi, unpublished data)

tensive patients when the cells are acid-loaded to a pH of 6.7. Using platelet swelling induced by acidification as a measure of Na^+/H^+ exchange, Livne et al. [42] and Schmouder et al. [43] demonstrated increased activity of the antiporter in platelets of hypertensives compared with control subjects.

We studied this mechanism using a procedure which measures platelet Na^+/H^+ exchanger activity with the intracellular pH indicator bis-(carboxyelhyl)carboxyfluorescein (BCECF) [44,45]. In hypertensive patients we found basal intracellular platelet pH higher than in a group of matched controls (Fig. 1). Despite the large overlap, the difference is statistically significant. These data are consistent with those reported by Astarie et al. [46]. By stimulating platelets with osmotically induced cell shrinkage, we observed a larger variation of pH_i in platelets of hypertensive patients.

On the basis of these data, it may be assumed that, following ADP stimulation, platelet response in hypertension is increased during the first stage of activation. In addition to enhanced Na^+/H^+ exchange activity, there may be a very early increase of Ca^{2+} influx, induced by ADP [27]. This could lead to enhanced activity of PLA_2 and the $AA-TXB_2$ pathway, thus triggering an amplification system capable of increasing cellular response. However, since we have not been able to demonstrate a significant increase of TXA_2 synthesis, at least in hypertensive patients without vascular complications, we believe this hypothesis is to be regarded with caution.

In other cell types, which do not recognize ADP as an agonist, but in which Na^+/H^+ exchange could be activated by other factors such as α-

adrenergic stimulation, subsequent events, although different from the AA-TXA_2 pathway, could still lead to increased cell activity.

It is tempting to speculate that the rise in $[Ca^{2+}]_i$ and pH_i in hypertension could be linked in some way. They might be accidental events related to an unknown variable, or the consequence of experimental procedure, as a result of in vitro activation. If this is not the case, the increase of both $[Ca^{2+}]_i$ and pH_i could be a consequence, rather than a cause, of hypertension, or could be the manifestation of a primitive phenomenon important from a pathophysiological point of view. In this case, Na^+/H^+ exchange could play a role in the pathogenesis of hypertension and its complications. In platelets it may enhance TXA_2-induced aggregability, while in the case of smooth muscle cells it may enhance cell proliferation. At this moment this is a stimulating hypothesis, deserving more detailed examination.

In view of the importance of platelets in the process of atherosclerosis and the possibility of their activation, it should be clear that various methods of enrolment of the hypertensive population are needed, depending on the objective of a study. If the aim is to investigate the existence of an anomaly determined, at least in part, genetically, it will be necessary to select patients who are homogeneous in regard to such environmental factors as diet and physical activity. It will also be advisable to study relatively young subjects free of vascular complications, in order to exclude as far as possible secondary alterations of platelet function. On the other hand, patients with more severe hypertension and vascular complications are more suitable for studying the role of platelets in the pathophysiology of these conditions and the possible impact of drug therapy. Failure to give due consideration to these factors may explain, at least in part, many discrepancies in the results reported in the literature.

References

1. De Clerck F (1986) Blood platelets in human hypertension. Agents Actions 18: 563–580
2. Nyrop M, Zweifler AJ (1988) Platelet aggregation in hypertension and the effects of antihypertensive treatment. J Hypertens 6:263–269
3. Lande K, Os I, Kjeldsen SE, Westheim A, Hjermann I, Eide I, Gjesdal K (1987) Increased platelet size and release reaction in essential hypertension. J Hypertens 5:401–406
4. Bruschi G, Minari M, Bruschi ME, Tacinelli L, Milani B, Cavatorta A, Borghetti A (1986) Similarities of essential and experimental hypertension: volume and number of blood cells. Hypertension 8:983–989
5. Gomi T, Ikeda T, Yuhara M, Sakurai J, Nakayama D, Ikegami F (1988) Plasma β-thromboglobulin to platelet factor 4 ratios as indices of vascular complications in essential hyprtension. J Hypertens 6:389–392
6. Kamal LA, Le-Quan Bui KH, Meyer P (1984) Decreased uptake of 3H-serotonin and endogenous content of serotonin in blood platelets of hypertensive patients. Hypertension 6:568–573

7. Guicheney P, Baudouin-Legros M, Valtier D, Meyer P (1987) Reduced serotonin content and uptake in platelets from patients with essential hypertension: is a ouabain-like factor involved? Thromb Res 45:289–297

8. Kjeldsen SE, Gjesdal K, Leren P, Eide IK (1988) Decreased platelet free dopamine and unchanged noradrenaline and adrenaline in essential hypertension. Thromb Haemost 60:251–254

9. Kjeldsen SE, Neubig RR, Weder AB, Zweifler AJ (1989) The hypertension-coronary heart disease dilemma: the catecholamine-blood platelet connection. J Hypertens 7:851–860

10. Erne P, Bolli P, Burgisser E, Buhler FR (1984) Correlation of platelet calcium with blood pressure: effect of antihypertensive therapy. N Engl J Med 310:1084–1088

11. Bruschi G, Bruschi ME, Caroppo M, Orlandini G, Spaggiari M, Cavatorta A (1985) Cytoplasmic free calcium is increased in the platelets of spontaneously hypertensive rats and essential hypertensive patients. Clin Sci 68:179–184

12. Lechi A, Lechi C, Bonadonna G, Sinigaglia D, Corradini P, Polignano R, Arosio E, Covi G, De Togni P (1987) Increased basal and thrombin-induced free calcium in platelets of essential hypertensive patients. Hypertension 9:230–235

13. Le Quan Sang KH, Devynck MA (1986) Increased platelet cytosolic free calcium concentration in essential hypertension. J Hypertens 4:567–574

14. Cooper RS, Shamsi N, Katz S (1987) Intracellular calcium and sodium in hypertensive patients. Hypertension 9:224–229

15. Ueno H, Mikawa M, Takata M, Asanoi H, Iida H, Sasayama S, Kagamimori S (1988) Does cytosolic free calcium concentration in platelets reflect tone and structural changes of resistance vessels? J Hypertens 6 [Suppl 4]:S255–S257

16. Hvarfner A, Larsson R, Morlin C, Rastad J, Wide L, Åkerström G, Ljunghall S (1988) Cytosolic free calcium in platelets: relationships to blood pressure and indices of systemic calcium metabolism. J Hypertens 6:71–77

17. Zimlichman R, Goldstein DS, Zimlichman S, Keiser HR (1986) Cytosolic calcium in platelets of spontaneously hypertensive rats. J Hypertens 4:283–287

18. Murakawa K, Kanayama Y, Kohno M, Kawarabayashi T, Yasunari K, Takeda T, Hyono A (1986) Cytoplasmic free Ca^{2+} is not increased in platelets of deoxycorticosterone-salt and spontaneously hypertensive rats. Clin Sci 71:121–123

19. Dominiczak AF, Morton JJ, Murray G, Semple PF (1989) Platelet cytosolic free calcium in essential hypertension: responses to vasopressin. Clin Sci 77:183–188

20. Pritchard K, Raine AEG, Ashley CC, Castell LM, Somers V, Osborn C, Ledingham JGG, Conway J (1989) Correlation of blood pressure in normotensive and hypertensive individuals with platelet but not lymphocyte intracellular free calcium concentrations. Clin Sci 76:631–635

21. Blaustein MP, Hamlyn JM (1984) Sodium transport inhibition, cell calcium and hypertension. The natriuretic hormone-Na^+/Ca^{2+} exchange-hypertension hypothesis. Am J Med 77(4A):45–59

22. Lechi C, Sinigaglia D, Delva P, Guzzo P, Arosio E, Steele A, Lechi A (1989) Platelet intracellular free Ca^{2+} after incubation with plasma from hypertensive patients. J Human Hypertens 3:17–20

23. Schaeffer J, Blaustein MP (1989) Platelet free calcium concentrations measured with fura-2 are influenced by the transmembrane sodium gradient. Cell Calcium 10:101–113

24. Robinson BF (1984) Altered calcium handling as a cause of primary hypertension. J Hypertens 2:453–460

25. Postnov YuV, Orlov SN (1985) Ion transport across plasma membrane in primary hypertension. Physiol Rev 65:904–945

26. Resink TJ, Dimitrov D, Zschauer A, Tkachuk TA, Buhler FR (1986) Platelet calcium-linked abnormalities in essential hypertension. Ann N Y Acad Sci 488:252–264

27. Rink TJ (1988) Cytosolic calcium in platelet activation. Experientia 44:97–100
28. Bonadonna G, Lechi C, Corradini P, Sinigaglia D, De Togni P, Grzeskowiak M (1986) Verapamil inhibits platelet aggregation by a calcium-independent mechanism. Thromb Haemost 56:308–310
29. Resink T, Tkachuk VA, Erne P, Buhler FR (1985) Platelet membrane calmodulin-stimulated calcium-adenosine triphosphatase: altered activity in essential hypertension. Hypertension 8:159–166
30. Marche P, Koutouzov S, Girard A, Elghozi JL, Meyer P, Ben-Ishay D (1985) Phosphoinositide turnover in erythrocyte membranes in human and experimental hypertension. J Hypertens 3:25–30
31. Heagerty AM, Ollerenshaw JD, Swales JD (1986) Abnormal vascular phosphoinositide hydrolysis in the spontaneously hypertensive rat. Br J Pharmacol 89:803–807
32. Koutouzov S, Limon I, Meyer P, Marche P (1988) Impaired phospholipase C activity is involved in the hyperreactivity of platelets in primary hypertension. J Hypertens 6 [Suppl 4]:S372–S374
33. Kroll MH, Schafer AI (1989) Biochemical mechanisms of platelet activation. Blood 74:1181–1195
34. Sweatt JD, Johnson SL, Cragoe EJ, Limbird LE (1985) Inhibitors of Na^+/H^+ exchange block stimulus-provoked arachidonic acid release in human platelets. J Biol Chem 260:12910–12919
35. Sweatt JD, Connolly TM, Cragoe EJ, Limbird LE (1986) Evidence that Na^+/H^+ exchange regulate receptor-mediated phospholipase A_2 activation in human platelets. J Biol Chem 261:8667–8673
36. Rouse IL, Beilin LJ (1983) Nutrition, blood pression and hypertension. Med J Aust 2 [Suppl I]:S10–S23
37. Naftilan AJ, Dzau VJ, Loscalzo J (1986) Preliminary observations on abnormalities of membrane structure and function in human hypertension. Hypertension 8 [Suppl 2]:II174–II179
38. Lagarde M (1988) Metabolism of fatty acids by platelets and the functions of various metabolites in mediating platelet function. Prog Lipid Res 27:135–142
39. Matsumoto M, Kusunoki M, Uyama O, Fujisawa A, Matsuyama T, Yoneda S, Kimura K, Abe H (1981) Platelet aggregation induced by arachidonic acid and thromboxane generation in patients with hypertension or cerebrovascular disease. Prostaglandins Med 7:553–562
40. Matsumoto M, Nukada T, Uyama O, Yoneda S, Imaizumi M, Miyamoto T, Kayama N (1980) Thromboxane generation in patients with essential hypertension or cerebrovascular disease and effect of oral aspirin. Thromb Haemost 44:16–22
41. Ng LL, Dudley C, Bomford J, Hawley D (1989) Leucocyte intracellular pH and Na^+/H^+ antiport activity in human hypertension. J Hypertens 7:471–475
42. Livne A, Balfe JW, Veitch R, Marquez-Julio A, Grinstein S, Rothstein A (1987) Increased platelet Na^+/H^+ exchange rates in essential hypertension: application of a novel test. Lancet I:533–536
43. Schmouder RL, Weder AB (1989) Platelet sodium-proton exchange is increased in essential hypertension. J Hypertens 7:325–330
44. Zavoico GB, Cragoe EJ, Feinstein MB (1986) Regulation of intracellular pH in human platelets. J Biol Chem 261:13160–13167
45. Livne A, Grinstein S, Rothstein A (1987) Characterization of Na^+/H^+ exchange in platelets. Thromb Haemost 58:971–977
46. Astarie C, Levenson J, Simon A, Meyer P, Devynck M-A (1989) Platelet cytosolic proton and free calcium concentrations in essential hypertension. J Hypertens 7:485–491

Aldosterone Receptors and Effector Mechanisms in Mononuclear Leukocytes in Different Forms of Hypertension.*

D. Armanini[1], M. Wehling[2], D. Bosson[1], U. Kuhnle[2], P.C. Weber[2], E. Milan[1], C. Pratesi[1], M.C. Zennaro[1], L. Martella[1], M. Scali[1], V. Zampollo[1], and F. Mantero

[1] Istituto di Semeiotica Medica, Università degli Studi di Padova, Via Ospedale Civile, 105, I-35100 Padova, Italy
[2] Medizinische Klinik Innenstadt und Kinderklinik, Universität München, W-8000 München, FRG

Introduction

One of the most important factors which regulate electrolytes, blood volume, and blood pressure is aldosterone. Mineralocorticoids by definition affect mineral metabolism, and probably all the factors related to hypertension usually studied at the cellular level are directly or indirectly influenced by the action of this hormone (sodium, calcium, electrolyte exchange systems, etc). The aim of the present study was to evaluate mineralocorticoid effector mechanisms in different types of hypertension. A change in plasma or urine content of aldosterone is not always an indicator of increased effect as demonstrated for example in pseudohypoaldosteronism. To have an effect aldosterone must act at the level of target tissues, and its effect is mediated by binding to mineralocorticoid receptors (type I corticosteroid receptors). The type I receptor shows high and equivalent affinity for aldosterone, deoxycorticosterone, and corticosterone. From these data it is difficult to explain why these receptors respond to aldosterone, since the plasma concentration of glucocorticoids is much higher and one would expect these receptors to be occupied by glucocorticoids. In vivo, however, the classical target tissues for aldosterone are highly specific for this hormone, probably as a consequence of intracellular specificity-conferring mechanisms. Some such mechanisms have been postulated or evaluated: presence in the tissue or in the cells of corticosteroid-binding globulin (CBG) [1], or a different intracellular metabolism of glucocorticoids (11-hydroxysteroid dehydrogenase and oxoreductase system and 5α- and 5β-reductase). These mechanisms would allow aldosterone to bind to its receptors by reducing the availability of free glucocorticoids in the cell. In other clinical situations it seems that cortisol binds to mineralocorticoid receptors, leading to a mineralocorticoid response. In this case reduction of 11-hydroxysteroid

* Partly supported by Nato grant 07 07 87.

G. Bruschi A. Borghetti (Eds.)
Cellular Aspects of Hypertension
© Springer-Verlag Berlin · Heidelberg 1991

dehydrogenase increases the concentration of free cortisol in the cells and the subsequent binding of the steroid to type I receptors [2,3].

In other tissues which are not classical targets for aldosterone, such as in the hippocampus, it seems that mineralocorticoid receptors act as high-affinity glucocorticoid receptors. The current concept of aldosterone action is that the protein synthesized after binding of the aldosterone-receptor complex to DNA increases the number of sodium channels at the level of the plasma membrane. Other mechanisms have been proposed: increased synthesis of citrate synthase, activation of Na, K-ATPase. It is likely that the initial effect of aldosterone is merely at the level of the sodium channels and that all the other mechanisms follow: in the classical target tissues with asymmetrical cells, the increase of intracellular sodium activates Na-K-ATPase, leading to sodium reabsorption and influx of potassium into the cells. The increase in potential difference due to these events produces a flux of positive electrolytes in tubular fluid, particularly of potassium, whose intracellular concentration is higher [4]. In the tissues which are not classical targets of aldosterone it is thought that mineralocorticoid receptors are occupied by glucocorticoids and give a glucocorticoid response.

It is very difficult to study mineralocorticoid receptors in humans, due to the low availability of classical target tissues; a few studies have been done in human kidney, confirming that the characteristics of mineralo-corticoid receptors in human kidney correspond to those found in rat kidney. In recent years several reports have been published characterizing type I receptors in other tissues (hippocampus, pituitary, gut, heart, arterial wall, mammary gland, and more recently, mononuclear leukocytes) [5,6].

The Mononuclear Leuckocyte Model

Mononuclear leukocytes are easily available for human studies and do not need culture. Preparation of these cells involves separating the mononuclear layer by an osmotic gradient and washing and resuspending the cells in medium. An aliquot of cells is incubated with increasing amounts of tritiated aldosterone plus an excess of a pure glucocorticoid (RU-26988) in order to rule out binding of tracer to glucocorticoid receptors. Incubations are also performed in parallel with addition to the system of an excess of cold aldosterone for calculating unspecific binding. It is thus possible to calculate the specific binding and the number of receptors per cell.

The number of mineralocorticoid receptors per cell in normal subjects ranges from 150 to 400 and the affinity of aldosterone for the receptors varies from 1.7 to 3 nmol/l [6]. Studies on the specificity, of steriods for the receptors have shown aldosterone, desoxycorticosterone, and corticosterone to have equal affinities. The affinity of cortisol is one-third that of aldosterone and of dexamethasone one-tenth [6]. This kind of hierarchy is consistent with preferential binding of glucocorticoids to these receptors. This theory has been proposed but not validated and in the present case we were not able to confirm it, at least in this model. We assessed a simple method for evaluating the effector mechanism of steroids in mononuclear leukocytes. Incubation of cells in steroid-free medium leads to loss of intracellular sodium and potassium, and this loss is prevented by adding to the system a physiological concentration of aldosterone ($1.4\,nM$). Incubation of the cells with cortisol does not prevent loss of electrolytes, while incubation with both aldosterone and cortisol gives the same effect as aldosterone alone [7]. From these studies we can conclude that aldosterone preferentially binds to type I receptors and produces a mineralocorticoid effect which is that of maintaining the electrolytes content of the cell. Since the flux of sodium is parallel to the flux of calcium, it is possible to conclude that immunological functions which are dependent on calcium content are in strict correlation to the effect of aldosterone.

Another interesting action of aldosterone in mononuclear leukocytes is to regulate the volume of these cells. Cell volume decreased by 16% when cells were incubated in medium alone, and the loss was prevented by addition to the system of aldosterone. The effect of aldosterone on both electrolyte content and volume regulation is blocked by addition to the system of canrenone, an aldosterone antagonist [8]. It is very difficult to explain why these cells preferentially respond to aldosterone in the form of a mineralocorticoid effect even in the presence of cortisol. It is possible, however, that in mononuclear leukocytes other mechanisms can prevent binding of glucocorticoids to mineralocorticoid receptors, as recently postulated by Funder and coworkers [4] for arterial wall. These authors have by in vivo and in vitro studies validated a mineralocorticoid effect in this tissue which binds steroids with a specificity parallel to that found in mononuclear leukocytes. This assumption comes from the finding that an 11-hydroxysteroid-dehydrogenase system is active in this tissue.

Applications of Measurement of Type I Receptors in Mononuclear Leukocytes (Fig. 1)

Pseudohypoaldosteronism [9]

Pseudohypoaldosteronism is a rare syndrome characterized by severe hyponatremia, hyperkalemia, volume loss, marked increase of plasma aldosterone and renin activity, and unresponsiveness to therapy with mineralocorticoids. We found that patients with this syndrome lack type I receptors in mononuclear leukocytes, which clarified the pathogenesis. These patients are treated with sodium supplementation and progressively all the most important clinical findings disappear, except for the elevated aldosterone and plasma renin activity and the lack of mineralocorticoid receptors. Probably there is a regulation of osmoreceptors at the cerebral level which is able to reverse the effect of absence or marked reduction of mineralocorticoid

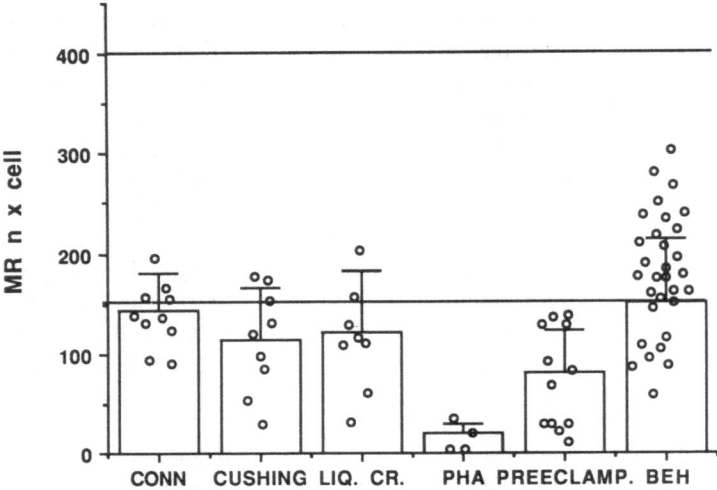

Fig. 1. Individual values and mean + SD of the number of mineralocorticoid receptors per cell found in patients with various forms of hypertension. All were significantly reduced in comparison with normal controls (p <0.05). *CONN*, Primary hyperaldosteronism (Conn's syndrome); *CUSHING*, Cushing's syndrome; *LIQ.CR*, long-term licorice treatment *PHA*, pseudohyperaldosteronism; *PREECLAMP*, preeclampsia; *BEH*, benign essential hypertension; *Horizontal line* defines normal range

effect. Studies on families of patients with pseudohypoaldosteronism have demonstrated that inheritance is sometimes autosomal dominant and sometimes autosomal recessive with different degrees of dominance.

Primary Hyperaldosteronism (Conn's Syndrome) [10]

Mineralocorticoid receptors are reduced both in patients with unilateral adenoma producing aldosterone and in those with idiopathic hyperaldosteronism, while they are normal in patients with dexamethasone-suppressible hyperaldosteronism. Reduction of mineralocorticoid receptors in primary hyperaldosteronism is important also in the determination of the escape phenomenon of target tissues to action of aldosterone on electrolyte exchange. In patients with unilateral adenoma the number of receptors is still reduced a short time after unilateral adrenectomy, while 3 months after surgery the number of the type I receptors becomes normal. These findings demonstrate that when plasma aldosterone returns to the normal range, type I receptors also normalize.

Secondary Hyperaldosteronism [11]

Mineralocorticoid receptors are down-regulated in high-renin hyperaldosteronism too. The down-regulation is present both in patients with hypertension (nephrogenic hypertension, renal artery stenosis, etc.) and in patients with low blood pressure (Bartter's syndrome, consumption of laxatives and diuretics, etc.). From these findings it seems that down-regulation of aldosterone receptors is not dependent, on volume expansion or hypertension.

Cushing's Syndrome [11]

In this disease too, where aldosterone and plasma renin activity are frequently suppressed, mineralocorticoid receptors are down-regulated while glucocorticoid receptors are always normal. It is possible that cortisol at the concentrations reached in this disease cross-reacts with mineralocorticoid receptors, leading to a mineralocorticoid effect.

Pseudohyperaldosteronism [12–15]

Hypertension from licorice derivatives is currently the subject of speculation. Initially it was demonstrated that glycyrrhetinic and glycyrrhizic acid and carbenoxolone bind to mineralocorticoid receptors, albeit with low affinity. Subsequently it was shown that the licorice derivatives can reduce the activity of 11-hydroxysteroid dehydrogenase, which is the enzyme that regulates the conversion of cortisol to cortisone at the cellular level. Measurement of the ratio of tetrahydrocortisol to tetrahydrocortisone (THF/THE ratio) in urine has demonstrated that during the first days of acute administration of licorice this is the main mechanism of action of the licorice. From these studies it could be ruled out that licorice acts via binding to mineralocorticoid receptors. From our studies the direct effect has been considered. We found that: (a) licorice derivatives bind to type I receptors in mononuclear leukocytes, (b) the effect of glycyrrhetinic acid in isolated mononuclear leukocytes is clearly a mineralocorticoid one in terms of electrolyte content of cells, and (c) the THF/THE ratio is not significantly different from normal after 7 days of consumption of high doses of licorice, in both normal volunteers and patients receiving long-term licorice treatment. In addition, we found that mineralocorticoid receptors are down-regulated during long-term ingestion of licorice but not during the first days of treatment. We conclude that enzymatic block only initially is the main mechanism of mineralocorticoid action of licorice; then, when the licorice concentration in plasma is consistent with mineralocorticoid receptor binding, the direct effect of licorice at the mineralocorticoid receptor level predominates.

Preeclampsia [16]

Preeclampsia is a syndrome characterized by hypertension and albuminuria in the third trimester of pregnancy. Blood pressure falls during pregnancy, and diastolic values of more than 90 mmHg are considered a sign of hypertension. Plasma aldosterone and plasma renin activity are equally elevated both in normal pregnancy and in preeclampsia. Several theories have been assessed to explain the difference. In normal pregnancy there is an increase of vasodilating prostaglandins and an insensitivity to pressor agents such as angiotensin II, epinephrine, etc. According to one theory, in patients with preeclampsia the prostaglandin concentration falls, restoring reactivity to angiotensin II.

Another theory claims the existence of immunological reactions which produce placental damage and subsequent production of pressor agents. We have examined the mineralocorticoid effector mechanism in preeclampsia and found that the numbers of mineralocorticoid receptors are low, while in normal pregnancy of the same gestational period they are normal. This reduced level of receptors is probably a marker of preeclampsia. More interestingly, we found that the mineralocorticoid effector mechanism in preeclampsia is increased as measured by subtraction potential difference (rectal potential difference minus buccal potential difference, which is a clear index of aldosterone effect), while in normal pregnancy it is normal. It is difficult to explain this; possibly the antagonism of progesterone at the mineralocorticoid receptor level is not effective in preeclampsia, due to placental or maternal factors. After delivery both the numbers of mineralocorticoid receptors and the subtraction potential difference become normal, supporting the hypothesis that the alterations are related to pregnancy and not to a preexisting factor.

Essential Hypertension [17]

We studied 40 patients with essential hypertension and found that a subpopulation of these patients (about 30%) had a reduced number of mineralocorticoid receptors. Taking all the patients into account, the mean number of mineralocorticoid receptors was significantly lower than in normotensive controls. It is interesting to note that we were not able to find a correlation between plasma aldosterone and number of receptors. This finding is not surprising, since a conditon of potential hypermineralocorticoidism can be associated with either high or low values of aldosterone. These data support the hypothesis that in a subgroup of patients with essential hypertension there is either an increase in the levels of some mineralocorticoids other than aldosterone, or there is a potentiation of the effect of aldosterone as previously demonstrated for carbenoxolone. Alternatively it is possible that altered intracellular corticosteroid metabolism produces an increase in the levels of mineralocorticoid-like compounds. A marker for characterizing these patients might be the measurement of mineralocorticoid receptors in mononuclear leukocytes.

Conclusion

The mononuclear leukocyte model has opened up the possibility of new interesting studies on the regulation of mineralocorticoid receptors and related effector mechanisms in normal subjects and in patients in various clinical conditions. Since this model is at present the only way to study the regulation of mineralocorticoid receptors, it may permit further information to be gained about the physiological mechanisms which regulate corticosteroid metabolism and related effects.

References

1. Krozowski Z, Funder JW (1984) Mineralocorticoid receptors and hippocampal corticosterone binding species have identical steroid specificity. Proc Natl Acad Sci USA 80:6056–6060
2. Pressley A, Funder JW (1975) Glucocorticoid and mineralocorticoid receptors in gut mucosa. Endocrinology 97:588–591
3. Edwards CRW, Stewart PM, Burt D, Brett L, McIntire MA, Sutanto VS, De Kloet ER, Mondeer C (1988) Localization of 11-beta hydroxysteroid dehydrogenase: tissue specific protector of the mineralocorticoid receptor. Lancet ii:986–989
4. Funder JW, Pearce PT, Smith R, Smith AI (1989) Mineralocorticoid specificity is enzyme, not receptor mediated. Science 242:583–585
5. Marver D, Kokko JP (1983) Renal target sites and the mechanism of action of aldosterone. Miner Electrolyte Metab 9:1–18
6. Armanini D, Strasser T, Weber PC (1985) Characterization of aldosterone binding sites in circulating mononuclear leukocytes. Am J Physiol 248:E388–E390
7. Wehling M, Armanini D, Strasser T, Weber PC (1987) Effect of aldosterone on the sodium and potassium concentrations in human mononuclear leukocytes. Am J Physiol 252:E505–E508
8. Wehling M, Kuhls S, Armanini D (1989) Volume regulation of human mononuclear leukocytes by aldosterone in isotonic media. Am J Physiol 20:E170–E174
9. Armanini D, Kuhnle U, Strasser T, Dorr H, Weber C, Butenandt I, Stockigt JR. Pearce P, Funder JW (1985) Pseudohypoaldosteronism: demonstration of aldosterone receptor deficiency. N Engl J Med 313:1178–1181
10. Armanini D, Witzgall H, Wehling M, Kuhnle U, Weber PC (1985) Aldosterone receptors in different types of primary hyperaldosteronism. J Clin Endocrinol Metab 65:101–104
11. Armanini D, Wehling, Kuhnle U, Witzgall H, Strasser T, Weber PC (1987) Mineralocorticoid effector mechanism in different clinical situations. In: Mantero F, Vescei P (eds) Corticosteroid and peptide hormones in hypertension. Ares Serono Symposia, 1987. Raven, New York, pp 285–293
12. Armanini D, Krozowski Z, Karbowiak I, Funder JW (1982) The mechanism of mineralocorticoid action of carbenoxolone. Endocrinology 111:1682–1686
13. Armanini D, Karbowiak I, Funder JW (1983) Affinity of licorice derivatives for mineralocorticoid and glucocorticoid receptors. Clin Endocrinol 16:606–609
14. Armanini D, Scali M, Zennaro C, Karbowiak I, Lewicka S, Vecsei P, Mantero F (1989) The pathogenesis of pseudohyperaldosteronism from carbenoxolone. J Endocrinol Invest 12:337–341
15. Armanini D, Scali M, Sonino N, Karbowiak I, Wehling M, Weber PC (1988) Pseudohyperaldosteronism from licorice: further insight into pathogenesis. Endocrine Society, 70th annual meeting, New Orleans 1988, P 56

16. Armanini D, Zennaro C, Martella L, Milan E, Pratesi C, Scali M, karbowiak I, Grella V, Kuhnle U, Mantero F (1989) The number of mineralocorticoid receptors is reduced in mononuclear leukocytes of patients with EPH-gestosis. Endocrine Society, 71st annual meeting, Seattle, 1989, p 127
17. Armanini D, Bosson D, Scali M, Milan E, Pratesi C, Zennaro MC, Mantero F (1989) Mineralocorticoid receptors in mononuclear leukocytes of patients wityh essential hypertension. 4th meeting of the European Society of Hypertension, Milan 1989, p 721

Role of Adrenergic Receptors in Essential Hypertension*

E. Fritschka and T. Philipp

Abteilung für Nieren- und Hochdruckkrankheiten, Medizinische Klinik und Poliklinik, Universitätsklinikum Essen, Hufelandstraße 55, W-4300 Essen 1, FRG

Introduction

Both the pressor response to exogenous norepinephrine (predominantly mediated via α-adrenoceptors) and the chronotropic effect of isoproterenol (mediated via β-adrenoceptors) have been reported to be increased in normotensive offspring of parents with essential hypertension [1,2].

We studied platelet α_2- and lymphocyte β_2-adrenoceptor densities by [^{3}H]yohimbine and [^{125}I]iodo-cyanopindolol (CYP) binding, respectively, in normotensive volunteers and in patients with verified essential hypertension, with the aim of identifying a possible influence of a familial predisposition to essential hypertension.

In addition, β_2-adrenoceptor-independent forskolin-stimulated lymphocyte adenylate cyclase activity was measured in normotensive subjects grouped according to their family history of hypertension in order to further evaluate a possible role of genetic factors in the described β-adrenergic hypersensitivity.

Subjects and Methods

A positive family history of hypertension was assumed if at least one parent with essential hypertension requiring medical treatment before the age of 50 years was reported. All subjects were drug-free for at least 3 weeks prior to the study. Absence of diseases was confirmed by physical examination and routine laboratory work-up. In patients with essential hypertension, secondary arterial hypertension was excluded according to the diagnostic guidelines of the German Hypertension League.

---

* This study was supported by the Deutsche Forschungsgemeinschaft grant no. FR/497/5-3.

G. Bruschi A. Borghetti (Eds.)
Cellular Aspects of Hypertension
© Springer-Verlag Berlin · Heidelberg 1991

Blood pressures were measured between 08.00 and 10.00 a.m. after 5 min sitting, using a random zero sphygmomanometer (mean of three measurements).

Platelet α_2-adrenoceptor density was determined on fresh platelet membranes from [3H]yohimbine saturation curves as described elsewhere [3]. $(-)$-[^{125}I]-CYP was used for the measurement of β_2-adrenergic binding sites on whole lymphocytes as also reported elsewhere [4]. Lymphocyte membranes were prepared for the determination of the β_2-adrenergic high- and low-affinity binding sites. Briefly, isolated lymphocytes suspended in phosphate-buffered solution were lysed after centrifugation at 40 000 g (4°C, 15 min) in ice-cold distilled water, homogenized for 15 s with an Ultra Turrax (Janke and Kunkel, Jungingen, FRG), centrifuged again at 40 000 g (4°C, 15 min), resuspended in 4 ml incubation buffer (12 mmol/l Tris-HCL, 0.9 mmol/l EDTA, 12.5 mmol/l Mgcl$_2$, 0.5 mmol/l ascorbic acid, pH 7.4) and kept frozen at -80°C. Rethawed membranes were again centrifuged at 40 000 g (4°C, 15 min) and resuspended in incubation buffer. Samples of 100-μl membrane suspension were incubated in the binding assay for 60 min at 37°C at a total volume of 200 μl with [^{125}I]-CYP (40 pmol/l) and different concentrations of isoproterenol ($10^{-10} - 10^{-4}$ mol/l), where $(-)$-propranolol (5×10^{-6} mol/l) was used to determine nonspecific binding. The reaction was terminated by the addition of 10 ml TRIS-buffer (20 mmol/l TRIS, 154 mmol/l NaCl, pH 7.4). Subsequently vacuum filtration for the separation of bound and free [^{125}I]-CYP using Whatman GF/C filters and counting of the radioactivity of the dried filters were performed as previously described [4]. The Lowry method was used for the measurement of the protein content of the samples [5]. Computer-assisted nonlinear regression (modified Gauss-Newton iterative procedure) was used to determine the fractions of receptors representing a low and a high affinity state [6].

Table 1. Platelet α_2-adrenoceptor density (means $\pm$ SEM) in normotensive subjects with (FH$^+$) and without (FH$^-$) a family history of hypertension

	FH$^-$ ($n = 53$)	FH$^+$ ($n = 19$)
Age (years)	36.9 ± 1.7	33.3 ± 2.2
Body weight (kg)	69.7 ± 1.3	72.7 ± 2.8
Systolic blood pressure (mmHg)	115.1 ± 1.4	114.1 ± 3.1
Diastolic blood pressure (mmHg)	74.4 ± 1.2	72.8 ± 3.1
Mean arterial pressure (mmHg)	87.9 ± 1.1	86.5 ± 2.9
Platelet α_2-adrenoceptor density (fmol/mg protein)	192.1 ± 7.4	225.4 ± 17.5*
Affinity of α_2-adrenoceptors (K_d; nmol/l)	1.4 ± 0.1	1.5 ± 0.1
Plasma norepinephrine (pg/ml)	250.2 ± 19.1	242.9 ± 25.3
Plasma epinephrine (pg/ml)	31.8 ± 3.3	38.2 ± 4.0

*$p < 0.05$.

Forskolin-stimulated adenylate cyclase (AC) activity was measured in triplicate by conversion of $[^{32}p]$-ATP (Amersham, Braunschweig, FRG) to $[^{32}P]$-cAMP in the presence of a phosphodiesterase inhibitor [7]. Twenty microliters of lymphocyte lysates were incubated at 37°C at total volumes of $100 \mu l$ containing 50 mmol/l TEA, 10 mmol/l $MgCl_2$, 1 mmol/l 3-isobutyl-1-methylxanthine, 0.1 mmol/l EGTA, 0.05 mmol/l ATP, 0.04 mg creatine kinase, 5 mmol/l creatine phosphate, 0.1 mmol/l guanylyl imidodiphosphate, 0.1 mmol/l ($\pm$)-propranolol, phosphorus-32, and forskolin (Sigma Co., St. Louis, Mo., USA). Forskolin was dissolved in 50% ethanol, and six increasing concentrations of forskolin (from 10^{-8} to 2×10^{-4} mol/l) were used in the assays. Nucleotides not converted were precipitated by addition of ice-cold zinc carbonate after 10 min incubation. The supernatant obtained after centrifugation at 13 000 g was applied to neutral alumina columns (100 $-$ 200 mesh, Biorad no. 737 1222, Munich FRG). The activity of phosphorus 32 in the effluent was determined by the measurement of Cherenkov radiation. The protein content was measured as mentioned above [5]. Linearity of cAMP accumulation was determined during 20 min incubation.

Statistical Methods

Means $\pm$SEM are reported. Student's t test for independent samples was used for comparison where appropriate, otherwise the Mann-Whitney test was used. Nonlinear least square analysis was employed for the evaluation of dose-response curves.

Results

α_2-Adrenoceptor Densities

A significantly increased density of α_2-adrenoceptors ($p < 0.05$) was found in 19 predisposed (FH$^+$) normotensive subjects (mean age 33.3 $\pm$ 2.2 years) compared to 53 age-matched (36.9 $\pm$ 1.7 years) volunteers without predisposition to essential hypertension (FH$^-$; Table 1). When only the male subjects from this population were taken into account a similar difference in α_2-adrenoceptor density was observed: FH$^+$ ($n = 14$): 241.9 $\pm$ 21.2 vs FH$^-$ ($n = 33$): 208.4 $\pm$ 8.9 fmol/mg protein ($p < 0.05$), while affinity of the receptors was similar (1.7 $\pm$ 0.1 vs 1.5 $\pm$ 0.1 nmol/l).

Platelet α_2-adrenoceptor density was not significantly different between age-matched (42.9 vs 42.9 years) normotensive controls ($n = 42$, 21 males) and patients with established hypertension ($n = 42$, 26 males) of both sexes (193 $\pm$ 9.8 vs 178 $\pm$ 6.7 fmol/mg protein). There was also no difference in affinity (K_d) of the receptors (1.3 vs 1.4 nmol/l).

Table 2. Platelet α_2-adrenoceptor density (means ± SEM) in hypertensive subjects with (FH$^+$) and without (FH$^-$) a family history of hypertension

	FH$^-$ (n = 37)	FH$^+$ (n = 42)
Age (years)	48.0 ± 1.5	46.8 ± 1.5
Body weight (kg)	76.3 ± 2.2	76.4 ± 1.8
Systolic blood pressure (mmHg)	155.5 ± 3.7	156.4 ± 2.8
Diastolic blood pressure (mmHg)	103.4 ± 1.5	103.4 ± 1.0
Mean arterial pressure (mmHg)	120.7 ± 2.1	121.0 ± 1.3
Platelet α_2-adrenoceptor density (fmol/mg protein)	171.9 ± 6.9	190.0 ± 7.6*
Affinity of α_2-adrenoceptors (K_d; nmol/l)	1.4 ± 0.1	1.6 ± 0.1
Plasma norepinephrine (pg/ml)	316.2 ± 23.4	273.4 ± 18.9
Plasma epinephrine (pg/ml)	32.5 ± 3.5	31.2 ± 2.8

* $p < 0.05$.

Patients with essential hypertension and a positive family history of hypertension had higher platelet α_2-adrenoceptor densities ($p < 0.05$) than age-matched patients without familial predisposition (Table 2). When only male subjects from this hypertensive group were taken into account a similar difference ($p < 0.05$) was obtained. The FH$^+$ group ($n = 26$; 46.9 ± 1.6 years) had an α_2-adrenoceptor density of 195.9 ± 10.2 and the FH$^-$ group ($n = 23$; 47.5 ± 1.8 years) of 169.4 ± 8.5 fmol/mg protein.

In all subjects studied ($n = 160$) there was an inverse relationship between age and platelet α_2-adrenoceptor density ($r = -0.23$, $p < 0.05$).

β_2-adrenoceptor densities

By contrast, there was no difference in lymphocyte β_2-adrenoceptor density between the age-matched normotensive FH$^+$ and FH$^-$ groups (Table 3).

Table 3. Lymphocyte β_2-adrenoceptor densities in normotensive subjects with (FH$^+$) and without (FH$^-$) a positive family history of hypertension

	FH$^-$ (n = 23)	FH$^+$ (n = 7)
Age (years)	37.2 ± 2.7	36.3 ± 4.7
Systolic blood pressure (mmHg)	111.1 ± 2.3	115.9 ± 3.3
Diastolic blood pressure (mmHg)	71.5 ± 2.0	76.0 ± 3.0
Mean arterial pressure (mmHg)	84.7 ± 1.8	89.3 ± 2.7
Lymphocyte β_2-adrenoceptor density (sites/cell)	2470.0 ± 180.1	2137.0 ± 153.0
Affinity of β_2-adrenoceptors (pmol/l)	24.5 ± 1.8	28.3 ± 5.6
Plasma norepinephrine (pg/ml)	268.3 ± 32.1	182.4 ± 19.8
Plasma epinephrine (pg/ml)	35.9 ± 3.3	33.0 ± 5.5

Table 4. Lymphocyte β_2-adrenoceptor densities in hypertensive subjects with (FH$^+$) and without (FH$^-$) a positive family history of hypertension

	FH$^-$ ($n = 27$)	FH$^+$ ($n = 35$)
Age (years)	47.2 ± 1.8	44.7 ± 1.6
Systolic blood pressure (mmHg)	152.9 ± 3.6	152.4 ± 2.7
Diastolic blood pressure (mmHg)	102.8 ± 1.5	103.7 ± 1.2
Mean arterial pressure (mmHg)	119.5 ± 2.0	119.8 ± 1.5
Lymphocyte β_2-adrenoceptor density (sites/cell)	2753.0 ± 214.0	2972.0 ± 189.0
Affinity of β_2-adrenoceptors (pmol/l)	21.6 ± 1.4	22.0 ± 1.2
Plasma norepinephrine (pg/ml)	292.7 ± 26.0	272.2 ± 21.4
Plasma epinephrine (pg/ml)	33.0 ± 4.3	30.2 ± 2.8

The FH$^-$ group of patients with essential hypertension also exhibited a similar number of lymphocyte β-adrenoceptor binding sites to the FH$^+$ group of patients with essential hypertension, indicating a lack of genetic influence on β-adrenoceptor density (Table 4).

Age had no influence on β_2-adrenergic density ($r = -0.026$).

Lymphocyte [^{125}I]-CYP-β-adrenoceptor density in 66 patients of both sexes with essential hypertension (45.6 ± 1.1 years) was increased ($p < 0.05$) compared to in 34 normotensive controls (2856 ± 137 vs 2416 ± 129 binding sites/cell).

We found in addition that patients with essential hypertension not only exhibited an increased total number of lymphocyte β-adrenergic binding sites compared to normotensive controls, but that they also have a concomitantly increased fraction of the functionally important high-affinity binding sites as assessed by competition curves with isoproterenol (Table 5).

Table 5. Lymphocyte β_2-adrenergic high-affinity binding sites related to the presence or absence of essential hypertension

	Normotensives ($n = 8$)	Hypertensives ($n = 8$)
[^{125}I]-CYP-bound (fmol/mg protein)		
Total binding sites	35.3 ± 3.6	51.0 ± 6.4*
High-affinity binding sites	13.4 ± 1.1	23.2 ± 6.0*
Low-affinity binding sites	21.9 ± 3.1	27.7 ± 5.4
Affinity of high/low binding sites		
(K_{dh}, 10^{-9} mol/l)	10.4 ± 2.3	3.1 ± 2.0
(K_{dl}, 10^{-7} mol/l)	3.8 ± 1.2	1.5 ± 0.7

*$p < 0.05$.

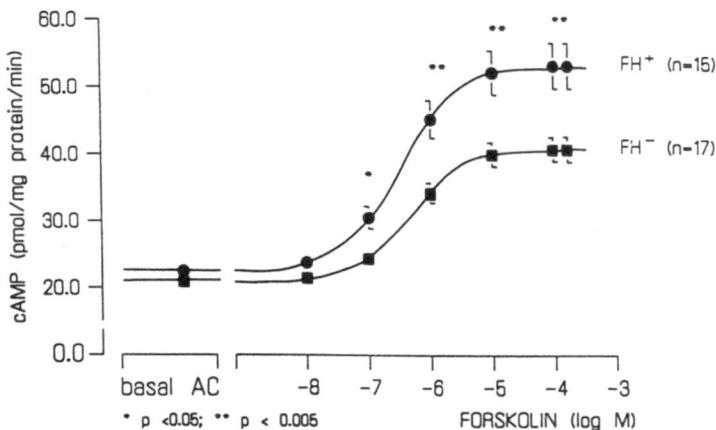

Fig. 1. Lymphocyte AC activity in normotensive males with (FH$^+$) and without (FH$^-$) a family history of hypertension in the absence or presence of increasing concentrations of forskolin in the incubation medium (means ±SEM). *$p < 0.05$; **$p < 0.005$

Forskolin-Stimulated Lymphocyte Adenylate Cyclase

The receptor-independent forskolin stimulation of lymphocyte membrane AC activity was increased ($p < 0.01$) in normotensive men ($n = 15$; mean age 27.8 ± 1.2 years) with a positive family history of hypertension as compared to controls ($n = 17$; 29.8 ± 0.9 years) without such a history and similar diastolic arterial pressure (FH$^+$: 75.6 ± 1.9 vs FH$^-$: 73.8 ± 1.5 mmHg). AC activity of the two groups in the absence of forskolin was similar (22.4 ± 0.9 vs 20.9 ± 0.6 pmol cAMP/mg protein min^{-1}). Stimulation of AC activity in the presence of forskolin demonstrated significantly higher AC activity at forskolin concentrations of 10^{-7} to 2×10^{-4} mol/l (Fig. 1).

Discussion

Significantly increased platelet alpha$_2$-adrenoceptor density was found in predisposed normotensive subjects compared to age-matched volunteers without predisposition for essential hypertension, confirming our previous preliminary report of increased platelet α_2-adrenoceptor density in patients with established hypertension with at least one hypertensive parent [3]. Our finding of a genetic influence on platelet α_2-adrenoceptor density has also more recently been confirmed in normotensive children [8], while others studying smaller samples did not observe such a difference (Table 6). Among the latter, however, Umemura et al. [11] confirmed that essentially hypertensive patients with a family history of hypertension have increased platelet α_2-adrenoceptor densities compared to patients without familial predisposition.

Table 6. Studies on familial predisposition towards hypertension and platelet α_2-adrenoceptor density (fmol/mg protein) in normotensive subjects

FH$^+$	n	FH$^-$	n	p	References
154.0 ± 34.0	12	182.0 ± 35.0	15	n.s.	Skrabal et al. [9]
225.4 ± 17.5	19	192.1 ± 7.4	53	<0.05	Fritschka et al. [10]
232.7 ± 18.5	12	214.7 ± 10.3	13	n.s.	Umemura et al. [11]
264.8 ± 17.5	34	208.0 ± 10.9	37	<0.01	Michel et al. [8]
241.9 ± 21.2	14	208.4 ± 8.9	33	<0.05	This study (males only)

In patients with established hypertension, platelet α_2-adrenoceptor density has not been consistently found to be increased compared to age- and sex-matched normotensive controls (Table 7). This might be in part due to the fact that, besides genetic factors, other variables such as age, concentration of plasma norepinephrine (which increases with age), and also the height of the chronically elevated blood pressure itself could play a role in the modulatation of α_2-adrenoceptor density as well.

By contrast to the findings concerning platelet α_2-adrenoceptors, there was no difference between the lymphocyte β_2-adrenoceptor densities in the four groups (normotensive and hypertensive FH$^+$- and FH$^-$ groups), suggesting that lymphocyte β_2-adrenoceptor density is not influenced by genetic factors related to essential hypertension. This finding is in keeping with the observation of Brodde et al. [19] indicating that the density of β_2-adrenoceptors in lymphocytes of children with normotensive parents was not significantly different from that in children with essentially hypertensive parents, although there was a trend towards a lower density of β_2-adrenoceptor density and, inversely, a trend towards a higher isoprenaline-

Table 7. Studies on platelet α_2-adrenoceptor density in patients with essential hypertension compared to normotensive controls

Author	Sex of subjects	Ligand	Substrate	Result
Kafka et al. [12]	Females	[^{3}H]DHE	Membranes	Increased
Kafka et al. [12]	Males	[^{3}H]DHE	Membranes	No difference
Motulsky et al. [13]	Males	[^{3}H]Yoh.	Cells	No difference
Boon et al. [14]	Males/fem.	[^{3}H]Yoh.	Cells	No difference
Continsouza−Blanc et al. [15]	Males/fem.	[^{3}H]Rauw.	Membranes	Decreased
Ashida et al. [16]	Males/fem.	[^{3}H]Rauw.	Membranes	No difference
Brodde et al. [17]	Males	[^{3}H]Yoh.	Membranes	Increased
Jones et al. [18]	Males/fem.	[^{3}H]Yoh.	Cells	Decreased
This study	Males/fem.	[^{3}H]Yoh.	Membranes	No difference

[^{3}H]DHE, [^{3}H]dihydroergocryptine; [^{3}H]Yoh., [^{3}H]yohimbine; [^{3}H]Rauw., [^{3}H]-rauwolscine.

Table 8. Studies on lymphocyte β_2-adrenoceptor density in patients with essential hypertension compared to normotensive controls

Author	Sex of subjects	Ligand	Substrate	Result
Kafka et al. [12]	Males/fem.	[³H]DHA	Membranes	No difference
Doyle et al. [20]	Males	[³H]DHA	Membranes	No difference
Brodde et al. [21]	Males	[¹²⁵I]CYP	Cells	Increased
Landmann et al. [22]	Males/fem.	[³H]DHA	Membranes	Increased
Middeke et al. [23]	Males/fem.	[¹²⁵I]CYP	Cells	Increased
Feldman et al. [24]	Not stated	[¹²⁵I]HYP	Membranes	No difference
This study	Males/fem.	[¹²⁵I]CYP	Cells	Increased

[³H]DHA, [³H]dihydroalprenolol; [¹²⁵I]HYP; ¹²⁵iodohydroxybenzylpindolol.

induced increase in lymphocyte cAMP content in children with hypertensive parents.

Since our patients with established hypertension had significantly higher lymphocyte β_2-adrenoceptor densities compared to corresponding normotensive controls, while age did not affect β-adrenoceptor density, it is probable that elevated blood pressure per se might influence β_2-adrenoceptor density, causing a secondary increase in β_2-adrenoceptor density.

A survey of available data shows that lymphocyte β_2-adrenoceptor density has been found by several investigators to be increased in patients with essential hypertension compared to normotensive controls, while none of these studies reported decreased β_2-adrenoceptor densities in patients with essential hypertension (Table 8).

We found additionally that not only do patients with essential hypertension exhibit an increased total number of lymphocyte β-adrenergic binding sites, but the fraction of the functionally important high affinity binding sites as assessed by competition curves with isoproterenol is concomitantly increased in patients with essential hypertension. The impact of a family history of hypertension on the fraction of high-affinity β-adrenergic binding sites was not determined.

Our finding of increased forskolin stimulation of lymphocyte AC activity in normotensive subjects with a familial predisposition to hypertension is compatible with the observation by Resinck et al. [25] of increased forskolin stimulation of AC activity in platelets obtained from patients with essential hypertension.

Previous studies by Stessmann et al. [26] in monozygotic twins demonstrated that a significant proportion of variance ($0.68-0.91$) of lymphocyte forskolin-stimulated AC activity was attributable to genetic factors. By contrast, neither basal nor isoproterenol-stimulated AC activity showed significant heritability. Increased isoproterenol stimulation of AC activity has been reported in cardiac membranes from prehypertensive spontaneously hypertensive rats (SHR) in the presence of normal myocardial β-adrenoceptor densities [27], and fluoride-stimulated AC activity was increased in parotid glands from young SHR [28], both compared to Wistar-Kyoto rats, suggesting that enhanced sympathetic nervous activity in rats with genetic hypertension may be a partial consequence of altered postreceptor events [28].

In summary, our results indicate that α_2-adrenoceptor density may be increased in normotensive and essentially hypertensive subjects with a familial predisposition to hypertension. If vascular α-adrenoceptor densities are increased as well in this population, a previously described increased reactivity to exogenous norepinephrine in normotensive subjects with a positive family history of hypertension could be in part explained by this mechanism [29]. As platelet α_2-adrenoceptor densities decrease with age, α-adrenoceptor densities may be not be consistently different between older patients with chronic hypertension and age-matched controls.

Genetic factors related to the development of arterial hypertension may not affect the number of lymphocyte β_2-adrenoceptor binding sites in normotensive or hypertensive groups. Lymphocyte β_2-adrenoceptor density may be secondarily increased in patients with established essential hypertension, as a possible consequence of factors associated with chronically elevated arterial pressure.

By contrast, preliminary evidence indicates that the receptor-independent stimulation of lymphocyte AC activity by forskolin is increased in normotensive subjects with at least one hypertensive parent as compared to appropriate controls without a familial predisposition to hypertension. The potential usefulness of this system as a genetic marker should be further examined.

The results obtained suggest that genetic factors related to essential hypertension affect platelet α_2-adrenoceptor density and lymphocyte membrane AC activity. These two observations could serve to explain α and β-adrenergic hypersensitivity in subjects predisposed to essential hypertension.

References

1. Bianchetti MG, Weidmann P, Beretta-Piccoli C, Rupp U, Boehringer K, Link L, Ferrier C (1984) Disturbed noradrenergic blood pressure control in normotensive members of hypertensive families. Br Heart J 51:306–311
2. Ferrara LA, Moscato TS, Pisanti N, Marotta T, Krogh V, Capone D, Mancini M (1988) Is the sympathetic nervous system altered in children with familial history of hypertension? Cardiology 75:200–205
3. Fritschka E, Kribben A, Haller H, Hoyer J, Thiede HM, Distler A, Philipp T (1987) Familial aggregation of altered adrenoceptor density and free intracellular calcium in patients with essential hypertension. J Cardiovasc Pharmacol 10 [Suppl 4]: S122–S125
4. Fritschka E, Kribben A, Harwig S et al. (1986) Effect of fludrocortisone on adrenoceptors and free intracellular calcium in man. In: Middecke M, Holzgreve

H (eds) New aspects of hypertension: adrenoceptors Heidelberg: Springer Berlin Heidelberg New York, pp 146–154

5. Lowry OH, Rosebrough NJ, Farr AL, Randall RJ (1951) Protein measurement with the Folin phenol reagent. J Biol Chem 193:265–275

6. Wiemer G, Wellstein A, Palm D, von Hattingberg HM, Brockmeier D (1982) Properties of agonist binding at the β-adrenoceptor of the rat reticulocyte. Naunyn Schmiedebergs Arch Pharmacol 321:11–19

7. Jacobs KH, Saur W, Schultz G (1976) Reduction of adenylate cyclase activity in lysates of human platelets by the alpha-adrenergic component of epinephrine. J Cyclic Nucleotide Res 2:381–392

8. Michel MC et al. (1989) Alpha- and beta- adrenoceptors in hypertension II. Platelet α_2- and lymphocyte β_2-adrenoceptors in children of parents with essential hypertension. A model for the pathogenesis of the genetically determined hypertension. J Cardiovasc Pharmacol 13:432–439

9. Skrabal F, Gruber G, Meister B, Ledochowski M, Doll P, Lang F, Cerny E (1985) Salt sensitivity in normotensives with family history of hypertension: studies of membrane transport, intracellular electrolytes and α_2-adrenergic receptors. J Hypertens. 3 [Suppl 3]:S25–S28

10. Fritschka E, Kribben A, Hoyer J, Thiede HM, Distler A, Philipp T (1987) Genetischer Einfluß auf α_2-adrenerge Rezeptoren und Adenylzyklaseaktivität bei Normotonikern und Hypertonikern. Klin Wochenschr 65 [Suppl IX]:170–171

11. Umemura S, Uchino K, Yasuda G, Ishikawa Y, Hatori Y, Tochikubo O, Ishii M, Kaneko Y (1988) Altered platelet alpha$_2$-adrenoceptors and adrenaline response in adolescents with borderline hypertension who have a family history of essential hypertension. J Hypertens 6 [Suppl 4]:S568–S571

12. Kafka MS, Lake CR, Gullner H-G, Tallman JF, Bartter FC, Fujita T (1979) Adrenergic receptor function is different in male and female patients with essential hypertension. Clin Exp Hypertens 1(5):613–627

13. Motulsky HJ, O'Connor DT, Insel PA (1983) Platelet alpha-adrenergic receptors in treated and untreated essential hypertension. Clin Sci 64:265–272

14. Boon NA, Elliot JM, Davies CL, Jones JV, Conway FJ, Graham-Smith DG, Sleight P (1983) Platelet α-adrenoceptors in borderline and established essential hypertension. Clin Sci 65:297–298

15. Continsouza-Blanc D, Elghozi J-L, Dausse J-P (1984) Alterations of platelet alpha-adrenoceptors in human hypertension. J Hypertens 2 [Suppl 3]:155–157

16. Ashida T, Tanaka T, Yokouchi M, Kuramochi M, Deguchi F, Kimura G, Kojima G, Ito K, Ikeda M (1985) Effect of dietary sodium on platelet α_2-adrenergic receptors in essential hypertension. Hypertension 7:972–978

17. Brodde OE, Daul AE, O'Hara N, Khalifa AM (1985) Properties of alpha- and beta-adrenoceptors in circulating blood cells of patients with essential hypertension. J Cardiovasc Pharmacol 7 [Suppl 6]:S162–S167

18. Jones CR, Elliott HL, Deighton N, Howie CA, Reid JL (1985) Alpha-adrenoceptor number and function in platelets from treated and untreated patients with essential hypertension and age and sex matched controls. J Hypertens 3 [Suppl 3]:S153–S155

19. Brodde OE, Michel MC, Beckeringh JJ (1987) Alterations of β-adrenoceptor function in essential hypertension. J Cardiovasc Pharmacol 10 [Suppl 4]:S50–54

20. Doyle V, O'Malley K, Kelly JG (1982) Human lymphocyte β_2-adrenoceptor density in relation to age and hypertension. J Cardiovasc Pharmacol 4:738–740

21. Brodde OE, Prywarra A, Anlauf M, Daul A, Bock KD (1983) Increased number of β_2-adrenoceptors in circulating lymphocytes of patients with essential hypertension. J Hypertens 1 [Suppl 2]:263–266

22. Landmann R, Bürgisser E, Bühler FR (1983) Human lymphocytes as a model for beta-adrenergic receptors in clinical investigation. J Recept Res 3:71–88

23. Middeke M, Remien J, Block LH, Kirzinger S, Landrock A, Holzgreve H (1983) Beta$_2$-adrenoceptor density on membranes and on intact mononuclear cells in essential hypertension. Res Exp Med 183:227–232
24. Feldman RD, Limbird LE, Nadeau J, Robertson D, Wood AJJ (1984) Leukocyte beta-receptor alterations in hypertensive subjects. J Clin Invest 73:648–653
25. Resink TJ, Bürgisser E, Bühler FR (1986) Enhanced platelet cyclic AMP response to prostaglandin E$_1$ in essential hypertension. Hypertension 8:662–668
26. Stessmann J, Mintzer Z, Lipschitz Y, Shemesh Z, Goldin LR, Ebstein RP (1985) Heritability of forskolin and hormone-stimulated adenylate cyclase activity in human lymphocytes. J Cyclic Nucleotide Protein Phosphor Res 10:317–326
27. Blumenthal SJ, McConnaughey MM, Iams SG (1982) Myocardial adrenergic receptors and adenylate cyclase in the developing spontaneously hypertensive rat. Clin Exp Hypertens [A] 4:883–901
28. Schmid G, Geiger H, Bahner U, Heidland A (1988) Glandular adenylate cyclase system in genetic hypertension: age-dependent response to catecholamines. Eur J Pharmacol 147:397–402
29. Fritschka E, Kribben A, Hoyer J, Rothschild M, Thiede M, Distler A, Philipp Th (1987) Regulation of alpha$_2$-adrenoceptor density in normotensive and hypertensive man. J Hypertens 5 [Suppl 5]:S349–S352

Na$^+$,K$^+$,Cl$^-$ Cotransport System in Primary Hypertension: Studies in Red Cells and in the Choroid Plexus of Spontaneously Hypertensive Rats

C. Rosati[1], M.-M. Ruchoux[2], K. Gruber[3], and R. Garay[1]

[1] Unité de Recherches de Physiologie et Pharmacologie Vasculaire et Rénale, Inserm U7, Hôpital Necker, 161, rue de Sèvres, F-75015 Paris, France
[2] CHU Bretonneaux, F-Tours, France
[3] University of Puerto Rico, Medical Sciences Campus, San Juan, Puerto Rico

Introduction

Apart from high blood pressure, there is no other clinical sign or laboratory abnormality common to all essential hypertensive patients. Thus, in only half of all hypertensive subjects does blood pressure rise after excess sodium intake. Some hypertensives are obese, others are hyperreactive to stress; some hypertensives have increased plasma catecholamines, others have increased plasma renin activity, or even a decreased urinary excretion of prostacyclin metabolites. We can be certain that this heterogeneity of essential hypertension is involved in the variability of therapeutic response to weight reduction, low-salt diet, or antihypertensive drugs.

In an attempt to provide a biochemical basis to the above-described heterogeneity, Laragh [3] has proposed classifying essential hypertensive patients according to their plasma renin activity. A further molecular insight has been recently provided by ion flux studies showing a large number of ion transport abnormalities in red blood cells from essential hypertensive patients and genetically hypertensive rats (for references see [2]). Table 1 summarizes most of these membrane abnormalities.

The question of whether or not the above abnormalities are etiological factors in primary hypertension is a matter under current investigation. We tried to elucidate physiological and pathological aspects of the Na$^+$,K$^+$,Cl$^-$ cotransport system in organs relevant to the regulation of blood pressure.

The Na$^+$,K$^+$,Cl$^-$ cotransport system was found in almost all cells investigated, and its physiological role in the kidney and other epithelial organs is well defined (for review see [1,6]). In Henle's loop and in several other absorptive epithelia, the Na$^+$,K$^+$,Cl$^-$ cotransport system catalyzes NaCl entry across *luminal* membranes. Conversely, in the choroid plexus and other *secretory* epithelia, the cotransport system catalyzes NaCl entry across *basolateral* membranes (Fig. 1). Unlike in epithelial cells, the physiological role of the Na$^+$,K$^+$,Cl$^-$ cotransport system in vascular and other nonepithelial cells is unclear. However, some indirect arguments suggest

G. Bruschi A. Borghetti (Eds.)
Cellular Aspects of Hypertension
© Springer-Verlag Berlin · Heidelberg 1991

Table 1. Ion transport abnormalities in red blood cells from essential hypertensive patients and genetically hypertensive rats

Transport pathway	Abnormality	Human hypertensives (white) (%)	Rat
Na^+,Li^+ countertransport	$V(+)$	20–50	?
Na^+,K^+,Cl^- cotransport	$V(+)$	30–50	SHR-HPS-MSH
	$R(-)$	20–40	SHR
Na^+,K^+ pump	$R(-)V(+)$	5–15	?
	$V(-)$	<10	RRM
Na^+,H^+ exchange	$V(+)$	30–50	?
Ca^{2+} pump	$R(-)V(+)$	20–30	?
Na^+ leak	Increased	10–30	HPS

$V(+)$, increased maximum rate; $V(-)$, decreased maximum rate; $R(-)$, decreased regulatory properties; *HPS*, hypertension-prone Sabra rats; *MSH*, Milano spontaneously hypertensive rats (for references see [2]).

that in vascular smooth muscle this carrier counterbalances the KCl loss associated with cell activation [7].

In regard to essential hypertension, the Na^+,K^+,Cl^- cotransport system may exhibit at least two independent abnormalities: (i) an increased maximum cotransport rate [$V(+)$ abnormality in Table 1] found in 30%–50% of white hypertensives and (ii) a decreased apparent affinity for internal Na^+ [$R(-)$ abnormality in Table 1] observed in 20%–40% of white hypertensives (for references see [2]). Interestingly, predisposition to hypertension in blacks appears to be correlated with a high frequency of the $R(-)$ cotransport abnormality.

In experimental hypertension, the $V(+)$ cotransport abnormality has been frequently found in red blood cells from several strains of genetically hypertensive rats. In particular, the spontaneously hypertensive rats of the Okamoto strain (SHR) have both $V(+)$ and $R(-)$ abnormalities [8].

The $V(+)$ and $R(-)$ abnormalities may have similar pathological consequences for epithelial tissues, i.e., increased cell NaCl entry by the Na^+, K^+,Cl^- cotransport system. However, it is important to note that while a $V(+)$ abnormality increases *basal* NaCl entry, the $R(-)$ abnormality only increases NaCl entry in *Na^+-loaded* cells (due to a decreased ability of internal Na^+ to exert a negative feedback on excess Na^+ entry).

Generally speaking, to ascertain the mechanistic relevance of the $V(+)$ and/or the $R(-)$ cotransport abnormalities, it is necessary first to identify the target organ where these abnormalities induce high blood pressure. Several arguments suggested to us that one of these target organs could be the choroid plexus: (i) the cotransport system is particularly active in the choroid plexus, (ii) the rate of CSF formation (and the intracranial pressure) influences brain function, and (iii) in SHR (and in human hypertensives)

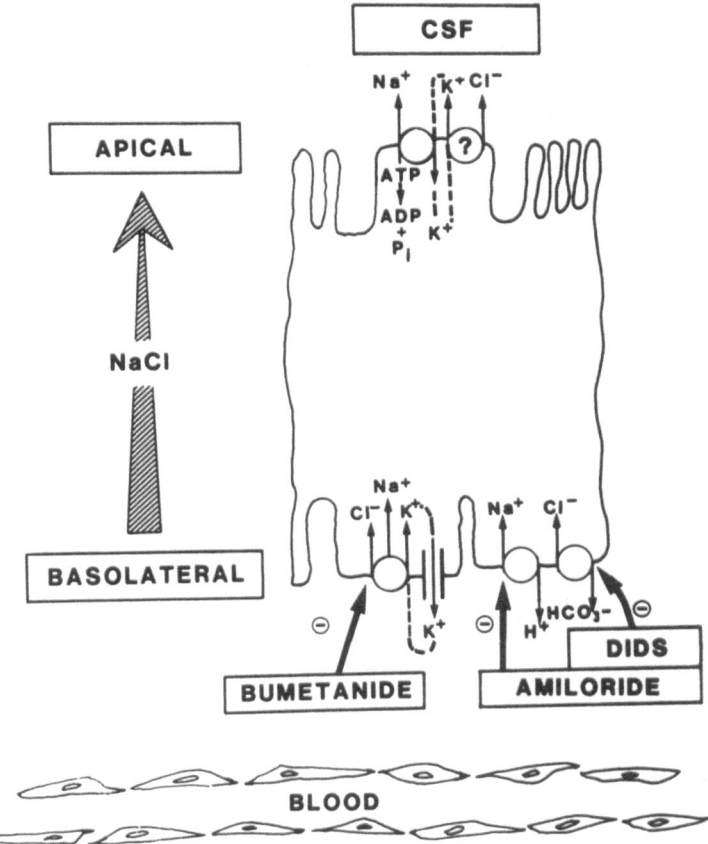

Fig. 1. Ion transport across epithelial cells of the choroid plexus. A bumetanide-sensitive Na$^+$,K$^+$,Cl$^-$ cotransport system is located in basolateral membranes, where it catalyzes net NaCl entry. Similarly to what is seen in Henle's loop, a K$^+$ channel operating in parallel may cycle K$^+$ across the basolateral membrane. In addition, an amiloride-sensitive Na$^+$,H$^+$ exchange, working in parallel with a DIDS-sensitive Cl$^-$,HCO$_3^-$ exchanger, seems also to catalyze NaCl entry. In apical membranes, Na$^+$ extrusion is only mediated by a Na$^+$,K$^+$ pump. As in Henle's loop, a K$^+$,Cl$^-$ cotransport system operating in parallel may cycle K$^+$ across the apical membrane

the presence of cotransport abnormalities is associated with increased noradrenergic activity (see for instance; [4]).

We therefore decided to investigate ion transport function in the choroid plexus of SHR.

Methods

Male Okamoto SHR and SHR-SP (stroke prone) rats and their normotensive controls, Wistar-Kyoto (WKY; derived from the original Okamoto stock and supplied by Centre d'Elevage Roger Janvier, Le Genest and by Iffa-Credo, L'Arbresle, France), were studied at 4–6 weeks of age.

The whole choroid plexus (from the third and fourth ventricles) was homogenized at room temperature in Ringer's medium containing type IV collagenase 45 units/ml. The Ringer's medium contained (mM): 145 NaCl, 5 KCl, 1 CaCl$_2$, 1 MgCl$_2$, 10 MOPS-TRIS buffer (pH 7.4 at room temperature) and 5 glucose. The cells were mechanically separated by simple aspiration with a syringe, then washed three times in cold Ringer's medium (without collagenase) and resuspended in the same medium.

Aliquots of the cell suspension were added in the cold to duplicate tubes containing Ringer's medium in which KCl was replaced by an equivalent amount of RbCl. Some duplicates contained 2 mM ouabain and some others contained 50 μM bumetanide. The cell suspensions were incubated at 37°C for 0, 5, 15, and 30 min, and then transferred to the cold for 1 min and washed three times with cold MgCl$_2$ 110 mM.

The cell pellet was resuspended in distilled water containing 0.02% Acationox (American Scientific Products, McGraw Park, Ill.). The suspensions were frozen and thawed two times and their Rb$^+$ and K$^+$ contents measured in an IL 457 atomic absorption spectrophotometer (Instrumentation Laboratory, Lexington, Mass., USA).

The initial rate of bumetanide-sensitive Rb$^+$ entry was equated to the inward Na$^+$,K$^+$,Cl$^-$ cotransport and expressed in μmol/internal K$^+$ per hour. The initial rate of ouabain-sensitive Rb$^+$ entry was taken as an indicator of Na$^+$,K$^+$ pump activity.

Results

Bumetanide-sensitive Rb$^+$ entry was significantly increased in the choroid plexus of SHR (2.20 ± 0.88 vs 1.07 ± 0.31 μmol/internal K$^+$ per hour in WKY rats; mean ± SD; $n = 5$; $p < 0.05$) but not in SHR-SP (1.37 ± 0.72 μmol/internal K$^+$ per hour).

Ouabain-sensitive Rb$^+$ entry was increased in SHR, but the statistical signficance was borderline [2.55 ± 1.25 vs 1.28 ± 0.39 μmol/internal K$^+$ per hour in WKY ($n = 5$; $p \approx 0.07$) and 1.65 ± 0.73 μmol internal K$^+$ per hour in SHR-SP].

Electronic microscopy in SHR revealed (i) partial loss of both the brush border in apical membranes and the invaginations of basolateral membranes, (ii) a high number of Golgi apparatus, vesicles, and mitochondria, and (iii) activated nuclei. More pronounced changes were observed in SHR-SP, particularly the appearance of degenerative processes such as replacement of brush border by ciliary bodies and mitochondrial swelling.

Discussion

The young SHR presented marked morphological alterations in the choroid plexus (high number of Golgi apparatus, vesicles, and mitochondria) which suggested increased secretory activity. This was further suggested by the flux experiments showing increased cotransport and pump activities. However, further experiments are required to see whether these abnormalities result in increased formation of CSF (and increased noradrenergic activity).

The most striking morphological abnormality in SHR was the partial loss of the apical and basolateral membranes, a fact that contrasted with the increased transport activity. One possible explanation could be that the increased secretory (and transport) activity reflects a compensatory phenomenon for some primary alteration in the cell membranes (perhaps in the cytoskeleton).

In SHR-SP, the morphological changes were much more pronounced than in SHR, perhaps explaining the reversal of the compensatory transport hyperactivation.

It is important to note that the cotransport abnormalities may disturb cell function in other tissues besides the choroid plexus. For instance, O'Donnell and Owen [5] have found a decreased number of cotransport units in vascular smooth muscle cells from SHR. This abnormality is compatible with decreased recovery of KCl after vascular cell activation. Another potential target organ that requires investigation is the kidney. In this regard, it is important to note that the loop of Henle is more adapted to regulating water than Na$^+$ homeostasis (which is finally regulated by the distal and collecting tubules).

To conclude: the choroid plexus from young SHR exhibits transport and morphological alterations compatible with increased formation of CSF. Whether this could be involved in the increased noradrenergic activity and/or in the hypertensive mechanism deserves further investigation.

References

1. Chipperfield AR (1986) The [Na$^+$,K$^+$,Cl$^-$]-cotransport system. Clin Sci 71:465–476
2. Garay RP (1990) Red blood cell Na$^+$ content is poorly related to essential hypertension and to membrane Na$^+$ transport abnormalities. Hypertension 15:225–226
3. Laragh J (1983) Personal views on the mechanism of hypertension. In: Genest J, Kuchel O, Hamet P, Cantin M (eds) Hypertension. McGraw-Hill, New York, pp 615–631
4. Mongeau JG (1985) Erythrocyte cation fluxes in essential hypertension of children and adolescents. Int J Pediatric Nephrol 6:41–46
5. O'Donnell ME, Owen NE (1988) Reduced Na-K-Cl cotransport in vascular smooth muscle cell from spontaneously hypertensive rats. Am J Physiol 255:C169–C180
6. Palfrey HC, Rao MC (1983) Na/K/Cl co-transport and its regulation. J Exp Biol 106: 43–54
7. Rosati C, Jeanclos E, Nazaret C, Chabrier PE, Hannaert P, Braquet P, Garay RP (1989) Stimulation of Na$^+$-H$^+$ exchange, the Na$^+$-K$^+$ pump and Na$^+$,K$^+$,Cl$^-$

cotransport by endothelin-1 in cultured vascular smooth muscle cells. J Hypertens [Suppl 6]:138–139

8. Saitta M, Hannaert P, Rosati C, Meyer P, Garay RP (1987) A kinetic analysis of inward Na$^+$,K$^+$ cotransport in erythrocytes from spontaneously hypertensive rats. J Hypertens 5 [Suppl 5]:285–286

Subject Index